Marine Engineering Economics and Cost Analysis

MARINE ENGINEERING ECONOMICS AND COST ANALYSIS

EVERETT C. HUNT
Consulting Engineer

BORIS S. BUTMAN
United States Merchant Marine Academy

CORNELL MARITIME PRESS
Centreville, Maryland

SOFTWARE LICENSE AGREEMENT

Library of Congress Cataloging-in-Publication Data

Hunt, Everett C.
Marine engineering economics and cost analysis / Everett C. Hunt, Boris S. Butman. — 1st ed.
p. cm.
Includes bibliographical references (p.) and index.
ISBN 0-87033-458-1 (hardcover)
1. Marine engineering—Costs. 2. Marine engineering—Decision making. 3. Ships—Cost of operation. 4. Shipbuilding—Costs. 5. Shipping—Costs. 6. Engineering economy. I. Butman, Boris S. . II. Title.
VM600.H93 1994
623.8\068\1—dc20 94-35353

Manufactured in the United States of America
First edition

For students of ship design, shipbuilding, and ship operations, with most particular appreciation of their tasks

Contents

Foreword

To my knowledge, this is the first text published during my 50-year career in the maritime industry which deals with the methods of economic evaluation of maritime decision alternatives from an engineering viewpoint. This text addresses two areas which have lacked a comprehensive and practical approach for the student as well as those in the industry. These areas are shipbuilding and shipping economics and their relation to one another. For too long, we have considered that shipping economics and shipbuilding are independent. This book applies engineering economics and cost analysis to the maritime industry and sets forth in a logical sequence the method to reach the most efficient vessel from both cost and capacity-required approaches.

Based on my experience in the maritime industry and government, I believe this book will help solve some of the persistent problems that have plagued our industry. During the 1960s, the American shipping industry witnessed the beginning of a change in cargo-carrying methods. A slow but increasing rate of change from break-bulk liner vessels to containerships was underway. However, the process has been inhibited by the absence of adequate evaluation of the validity of this change by either industry or government. As a result, construction of the technologically obsolete break-bulk vessels continued, increasing vessel-operating and cargo-handling costs which finally forced most owners to recognize the need for conversion of existing tonnage and the construction of new tonnage for container transport. It has been 30 years since the first containership entered service.

My concern for the future of the American flag merchant marine leads me to look forward for a better, more efficient, and less costly system to replace or modify current containerization in the next 10 to 20 years. The methods of engineering economics and cost analysis presented in this text are necessary for the logical evaluation and validation of such future systems. I am sure that other authors will build on the methods and software of this text.

Engineers in the American shipbuilding industry, which is in decline and struggling to become competitive in the world market, should benefit from the methods detailed in this book. After 15 years of primarily naval construction, the shipbuilders are now trying to turn to commercial shipbuilding. They face tremendous challenges despite their competitiveness on a labor-hour cost basis. Shipbuilders must change their management techniques and philosophies. Production methods must be modified and improved if they intend to gain a portion of the market. These tasks cannot be taken lightly.

In addition, the question of financing must be addressed, for if an adequate financing scheme is not available to the shipbuilders, they will fail in their goal of being competitive in the world market.

I am confident that given a rational approach to our declining American shipbuilding we can reverse the current trend. We must objectively review our past and retain only what is applicable for today and the near future. Past operation criteria, design, and construction methods are not the answer. Taxpayer funding of current promotional programs such as operating differential subsidies and construction differential subsidies is not the answer. Some type of government support may be needed although I do not know what type of support would be effective.

I am privileged to be able to lend my experience and background to set forth briefly the need for this book and for others which are sure to follow. This text, written primarily for the engineering student, will serve as a refresher and resource for those in the maritime industry. We need new thinking, new approaches, and open minds if we are to solve our problems and be successful.

Captain Warren G. Leback
Former Maritime Administrator

Princeton, N.J.
October 14, 1993

Preface

This book is written for students and practitioners of ship design, shipbuilding, and ship operations who must understand and apply the concepts of engineering economics to their routine engineering decisions. Decisions made without adequate consideration of the economic component frequently result in the wrong choice. Naval architects, marine engineers, industrial engineers, transportation engineers, ocean engineers, and operations management are among the disciplines for which this book will be useful.

Since decisions in the maritime environment entail future uncertainties with significant risks to shipowners and shipbuilders, this text includes methods for quantifying such risks and making rational choices in the face of the defined uncertainties. The necessary calculations are sometimes tedious for the analyst. Therefore, some elementary computer software is provided with the text to reduce the effort of analysis necessary for an optimum decision and to encourage the student to consider the preparation of more comprehensive and sophisticated software in the future.

Shipping and shipbuilding are international businesses. Since students and engineers must be familiar with metric and U.S. customary units of measurement, both are used throughout this text. The reader acquainted with undergraduate engineering mathematics including probability and statistics and with access to an IBM-compatible personal computer will be comfortable using this text. The authors have included brief reviews of mathematical methods when necessary for the understanding of the decision methods employed. This text is suitable for a one-semester upper level undergraduate course or an introductory graduate course. The engineer in industry will find it a useful reference.

The text begins with an overview of the ocean transport and related industries including a historical perspective and some predictions concerning trading patterns and industry growth. The roles of the service industries and government are discussed. Chapter 1 also introduces the definition of engineering economics. The principles of engineering economics, including the time value of money concept, compound interest, and present worth are explored in Chapter 2. Chapter 3 employs these principles to evaluate economic alternatives assuming a deterministic scenario. Methods for dealing with decisions under uncertainty and risk are presented in Chapter 4. Chapters 6 and 7 employ these methods for the analysis of a fast ferry system and required freight rate calculations for ocean transport vessels. Both deterministic and stochastic methods of engineering economics are applied to ship design in Chapter 8. In Chapter 9, the student is introduced to the details of a shipbuilding cost estimate and the structure of a cost-accounting system for collecting shipbuilding costs. Chapters 10 through 13 present applications of engineering economics to shipyard operations, ship construction, ship repair, and ship operations. Chapter 14 is a review of the methods and sources of ship financing. Finally, the last chapter presents a method for multicriteria decision making for those who wish to extend decision criteria beyond the single measure-of-merit usually employed in engineering economics.

Most chapters include a reference and bibliography section to permit the reader to study sources and explore the subject beyond the limits of this text. A section titled "Questions and Projects" following each chapter is designed to assist in a review of the chapter, to raise pertinent issues not included in the chapter, and to provide projects for assignment to groups of students. Most chapters also include a set of numerical problems for assignment to students. A solutions manual is available to instructors upon request.

This text represents the experience of the authors in industry and in teaching to engineering students. Some of the methods described were very successfully employed in industry.

The text has benefited from the advice and counsel of many colleagues, friends, and students. The authors are particularly grateful to Theodore A. Ulrich who wrote Chapter 14, "Ship Financing," and to Daniel J. Mulcahy who reviewed and contributed to Chapter 5, "Depreciation and Taxes." Messrs. Ulrich and Mulcahy are with Cadwalader, Wickersham & Taff of New York. The contribution of Geoffery Uttmark of Transtech Marine Company, Brooklyn, N.Y., who wrote the shipping industry portion of Chapter 1 and reviewed several of the chapters and the software of this book, is acknowledged with gratitude. We thank Captain Warren G. Leback for his kindness in agreeing to write the Foreword.

Marine Engineering Economics and Cost Analysis

Chapter 1

Maritime Industries

Six or eight men, therefore, by the help of water-carriage, can carry and bring back in the same time the same quantity of goods between London and Edinburgh, as fifty broad-wheeled wagons, attended by a hundred men, and drawn by four hundred horses.
—*Adam Smith,* The Wealth of Nations, *Book One, Chapter III*

INTRODUCTION

Historically, making decisions under conditions of risk and uncertainty has always been the shipowner's challenge. This business environment continues today. In the last decade of the twentieth century shipowners must consider the following factors:

1. Ships cost more each year in absolute terms due to larger size, greater complexity, and improved and higher technology.
2. Most ships are less flexible in employment as trades become more segmented by specialty tonnage.
3. Economic life cycles of assets are shorter due to the increasing pace of technological innovation and changing requirements.
4. Competition is intense as the information age blurs distinctions between potential customers and competitors.
5. Financial resources, the lifeblood of capital-intensive shipping, have become more difficult and costly to obtain.

Engineers involved in analysis and decision making in this business environment must be prepared to address quantification of risk and uncertainty, to make optimal use of resources including labor, material, capital, and technology, and to apply effectively the techniques of engineering economics and cost analysis.

The global economy of the 1990s could not exist without ocean shipping. Intercontinental economic integration supported by ocean transport has been developing for centuries. Ships now transport more than 4 billion tonnes of cargo each year and the total annual tonne-miles exceeds 19 trillion. Ships facilitate trade and are a crucial component of a nation's competitiveness in international markets. While the contribution of mechanized farming and automated factories to lower consumer prices is generally understood, the contribution of modern global transportation and distribution systems is not fully appreciated. During the decade of the eighties, a silent revolution in transport economics has occurred, mostly out of sight of land.

The ocean shipping industry is a diverse group of businesses with little in common except that they share the same oceans, lakes, and rivers. In 1992 the world's deep-sea cargo-carrying fleet included 23,943 vessels. The dominant ship types are general cargo, tankers, and dry bulk carriers. When unpowered vessels, such as barges, and vessels of less than 1,000 grt, such as tugs, fishing vessels, coasters, excursion vessels, etc., are included, the worldwide fleet of commercial vessels exceeds several hundred thousand ships. Each must be designed, financed, built, maintained, and operated profitably.

There are significant differences among the economic and cost structures of the many shipping industry market sectors. Large differences in the economics also exist among vessel sizes in the same market sector. Figure 1-1, a summary of the world fleet, defines the types of seagoing self-propelled commercial vessels. Some of these general ship types have been further classified by size as shown in Figure 1-2. A typical size classification is the popular 125,000-cubic-meter liquified natural gas carrier (LNG). Ships which are designed and constructed to operate in a specific trade during their lifetime are optimized for the unique characteristics and constraints of the trade. Typical characteristics and constraints may include type and amount of cargo, distance between ports, service frequency, port limitations, cargo-handling facilities, terminal size, and bunkering schedule. A vessel which is specialized for a certain trade is referred to as a "one off" design, though several vessels may ultimately be built to the design. Generalized designs are built for unspecified trading such as the breakbulk cargo vessel which is disappearing from international trade. It is not uncommon for a generalized trading vessel to be engaged under a long-term charter in a specific trade. Comparable cargo-carrying capacity is the only criterion for grouping ships in a size class. Design variations among a class can be substantial as designers and owners seek characteristics which will improve the economics of vessel operation.

Vessel Type[1]	*Number*	*Percent*	*Grt, (000)*	*Percent*	*Dwt, (000)*	*Percent*
Liquid Cargo[2]	5,542	23.1	153,309	38.6	280,572	43.0
Dry cargo[3]	5,473	22.8	139,295	35.0	247,679	38.0
Combo carriers[4]	347	1.5	4,690	1.2	1,425	0.2
General cargo[5]	10,123	42.3	59,137	14.9	79,479	12.2
Containerships	1,290	5.4	26,844	6.8	29,686	4.5
Partial containerships[6]	162	0.7	1,874	0.5	2,379	0.4
Roll-on/roll-off[7]	966	4.0	11,096	2.8	9,448	1.5
Barge carrier	40	0.2	980	0.2	1,257	0.2
Totals	23,943	100.0	397,225	100.0	651,925	100.0

[1] Vessels over 1,000 grt, excluding Great Lakes, as of January 1992.

[2] Includes gas carriers, crude carriers, parcel, product, and specialty tankers of single skin, double bottom, and double hull construction.

[3] Includes geared and gearless specialty bulk carriers.

[4] Includes ore-oil (O/O), bulk-oil (B/O), and ore-bulk-oil (O/B/O) carriers.

[5] Includes single deckers, 'tween deckers, breakbulk, reefer, warehouse (pallet), semisubmersible, heavy-lift, and specialty vessels.

[6] Includes container/car carriers, container/rail car carriers, and container/roll-on/roll-off vessels.

[7] Includes wheeled containers and oceangoing passenger-vehicle ferries.

Figure 1-1. World cargo fleet summary. Courtesy MarAd

Shipping markets comprise tonnage, i.e. ships, that are both old and new, large and small, generalized and specialized, fast and slow, with or without cargo-handling capacity, and luxurious and austere. Competition among owners to attract cargoes and earn profits by anticipating and preparing for uncertain futures creates the dynamic environment where success may be richly rewarded and error may lead to financial ruin. The consuming public, unaware of the search for competitive advantage on the oceans, is the ultimate beneficiary of these activities.

TANKERS: A SPECIALIZED TRADE

Tankers are the single largest component of the world merchant fleet in deadweight capacity and annual tonnage transported. This is not surprising, because demand for hydrocarbons is ever increasing and the producing centers are located far from the consumption areas. Demand for oil, used mainly for energy, approached 3.5 billion tonnes annually during the early 1990s. Japan is totally dependent on imported energy; Europe imports 70 percent of its oil needs; North America imports 50 percent. The main oil-producing states border the Persian Gulf which accounts for more than 50 percent of the world oil exports. Smaller exporting areas include North and

West Africa, South America, the North Sea, and Eastern Europe. Vast largely undeveloped oil reserves exist in Russia, China, and some former Soviet Union republics such as Kazakhstan. Exploration for new oil-producing fields continues in these and other places, sometimes fostering the development of new ship types and associated technology, for example, the high-powered double-hull ice-breaking very large crude oil carrier.

Tankers for the transport of crude oil and petroleum products are classed according to function and size. Tankers have also been developed for transport of liquid chemicals, latex, molasses, wine, vegetable oil, orange juice, molten sulphur, fresh water, and other liquids. Specialized tankers have special characteristics: a wine tanker has stainless steel tanks; a concen-

Deadweight[1]	*Tanker*	*Bulk Carrier*	*General Cargo*
500			
			Inter-island coaster
	Coastal	By vessel dwt	River-sea ship feeder vessel
10,000			
16,000	T-2 (WW II)		
20,000	Handysize	Handysize	By vessel TEU, dwt, grt, lane meter capacity
30,000	50,000 cubic meter LPG		
45,000	Handymax 75,000 cubic meter LPG	Handymax	
55,000	125,000 cubic meter LNG		
75,000	Panamax	Panamax	Panamax
80,000	Aframax		Post-Panamax[2]
100,000	Supertanker	Capesize	
128,000		Hamptonmax	
140,000	Suezmax		
160,000	VLCC	VLBC	
200,000			
320,000	ULCC	ULBC	
350,000			
500,000	Mammoth		

[1]Not to scale.

[2]Beam (not necessarily dwt) exceeds Panama Canal limit of 105′ 6″.

Figure 1-2. Vessel standard size categories

trated orange juice carrier has complicated refrigeration capabilities; parcel tankers have elaborate tank, piping system, and pump segregation to permit the simultaneous accommodation of many different cargoes.

Parcel Tankers

Tankers for the transport of oil are classified as product carriers or crude carriers. Product carriers are designed to transport the products of the oil-refining process. Products cover a large range of viscosities, requiring different handling methods for gasoline, diesel fuel, bunker C fuel, and asphalt. Early product carriers were built in the range of 15,000 to 30,000 dwt. As the distance between refineries and consumers increased, the size of product carriers increased to take advantage of economies of scale. Product carriers of 100,000 dwt are now in service on the longer ocean trade routes. Such ships often have eight to ten cargo segregations in specially coated tanks. Additional flexibility is sometimes provided by fitting these ships with tank-heating capability and high viscosity cargo pumps to permit them to transport a wide range of cargoes. Asphalt is a very special cargo requiring special hull strengthening to accommodate cargo heating to 150°C.

Crude Oil Carriers

Crude oil carriers are single-purpose vessels designed to transport crude oil from the source to the refinery. The worldwide capacity of crude oil carriers is often expressed in cargo-carrying capacity equivalents of the 16,500-dwt T-2 tanker. The T-2 tankers, developed in the early 1940s, built and used during World War II, were very successful. Some were jumboized after the war and remained in service for many years. Technically the T-2 was a products carrier, but it was eventually used for many purposes. As suggested in Table 1-1, economies of scale have led to the design and construction of ever-larger crude oil carriers. Port limitations on vessel draft have been the only restraining factor on the maximum size of crude carriers. Despite the concerns for the environment in the event of a spill, the size of crude carriers is driven by the major costs to the owner, i.e., capital cost per deadweight capacity, fuel consumption, and labor, all of which are reduced per unit of cargo transported by increasing ship size.

TABLE 1-1. Typical Tanker Dimensions

dwt	*Length*	*Beam*	*Draft*
16,500	532′	70′	30.6′
100,000	861′	125′	49.6′
250,000	1,141′	170′	65.4′
500,000	1,130′	233′	82.0′

The size advantage is best illustrated by considering that, as cubic capacity increases with larger size tankers, the principal dimensions increase as the cube root of the capacity. Therefore, a 250,000-dwt tanker is five times larger than a 50,000-dwt vessel but the linear dimensions ratio of the larger vessel to the smaller vessel is the cube root of 5 or 1.7. Fuel economy per cargo unit is improved for larger tankers since the resistance of the vessel does not increase directly with the increase in cubic size for properly designed hulls. The admiralty formula (which states that the power requirements of similarly proportioned vessels is proportional to the ratio of displacements to the ⅔ power times the ratio of speeds cubed) can be used to make rough comparisons of power requirements for different size tankers. The approximate power required to propel a VLCC at the 15.7-knot speed would be slightly more than twice the power of the smaller vessel.

Crew costs are most apparent in considering economies of scale. Minimum crewing standards are established by the maritime regulatory authority of the registry country. Nevertheless, there is little or no difference among nations in crewing requirements for deep-sea vessels on the basis of ship size. The same number of crew members are required to operate a VLCC as a small vessel, hence the cost of labor per unit cargo decreases with vessel size. Vessel automation is another factor which contributes to reduction of crew size on large vessels where the substantial capital cost increase can be justified. The large size of crude carriers is one reason that remote Persian Gulf crude oil is a low-cost source of hydrocarbons today.

Double Hull Tankers

In 1990, the U.S. Congress passed the Oil Pollution Act which has become know as OPA 90. The act mandates double hulls for all new tankers operating within the 200-mile U.S. exclusive economic zone. Since in 1990 most tankers were still single hull, the act provides time for the industry to adjust to the new requirement by establishing a phase-out schedule for single-hull tankers. This schedule is shown in Figure 1-3. Tankers serving the Louisiana offshore oil port (LOOP) and others which offload to shuttle tankers more than 60 miles off the coast are exempted from the act. Predictably, OPA 90 has stimulated new tanker construction, since retrofitting double hulls to existing tankers is not a cost-effective option. For tankers exceeding 100,000 dwt the double-hull requirement adds about 12 percent to the hull steel weight. For smaller tankers, the steel weight increase is over 20 percent. Thus, for crude oil carriers, the higher cost of construction may increase the price of a ship by 10 to 20 percent.

DRY BULK CARRIERS

Seaborne dry bulk trade is measured in tonnes of cargo lifted and in tonne-miles (which is the product of tonnes lifted times the distance transported).

By either measure, the dry bulk trades have had a steady average growth of about 4 percent per year during the two decades from 1970 to 1990.

The major bulk commodities are iron ore, steam and coking coal, grain, bauxite, and phosphate rock. Iron ore has dominated the group but is expected to be surpassed by coal tonnage by the year 2000. Iron ore demonstrates the importance of employing tonne-miles as the most useful demand measure for bulk shipping capacity. In the period 1970 to 1990, annual iron ore sea transport tonnage increased from 240 to 360 million tonnes, a 50 percent increase, while for the same cargo movement, the

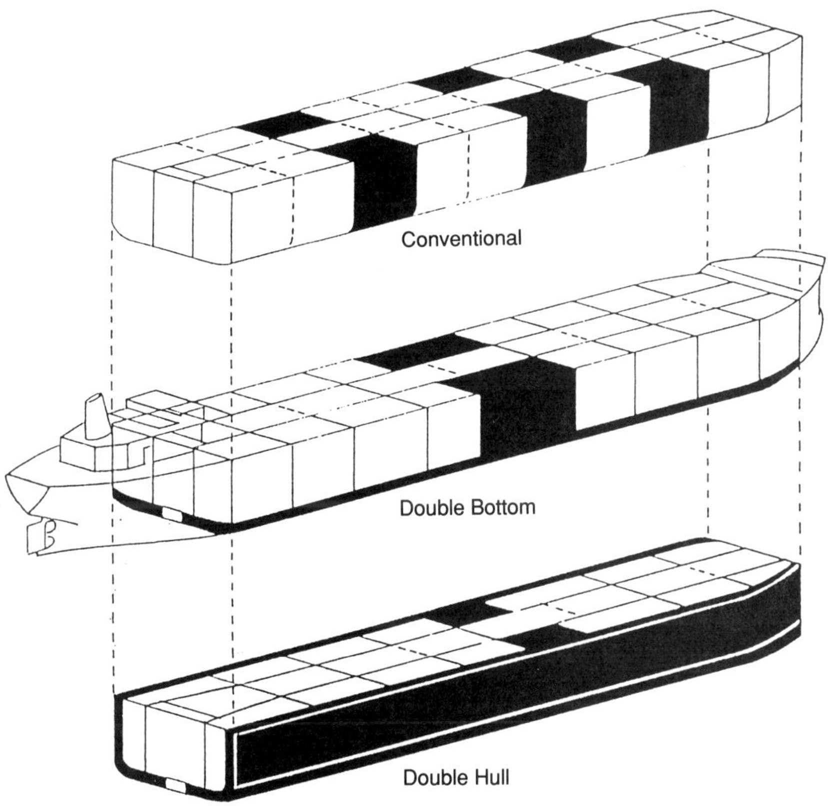

Single-skin tankers use cargo space efficiently; the entire enclosed hull volume holds oil, except for the segregated ballast located in wing tanks. Double-bottom hulls typically add tankage across the bottom of the hull. Conversely, double-side hulls (not shown) employ a single bottom skin and full-length wing tanks.

The full double-skin hull is configured with full-length ballast wing tanks and a double bottom; its attraction is that it offers a barrier totally enveloping the cargo volume.

The black areas in these drawings show the typical location of ballast spaces in the three types of tanker.

Figure 1-3a. Comparison of conventional, double-bottom, and double-hull tankers. Courtesy of American Bureau of Shipping

Year	*Vessels Under 5,000 gt*	*Vessels of 5,000 gt to 14,999 gt*	*Vessels of 15,000 gt to 29,999 gt*	*Vessels of 30,000 gt and Over*
		# Yrs or Older	*# Yrs or Older*	*# Yrs or Older*
1995		40	40	28
1996		39	38	27
1997		38	36	26
1998		37	34	25
1999		36	32	24
2000		35	30	23
2001		35	29	23
2002		35	28	23
2003		35	27	23
2004	Exempt until 2015	35	26	23
2005		25	25	23
2006		25	25	23
2007		25	25	23
2008		25	25	23
2009		25	25	23
2010				
2011			Must have Double Hull	
2012				
2013				
2014				
2015	Must have Double Containment System			

How to Read the Calendar

In all of the above cases, vessels with double bottoms or double sides are allowed an extra five years.

For all of the above, a vessel's age is determined from the date on which it was delivered after original construction, after completion of a major conversion, or qualification for documentation under the "Wrecked Vessel Act," whichever is later.

The delivering vessel in a lightering operation (the larger vessel bringing oil from another location) is exempt from the double hull requirement until 2015.

Vessels calling at LOOP (Louisiana Offshore Oil Port) are exempt until 2015.

All vessels 5,000 gross tons and above, for which a building contract or contract for major conversion was placed before 30 June 1990 and that are delivered under that contract before 1 January 1994, must adhere to this phase-out schedule. Vessels that have appraised salvage values determined by the United States Coast Guard before 30 June and qualify for documentation under the Wrecked Vessel Act before 1 January 1994 must also adhere to the schedule. That is, such a vessel may operate with a single hull until the deadline applicable to its respective size and age.

All vessels 5,000 gross tons and above, for which a building contract or contract for major conversion is placed on or after 30 June 1990 and that are delivered under that contract on or after 1 January 1994 must be equipped with double hulls. Vessels that have appraised salvage values determined by the United States Coast Guard on or after 30 June 1990 and qualify for documentation under the Wrecked Vessel Act on or after 1 January 1994 must also be fitted with double hulls.

Figure 1-3b. Schedule for phase-out of single-hull tankers. Courtesy American Bureau of Shipping

tonne-miles increased from 975 billion to about 1,950 billion, a 100 percent increase. Though lower in magnitude, other commodities experienced similar trends as users of raw materials have implemented strategies for diversifying sources of supply.

Table 1-2 shows the annual growth rate in dry bulk shipments.

TABLE 1-2

Annual Growth in Dry Bulk Shipments

	Major Bulks		*Minor Bulks*		*Total*	
	Tonnes	*Tonne-Miles*	*Tonnes*	*Tonne-Miles*	*Tonnes*	*Tonne-Miles*
1970	500	2.1	195	1.0	695	3.1
1990	950	5.1	600	2.6	1,550	7.7
Annual growth	3.3%	4.5%	5.8%	4.9%	4.1%	4.6%

Steel mills in Japan now obtain ore from Sweden, South America, and Australia. Electric utilities, such as Taiwan Power Corporation, obtain steam coal from the Americas, South Africa, and Australia. The economies of large-size bulk carriers make such source diversification strategies feasible. The average cost per tonne of raw material is higher, but the risk-adjusted cost per tonne is lower than would be experienced when depending on single source materials. For steel mills and power plants, disruption of production is far more costly than the slightly higher prices from supply diversification. There is also the price advantage from competitive bidding when multiple sources are utilized.

Tonnage of other major bulk commodities has also increased. As a group, grain, bauxite, and phosphate rock composed about 275 million tonnes in 1990 compared to 150 million tonnes in 1970. Tonne-miles have also increased in the same period from 600 to 1,400 billion tonne-miles. Grain shipments are subject to unpredictable factors such as weather effect on harvests and political decisions affecting the availability of export credits.

The minor bulk trades include nonferrous ores, limestone, gypsum, cement, sulfur, potash, salt, sugar, rice, tapioca, and numerous others. Individual commodity shipping volumes are modest, but the group comprises nearly 40 percent of all seaborne dry bulk trade. Growth rates of the minor bulk trades are not precisely known because the cargo base is not static. There is a natural tendency to dedicate specialized shipping capacity to individual cargoes when volume levels become large enough to support ship-size lots. There is also a trend to replace multipurpose general cargo vessels with container vessels in most liner trades. As a result, the handysize dry bulk carrier with cargo gear has assumed the former role of break bulk general cargo vessels for the shipment of small consignments of dry bulk cargoes. Table 1-3 illustrates the composition of dry bulk seaborne trades.

TABLE 1-3

Dry Bulk Seaborne Trade Cargoes

Cargo	*Percent of trade*
Phosphate rock	2.30
Bauxite	3.10
Iron ore	22.30
Grain	13.30
Coal	20.70
Minor bulks	38.30
Total	100.00

The principal types of dry bulk carriers are handysize, panamax, and capesize, shown in Figure 1-4. Handysize are so designated because of the versatility provided by the onboard cargo-handling gear and the shallow draft which permits entry into most smaller ports. Handysize carriers operate predominantly in the minor bulk trades. When these ships are fitted with specially strengthened hatch covers, they can carry containers on the deck.

Panamax-class bulk carriers are capable of carrying up to 80,000 deadweight tons of cargo. However, this deadweight requires the full Panama Canal depth to accommodate a draft of 41 feet. In summer months, the canal transit vessel draft may be limited to 37 feet which reduces the vessel deadweight capacity to 72,000 dwt. Panamax bulk carriers are used extensively to carry coal from U.S. East and Gulf ports to Asia via the Panama Canal, although many are employed in trades which do not require transit of the Panama Canal. Panamax size vessels do not normally have onboard cargo-handling gear hence they are called gearless. As ship size increases, the number of cargoes and/or ports which can be served decreases. This makes it economical to position cargo-handling equipment in the shore terminal, resulting in an increase in the utilization of the cargo-handling gear and the ship's deadweight capacity.

Capesize ships are so named because they are commonly used in trades where they transit between oceans via the Cape of Good Hope. The capesize ships are used almost exclusively in the iron ore and coal trades, although some cargoes of grain and tapioca are also lifted, often to defray vessel repositioning costs. Of the 5,000 dry bulk carriers in the world fleet, less than 500 exceed panamax size. Nevertheless, these capesize and larger vessels transport one fourth of all dry bulk cargo. For some dedicated long haul trades such as the transportation of iron ore from Brazil to Japan, dry bulk carrier ship designs have exceeded capesize to an extent that a new vessel class will be defined as a very large or ultra large dry bulk carrier.

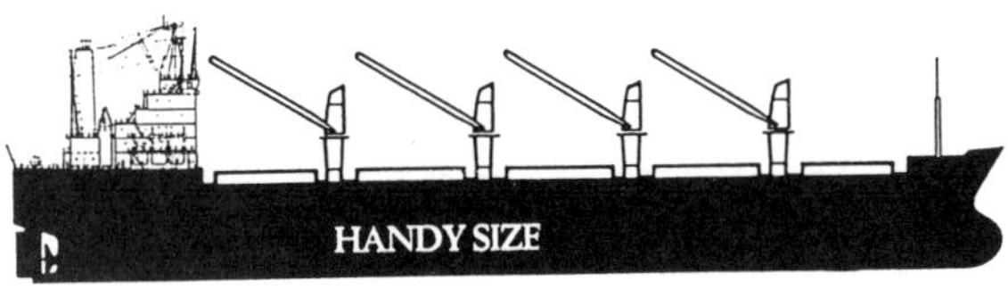

10,000 – 50,000 dwt

LOA 155 M, Beam 24.5 M, Draft 10.7 M
BHP 9,000
Speed 15.0
DWT 30,000

50,000 – 80,000 dwt

LOA 225 M, Beam 32.2 M, Draft 12.5 M
BHP 12,500
Speed 14.5
DWT 65,000

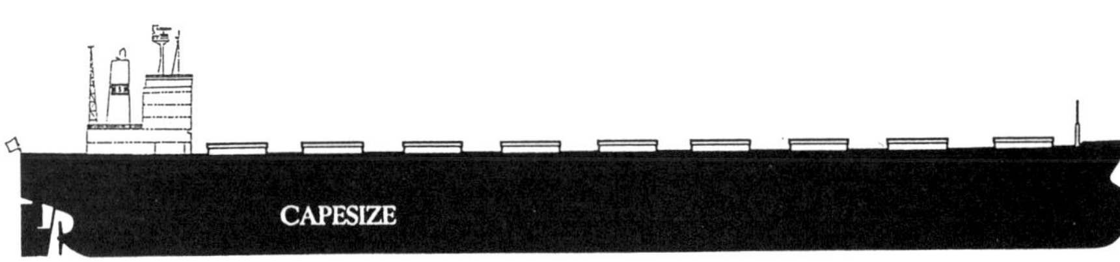

80,000 – 150,000 dwt

LOA 270 M, Beam 43.0 M, Draft 17.3 M
BHP 16,900
Speed 14.0
DWT 149,000

Figure 1-4. Typical designs for classes of dry bulk carriers

Aging Capesize Fleet Replacement. Renewal of the world's merchant fleet in general, and specific vessel types in particular, is unevenly distributed over time. This is the result of the interaction of complex forces generated by shipowners' expectations of future economic activity, availability of shipyard building berths, interest rates, credit terms, changes in regulatory requirements, vessel scrapping rates, and the actions of shipping competitors. Classes of vessels may experience building cycles which skew the age of the fleet. For example, illustrating shipowners' anticipation of very large increases in the coal trades in the early 1980s, Figure 1-5 shows that many orders for capesize vessels were placed in the late 1970s. This ageing tonnage in 1992 formed a disproportionate share of the global capesize fleet. During the early 1990s, a high incidence of capesize vessel structural failures prompted the International Association of Classification Societies to publish information on specific ship structure which requires the special attention of ship surveyors as shown in Figure 1-6.

Anticipating *en bloc* obsolescence of a large portion of the capesize fleet coupled with improving economic conditions in North America and Europe, owners in the early 1990s placed many orders for capesize vessels. Figure 1-7 illustrates the typical specifications for a popular capesize vessel for trades with the Asian market. Another buildup cycle has begun and global shipbuilding capacity which has been shrinking since the mid-1970s is projected to be in short supply by 1995. In the typically cyclical nature of international shipping, the possibility exists that shipbuilding capacity will expand to build an excessive number of speculative capesize vessels which will ensure future poor rates of return for all owners. Engineers employed in shipbuilding and ship operations who are skilled in measuring and predicting financial risks and rewards can help to develop more rational strategies for the future.

Specialty Tonnage

The efforts to improve shipping economics is neverending. The OBO (ore/bulk/oil) vessel, also called a combination carrier, and the dry bulk carrier with hoistable vehicle decks shown in Figure 1-8 are examples of novel designs for increasing revenue by decreasing time underway in ballast. The log carrier with cargo gear and the square box section for low density wood chips also shown in Figure 1-8 are special designs for increasing effectiveness and revenue in specialty trades. Such vessels would have serious competitive disadvantages if operated outside the trades for which they have been specifically designed. Self-unloading dry bulk carriers illustrated in Figure 1-9 are used in trades where faster vessel turnaround is advantageous despite the burden of higher vessel construction costs and the reduced cargo capacity due to the deadweight of the cargo-handling gear. Self-unloading capability is most attractive for short trade routes where the gear has high utilization and the reduction in port time permits

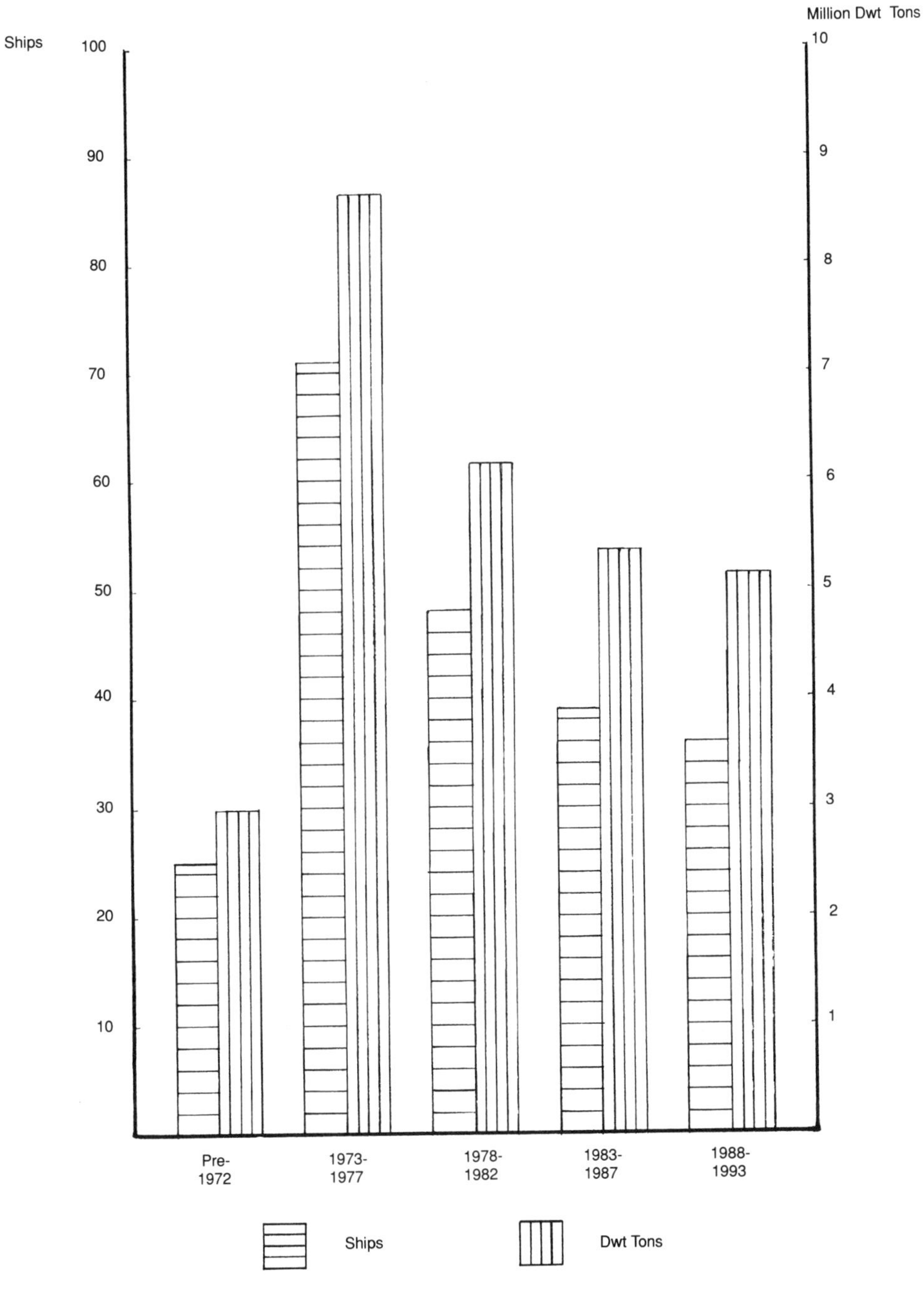

Figure 1-5. World capesize fleet tonnage (100,000 – 150,000 dwt) and age distribution. Courtesy Oak Steamship Company

more voyages per year. Short distances and an abbreviated shipping season are the reasons that self-unloaders are popular in the Great Lakes iron ore, coal, and cement trades. Large seagoing self-unloaders are also employed in dedicated closed-loop services.

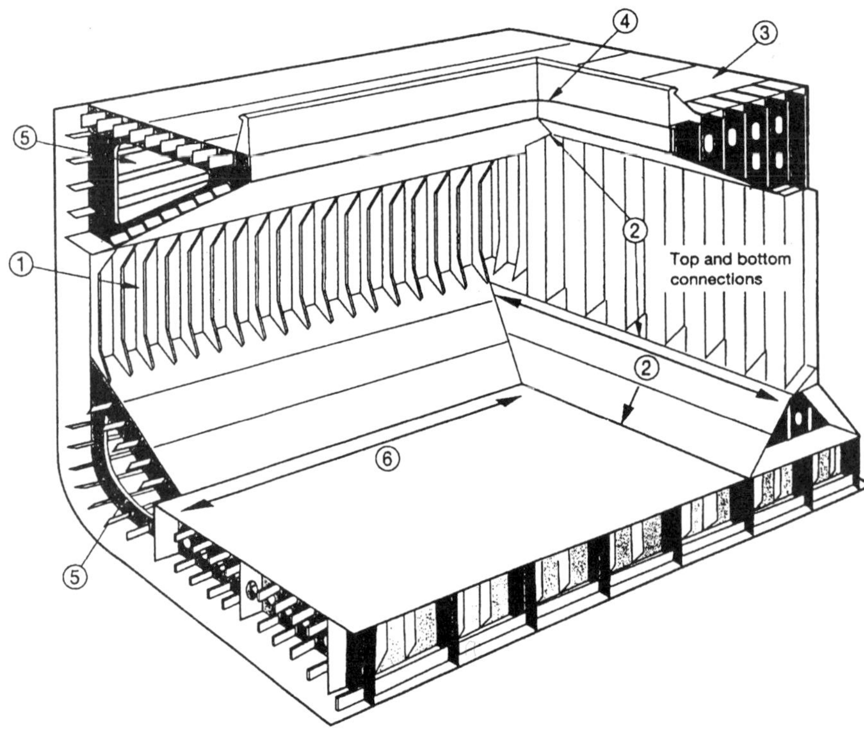

1. Hold frame — connection to upper and lower wing tanks and side shell.
2. Boundaries of transverse bulkheads and bulkhead stools.
3. Cross deck structure.
4. Hatch corners/hatch coaming brackets.
5. Localized cracking and buckling of web frames and breakdown of coatings in water ballast tanks.
6. Inner bottom plating/hopper plating intersection.

Figure 1-6. Problem areas of ageing capesize bulk carriers. Courtesy of I.A.C.S.

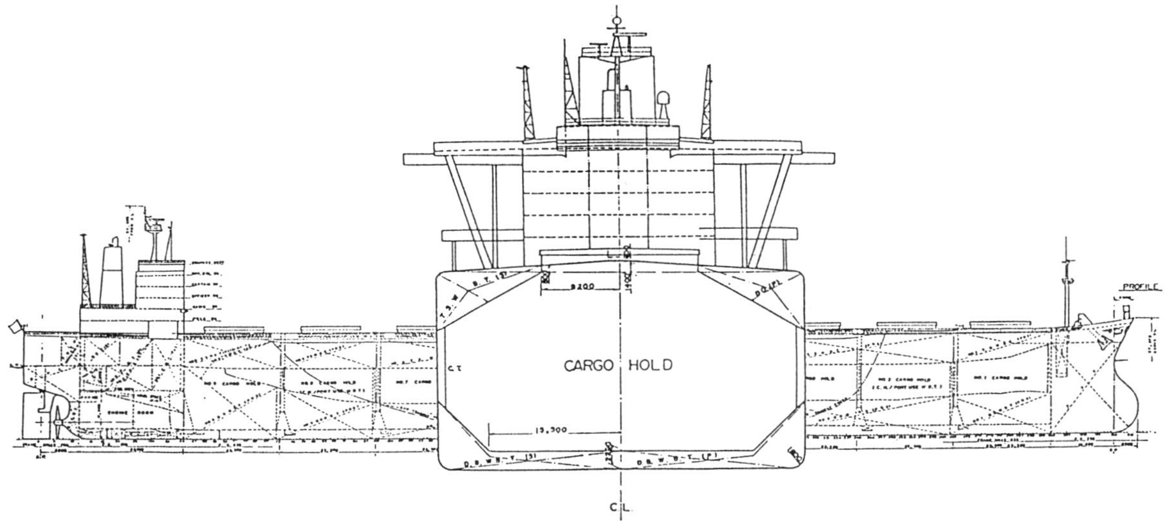

Length over-all	270.00 M
Length between perpendiculars	261.12 M
Breadth moulded	43.00 M
Depth moulded	23.90 M
Draft (1966 LLC. moulded)	17.30 M
Corresponding deadweight	149,000 M-tons
Gross tonnage	77,090
Type	Single screw, flush decker, self trimming bulk carrier with aft. engine
Cargo capacity	264,500 M^3

Main engine type	M.A.N. B&W 5L80MCE
Max. continuous rating	16,900 BHP at 88 RPM
Normal output rating	14,370 BHP at 83 RPM
Speed: Trial	15.90 Knots
Service	13.90 Knots
Cruising range	26,500 N. Miles
Classification:	Germanischer Lloyd GL+100A4 bulk carrier, strengthened for heavy cargo holds No. 0,2,4,6 and 8 holds may be empty+Mc Aut.

Figure 1-7. Capesize dry bulk cargo vessel specifications most popular for Asian service. Courtesy Oak Steamship Company

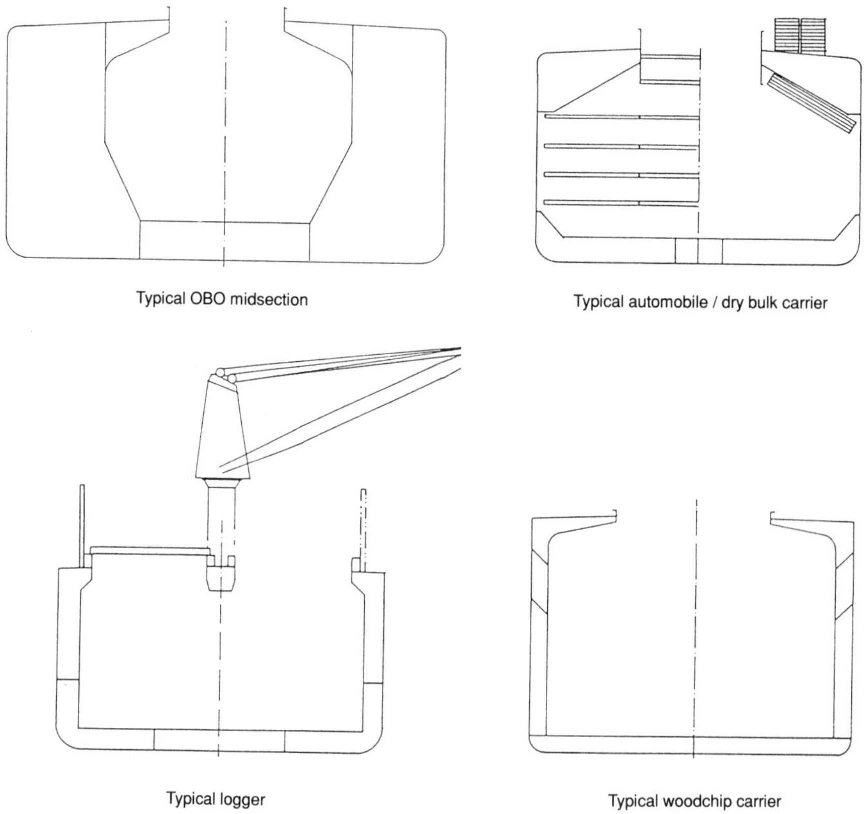

Figure 1-8. Midsections of specialized bulk carriers. Courtesy SNAME

GAS CARRIERS

The LNG Ship

Seaborne transport of natural gas began in 1957 with conversion of two American C-1 general cargo freighters to carry 5,000 cubic feet of LNG each. LNG is liquefied natural gas containing mostly methane, but also present are smaller amounts of ethane, propane, and butane. “Wet” or “associated” LNG contains larger quantities of butane and the heavier hydrocarbons. “Dry” or “unassociated” gas is predominantly methane. The properties of some liquified gases at atmospheric pressure are shown on page 1-18.

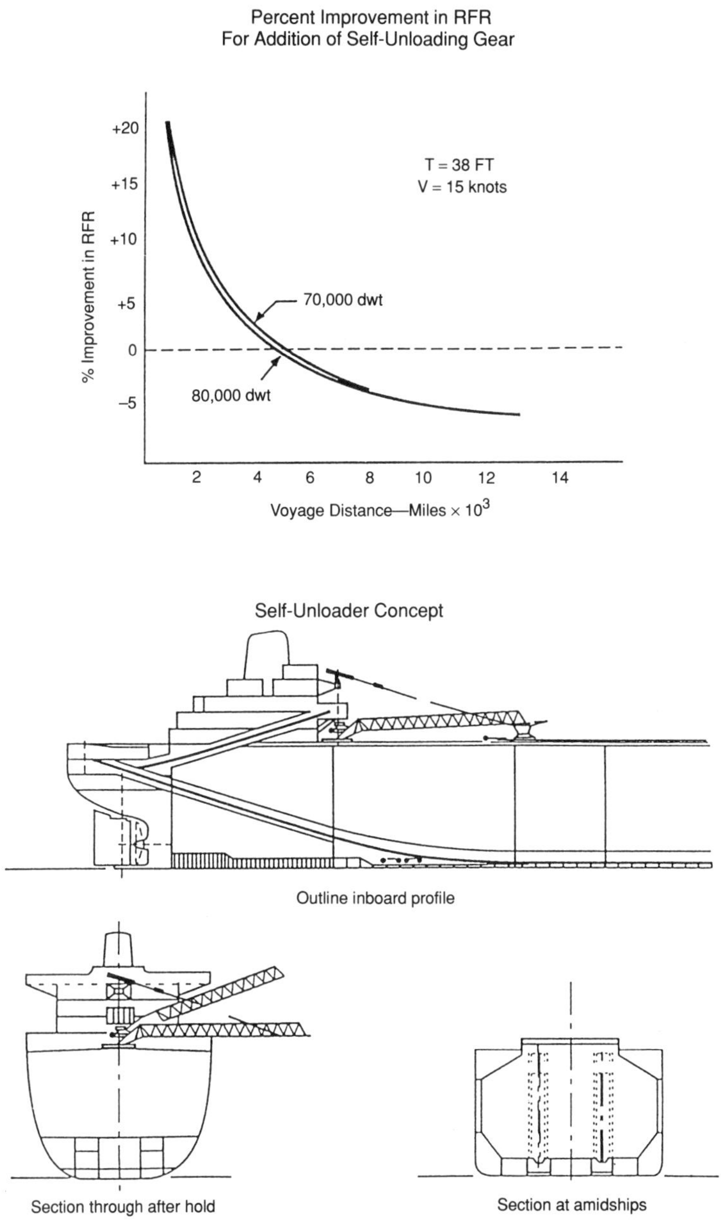

Figure 1-9. Self-unloading dry bulk vessel and percent improvement in RFR for addition of self-unloading gear. Courtesy David Taylor Research Center

TABLE 1-4

Properties of Some LNGs at Atmospheric Pressure

	Boiling point, °C	*Density*
Nitrogen	-196	.81
Methane	-162	.42
Ethylene	-104	.57
Ethane	-89	.56
Propane	-42	.59
Butane	-.5	.60

Liquefaction of natural gas is achieved by a cascade vapor-compression cycle which lowers the temperature to -165°C (-260°F) under atmospheric pressure, or to -82°C with increased pressure. LNG is a colorless, odorless, nontoxic liquid which occupies about $\frac{1}{620}$ of the volume of the gaseous equivalent mass. Depending on its constituents, LNG ranges between .42 and .48 of the density of seawater, which is slightly heavier than crude oil. Therefore, a 125,000-cubic-meter LNG carrier has a cargo deadweight of 55,000 tons but the principal hull dimensions will approximate those of a 120,000-dwt crude carrier. There will be considerable differences in the power of the propulsion systems also, since a .2 percent cargo boil-off every 24 hours provides the economic justification for a faster vessel. Steam turbine propulsion continues to be the plant of choice for LNG carriers since the cargo boil-off gases can be utilized as fuel in a properly designed dual-fuel boiler. While modern 120,000-dwt crude oil carriers typically have propulsion powers less than 30,000 SHP, LNG carriers of the same physical size have higher power to provide speed of 19 or more knots compared to the 15- to 16-knot speed of a crude oil carrier. LNG carriers have finer hulls, i.e. a block coefficient of less than .75, compared to higher than .8 for a crude carrier to accommodate the higher speed. The standard size 125,000-cubic-meter LNG carrier was defined by the largest terminal facilities for these vessels at the time they were developed in the early 1970s. Some LNG ships proposed in the early 1990s have capacities of up to 200,000 cubic meters to take advantage of the economies of scale which will be discussed in subsequent chapters.

Growth in the LNG trade has been rapid, expanding from 34 billion cubic meters per year in the early 1980s to 77 billion in 1992. Seventy percent of the market demand comes from Asia, with Japan as the principal trade destination, followed by Taiwan, Korea, and Hong Kong. The remaining market demand is mostly from Europe. The major suppliers are Indonesia, the Persian Gulf states, and North Africa. The United States has ample sources of natural gas in North America so that LNG shipments have not become a significant trade. However, due to the very high environmental acceptance of natural gas it is expected that this trade will continue to grow, supplying more than the 20 percent of world energy presently shipped to the Far East, Europe, and America.

Figure 1-10 is a list of anticipated future LNG trading routes. Seaborne trade in LNG is forecast to exceed 100 billion cubic meters annually by the year 2000 and to double the 1990 level by the year 2005. Such a trade will require the current world LNG fleet of 7.5 million cubic meters to be doubled. This will be a challenge to shipbuilders, owners, and finance organizations since LNG carriers are the most expensive cargo ships in the world—US$ 275 million to 300 million for a 125,000-cubic-meter vessel built in the Far East for delivery in 1993. A comparatively small number of shipyards are capable of building LNG carriers, including two in the United States.

LNG Shipping Distances on Actual / Possible Supply Routes

Producing Country	*Market*	*Distance (Nautical Miles)*	*Round Trip (Days)*
Alaska	Japan	3,280	18.5
Alaska	USA (West Coast)	2,160	13
Algeria (Arzew)	France (Le Harve)	1,210	9.5
Algeria (Arzew)	UK	1,540	10
Algeria (Skikda)	France (Fos)	400	4.5
Algeria (Skikda)	USA (East Coast)	3,470	19.5
Brunei	Japan	2,400	14
Indonesia	Japan	3,250	18.5
Libya	Italy (La Spezia)	990	7
Libya	Spain (Barcelona)	1,060	8
Libya	USA (East Coast)	4,540	25
Nigeria	Argentina	4,390	24
Nigeria	USA (East Coast)	5,080	27.5
Persian Gulf	Japan	6,770	36
Trinidad	USA (East Coast)	1,930	12
Venezuela	Argentina	4,860	26
Venezuela	USA (East Coast)	2,000	12

Figure 1-10. Future LNG trading routes

Cargo Containment Systems. The distinguishing feature of an LNG ship, and the most costly, is the cargo tank system. Because of the cryogenic temperatures, the tanks must be made of materials which require special technology to fabricate and install. Depending on the designer, materials such as aluminum alloy, stainless steel, and Invar have been employed. Two general approaches to design are offered: the free-standing or self-supporting type and the membrane, semimembrane, or integrated type. Free-standing tanks are independent of the vessel's hull and are capable of supporting their own weight as well as the static and dynamic loads of the liquid cargo. The free-standing tanks are supported by conventional ship structure. The Moss Rosenberg spherical type tanks and the Conch prismatic design tanks are examples of the free-standing tanks.

Membrane and semimembrane integral tanks are made of light gage metal alloy incapable of self-support. Gaz Transport, Technigaz, Hitachi Zosen, and McDonnell Douglas have offered variants of this technology. The tanks are supported by the ship's structure through load-bearing insulation. A secondary insulating barrier over the entire tank exterior surface protects the ship structure from the cargo cryogenic temperature. Figure 1-11 shows the semimembrane tank arrangement designed for liquid propane gas cargo.

LPG Ships

Liquid propane gas carrier is the name applied to a ship designed to carry propane and butane and, with some modification, other cargoes such as ammonia, butadiene, propylene, ethane, ethylene, and vinyl chloride. LPG ships, which usually have cargo capacity from 50,000 to 75,000 cubic meters, are far more numerous than LNG vessels. Since the boiling point of LPG is –42°C, much less insulation is required compared to the LNG carriers. The smaller size is a result of trading patterns. Propane and butane are produced in large quantities when crude oil is refined. This creates an intraregional transport demand, such as within Europe and between the Caribbean and North America. If oil-producing regions such as the Middle East continue to increase refining capacity, LPG routes may be expected to lengthen, leading to a demand for larger LPG carriers.

CARGO SHIPS

The term general cargo ship describes a design conceived to transport diverse cargoes, including some passengers, in trade routes ranging from shuttle, radial, or triangular, to tramping. Vessels maintaining regularly scheduled common carrier service with published tariffs are called liners. Vessels employed by their owners to transport single cargoes in ship-size lots in whichever trade offers the greatest earnings are called tramps. Distinction between liner and tramp services are sometimes blurred by the recent development of con-bulkers and semiliners. These ships typically transport dry bulk cargoes on one voyage leg and general cargo, often in containers, on another leg. The handysize dry bulk carrier fitted with cargo-handling gear, open box-shaped holds, and strengthened hatch covers to maximize container capacity, as shown in Figure 1-12, typifies

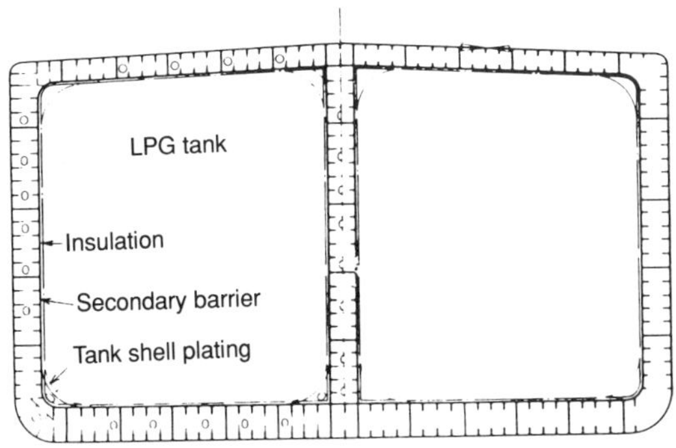

Midship section of a semi-membrane tank

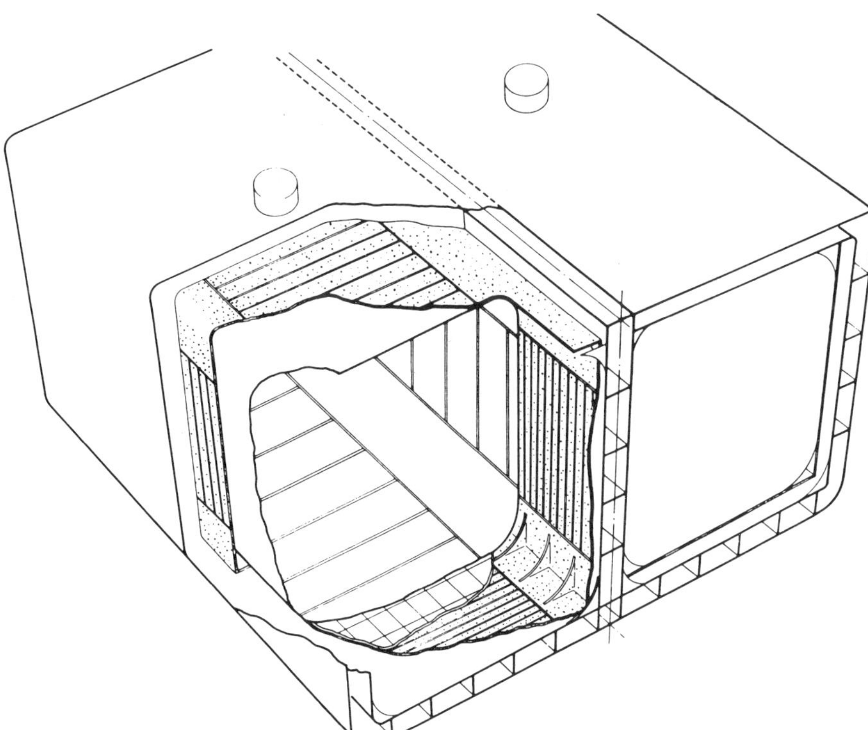

Inside view of a semi-membrane tank

Figure 1-11. Semimembrane cargo tank for liquid propane

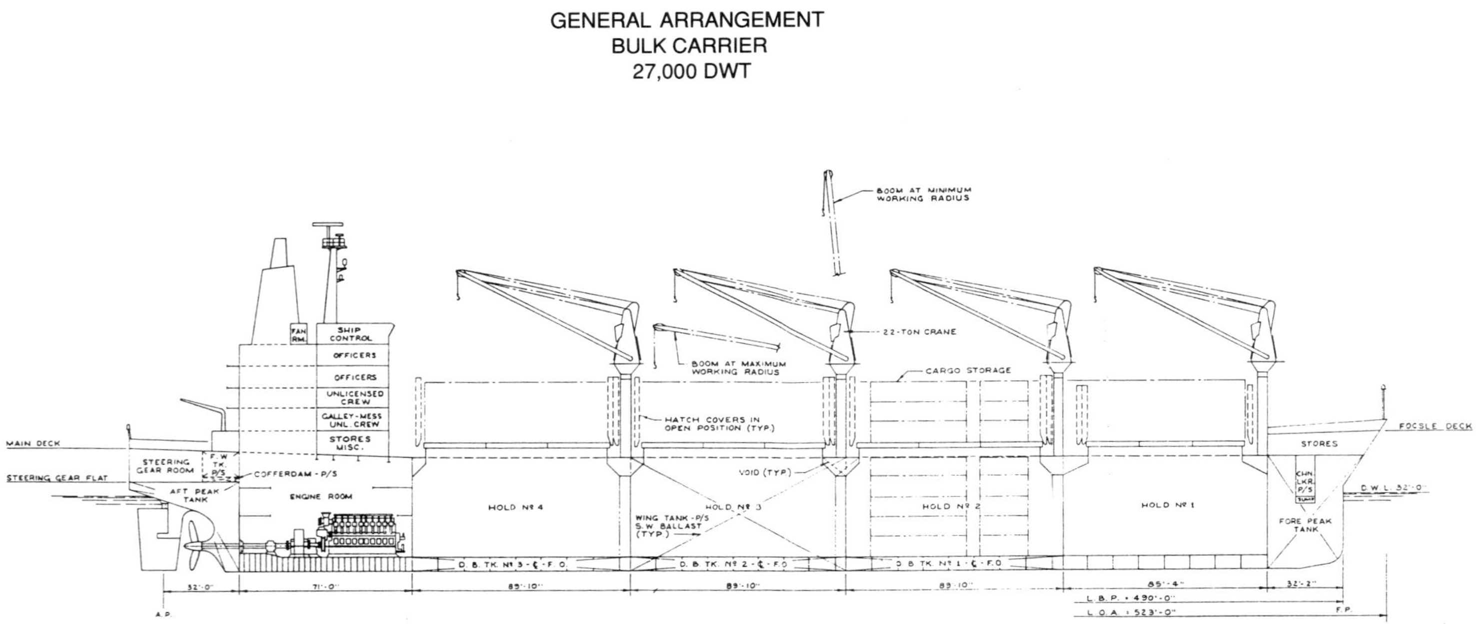

Figure 1-12. Typical handy-size dry bulk carrier. Courtesy SNAME (continued on facing page)

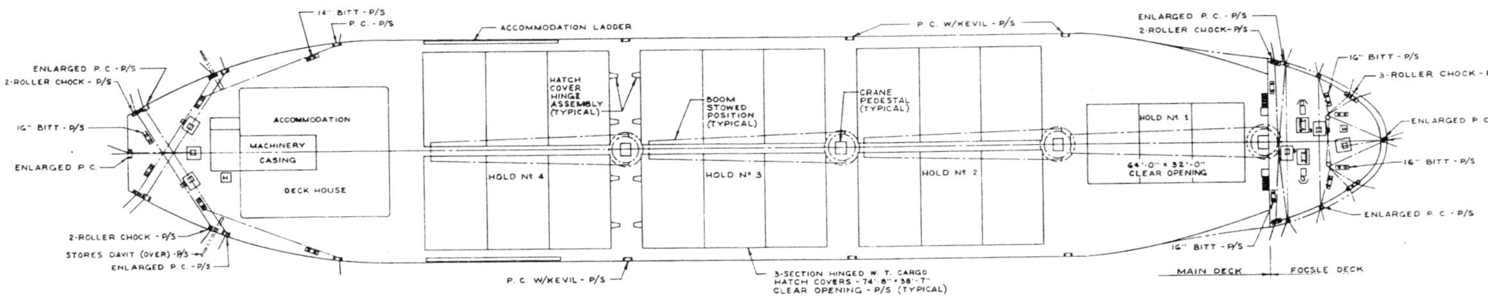

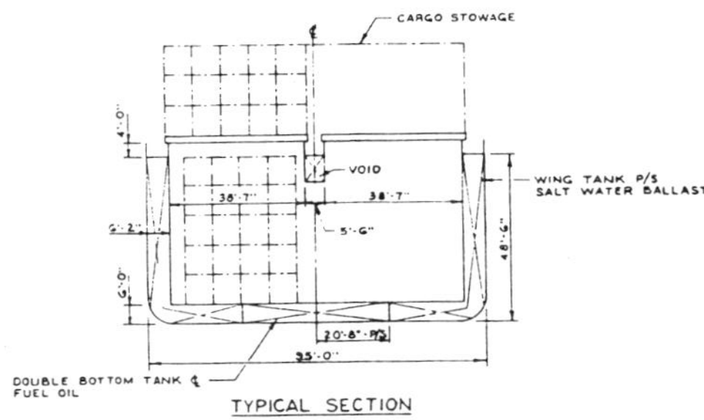

Capacities

Cargo, grain, CU FT	1,181,000
Ballast, 100% LT	10,280
Fuel oil, 98% LT	1,541
Diesel oil, 98% LT	360
Fresh water, 100% LT	200

Principal Characteristics

Length Overall	523.0
Length between perpendiculars	490.0
Beam, molded	95.0
Depth, molded	48.6
Draft, molded (design)	32.0
Deadweight @ 32.0 Draft, LT	24.233
Draft, molded, summer freeboard	35.0
Deadweight @ 35.0 Draft, LT	27.603
Horsepower MCR BHP	12.184
Trial Speed @ MCR	16.0

Figure 1-12 (continued).

vessels that can serve both bulk and general cargo markets. Emergence of neo-bulk cargoes further encouraged this dual capacity. (Neo-bulk are cargoes that did and still do move in smaller lots aboard liner vessels but which because of continuing growth in volume in some trades are now transported more economically in ship-size lots. Such cargoes include automobiles, semifinished steel products, unrefined rubber, and the range of forest products from logs to newsprint.)

Cargo in crates, casks, cartons, drums, or bags, etc., loaded aboard by slings or pallets using the ship's cargo-handling gear is collectively called breakbulk cargo. Stowage in the hold and between deck spaces is done manually. Consequently, loading and unloading is time consuming and labor intensive with emphasis. The emphasis is on maximizing the use of the vessel's internal volume and deadweight capacity. Modern general cargo ships encompass a range of designs that participate successfully in the 1.3 billion tons per year of general cargo movement. Different designs often operate in the same trade and compete for the same cargoes. Differences in applied technologies among the ship designs may be attributed to an owner's preference and experience, the targeted base cargo, or the strategic assessment of the directions and dimensions of future trade growth.

Owner specification of a new ship encompasses compromise between numerous and sometimes contradictory objectives. In any given trade or service one type of general cargo vessel ship may be more productive or profitable than another. Preliminary or parametric ship design investigations are simplified through use of sophisticated computer models that enable engineers, economists, and managers to study many more design and economic variables than was previously possible. A typical general cargo liner service where designs are ranked according to level of service provided to the shipper would include roll-on/roll-off, containerships, barge carriers, and multipurpose ships.

Roll-on/Roll-off

RO/RO ships offer shippers the most expeditious service. Cargoes transported over highways in trailers are driven directly aboard the ship via ramps which can be shore based or carried aboard ship. (Ships will normally carry their own ramps except in dedicated shuttle service where shore-based ramps are more productive.) A disadvantage of the RO/RO is the inefficient use of vessel cargo space. Truck trailers cannot be stowed effectively due to space required for maneuvering. Furthermore, the wheeled chassis require large amounts of the ship's cubic capacity. As a result, RO/RO ships usually attract high value freight which can obtain higher freight rates for a superior service including high speed and quick port turnaround.

Containerships

Containerships transport cargo in unitized vans without the trailer chassis. Van width is almost always 8 feet, while the height and length vary. Standard height is 8 feet 6 inches, however, for low-density cargoes 9-foot 6-inch-high containers are available and "half-height" containers are available for very high density cargoes. Standard lengths are 20 feet, 40 feet, and 45 feet. The original containers used by Sea-Land Service, Inc., were 35 feet since that was the largest trailer permitted on U.S. highways in the mid-1950s.

Matson Navigation Co. also pioneered containerization in the Hawaiian trade during the 1950s and selected a nonstandard 24-foot container as the optimum for their cargo mix. For convenience, containership capacity is measured in TEUs, 20-foot equivalent units. Average per-ship capacity on major trade routes has grown steadily from about 2,000 TEUs in the 1970s to 3,000 TEUs in the 1980s. In the 1990s, 4,000-TEU containerships have made an appearance. These vessels are too large to transit the Panama Canal. Maximum gross weight for standard 40-foot containers is about 67,000 pounds (8,000 tare plus 59,000 payload). Gross weight for the 20- foot container is 5,000 pound tare plus 40,000 pound payload. For specialty cargoes some containers have open tops, tanks, decks, or refrigeration, as the technology of intermodalism continues to expand, dominating the general cargo trades and penetrating into smaller special trades.

Containerization's attraction to the shipowner is a vast increase in productivity coupled with the near elimination of cargo damage and pilferage. The replacement of labor intensive breakbulk ships with capital-intensive intermodal transport systems has required massive investment: new ships, huge fleets of containers and land transport chassis, highly productive container cranes, vast terminals, sophisticated computer systems, and new management procedures. Transition on such a massive scale has forced smaller shipping lines to merge or form consortia. Those unable to achieve and sustain the economies of scale needed to compete have disappeared. Individual carrier expansion is driven by the need for ever larger cargo volumes to achieve superior economic performance and by the relatively small incremental cost to serve new markets once an intermodal transport system is in place. In some trades, owners have compromised between higher service RO/RO and the more productive containership service by introducing ships which combine both technologies as shown in Figure 1-13. One design approach carries containers on the weather deck and RO/RO cargo in the hold. Another approach divides the ship transversely; the forward section is fitted with cell guides to the tank top and the aft section is fitted for RO/RO trailers.

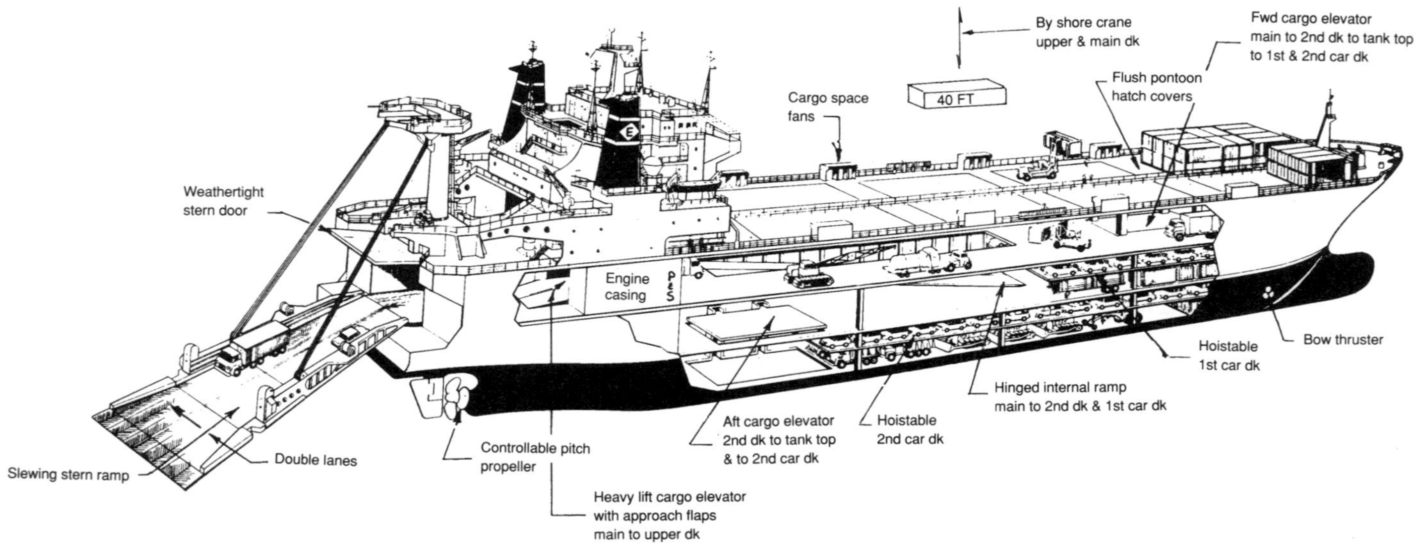

Figure 1-13. Roll-on/Roll-off vessel. Courtesy Avondale

Barge Carriers

Barge-carrying vessels (BCV) are hybrid designs that combine elements of unitization and break-bulk or neo-bulk shipping. The lighter aboard ship (LASH) system can transport up to 89 barges of 400 tons each. The Lykes Seabee design is capable of carrying 38 barges of 1,000 tons each. A similar Finnish-built ship transports 26 barges of 1,300 tons each. The LASH system loads barges over the stern using a rail-mounted gantry crane. Barges are stowed six high. Seabee systems lift barges over the stern by elevator to one of three decks for stowage.

Barge-carrying ships were developed to interface with extensive inland waterways in North America, Europe, and Asia. Dispersal of barges to multiple inland terminals and the ability to load/discharge the mothership in stream greatly expand geographic cargo draw area and speed up the port turnaround time. Barges typically carry breakbulk and neo-bulk cargoes. BCVs have not been successful in attracting high-value container cargo nor have they achieved the low cost levels required to compete as dry bulk carriers. This has tended to limit their employment, leading to the conversion of some LASH vessels to carry a partial load of containers. Notwithstanding, even larger BCVs have been proposed for the future. The proposed jumbo barge carrier would transport only six or eight barges of up to 30,000 dwt each configured to carry containers, RO-RO, or bulk cargo as shown in Figure 1-14. For this vessel, barges would be loaded and unloaded using the float-on/float-off method by ballasting down the carrier vessel to form a dock. Lower handling costs for fewer barges coupled with the economies of scale are the attractive features of this system, which has not as yet been tried in the marketplace.

Multi-purpose Ship

The cycle of trade expansion leading to cargo segmentation and vessel specialization is continuous. However, it is unlikely that the process will ever entirely eliminate the multipurpose breakbulk vessel fitted with cargo-handling gear. Productivity improvement features, including pedestal crane, side loaders, internal skids, etc., are continually being incorporated into break bulkvessels. It is estimated that breakbulk cargoes represent up to 20 percent of cargo moved in some liner trades. The high capital cost of an intermodal system and the abundance of low cost labor pools in many parts of the world seem to assure that the flexible multipurpose breakbulk cargo vessel will continue to trade for some time to come.

FREIGHT RATES

During the period 1979 to 1992, the average increase in freight rates for all types of ocean transport service was 3.096 percent per year. If we accept

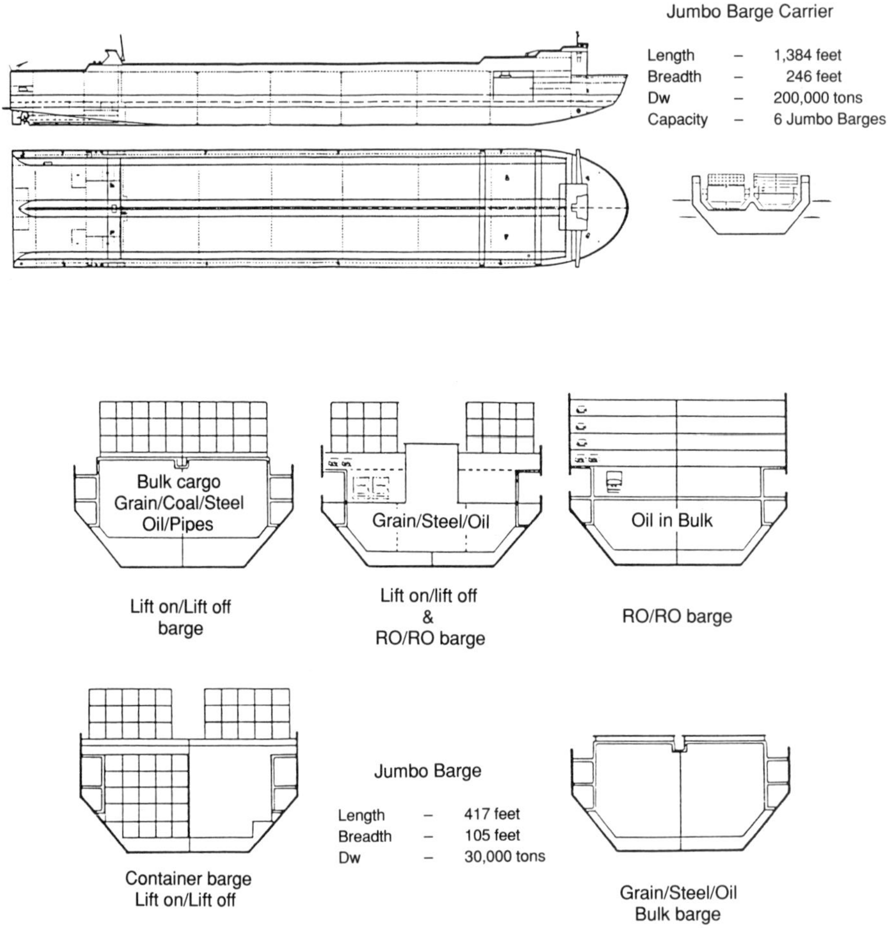

Figure 1-14. Arrangement of jumbo barge carrier and five basic barge types providing multiple cargo combinations. Courtesy MarAd

the assumption that this average rate will continue in the near future, it is a useful quantity for estimating future freight rates.

MAJOR MERCHANT FLEETS

In terms of total deadweight tonnage, the fleets of Cyprus, Greece, Japan, Liberia, and Panama were the major ocean transport fleets of the world during the period 1987 to 1992. It is likely that these fleets will continue to dominate ocean shipping. Table 1-5 is a listing of the total deadweight tonnage of these major fleets compared to the U.S. flag private fleet.

TABLE 1-5
Major Merchant Fleets of the World
Compared to the USA Fleet
(Total Deadweight, Thousands of Tons)

Flag	*1987*	*1988*	*1989*	*1990*	*1991*	*1992*
Cyprus	21,481	30,318	31,832	29,729	34,171	35,746
Greece	45,910	38,387	37,130	36,537	40,653	43,184
Japan	51,523	44,819	39,699	36,237	35,232	33,376
Liberia	94,406	93,537	89,200	88,275	92,057	95,552
Panama	66,119	68,884	72,977	70,537	67,938	74,905
U.S.A.	20,727	21,200	21,601	20,439	20,771	19,719

SERVICE INDUSTRY

There are a variety of organizations which exist solely to serve and/or support the water transport industry. These companies range in size from major international firms to small local companies. In the economic analysis of shipping ventures the costs of these service industries must be considered.

Insurance

Insurance coverage for loss of ship and cargo has existed from the beginning of water transport. Most maritime nations have enacted into law the division of responsibilities among the ship operators, the ship, and the cargo owners. The hull policy describes in detail the provisions to safeguard the vessel in colorful and archaic language. Such language which has been tested and defined by litigation includes protection from "adventures and perils of the seas, men-of-war, fire, enemies, pirates, rovers, thieves, jettisons, letters of marque, surprisals, takings at sea, arrests" and numerous other modern and ancient risks of a vessel at sea. The owner purchases the amount of insurance needed to protect his or her interest, being careful to avoid the potential problems of underinsurance and the cost problems of overinsurance. Protection and indemnity (P & I) insurance is purchased to protect the owner against liability for loss of life, personal injury, and damage to cargo. This is a more modern type of insurance based on the owner's membership in an insurance club which pools the liability claims among the club members. Thus the cost of P & I insurance reflects the pool members' recent experience in claims. Classification societies, which will be discussed later, play an important role in determining the risk of insuring a vessel by performing surveys to rate a ship regarding its seaworthiness, construction, and suitability for the trade in which it is employed. In recent times special war risk insurance has been purchased for vessels intentionally trading in a war zone.

Ship Agencies

Small to medium size owners or large ship owners calling at seldom visited ports cannot afford to have a shore staff in every port. A shore staff is needed to arrange for spare parts delivery, ship chandler service, port entrance and exit documentation, medical problems, bunkering, cargo transfer, legal service, port fees, and numerous other services needed to support a large modern vessel. The port call cost of a large vessel amounts to many thousands of dollars even for a brief visit. Excluding cargo-handling fees, port call fees vary from 2 to 8 percent of the annual operating costs. A large industry called ship agents has grown up to provide these needed services. Some ship agencies are large multiport companies while others are small single port organizations. Agency fees are based on the size of the ship.

Crewing Brokers

Some shipping companies, operating under flags of convenience such as Panama or Liberia, which do not require that the crew be citizens of the country, hire shipboard personnel from any country where qualified seamen can be retained at a low wage. In 1993, popular sources of officers and crew members included Russia, the Philippines, Taiwan, Italy, Korea, and India. Crewing brokers recruit such seamen in their native land and provide them to owners as required. The owners pay the brokers a percentage of the seamen's wages for this service. Typical wages for international seamen compared to typical American seamen are shown in Table 1-6.

TABLE 1-6

Typical Seaman's Monthly Wages

(Paid 12 Months a Year for 7 Months' Work, 1991)

Rank or Rating	*International, $*	*USA, $*
Master	2,766	6,600
Chief engineer	2,514	6,000
First officer or first engineer	1,786	3,260
Second officer or second engineer	1,430	2,880
Third officer or third engineer	1,378	2,500
Electrician	1,430	2,400
Radio officer	1,430	2,780
Chief steward	1,430	1,840
Boatswain, carpenter, chief cook, pumpman	917	1,500
Able seaman, motorman, oiler, 2nd steward	821	1,250
Second cook, messman	699	960
Ordinary seaman, wiper	611	1,140

Overtime pay for American crew members frequently ranges from 20 to 200 percent of base wages. Fringe benefits for American crew members is approximately 200 percent of base wages. Base wages vary for the type and size of the vessel.

Ship Managers

Small shipowners who do not wish to maintain the organizational infrastructure needed to support a ship in a trade may hire a ship manager who uses economies of scale to operate the ship in a manner which produces a profit for both the owner and the manager. The ship manager frequently provides nearly everything, including maintenance, fuel, insurance, supplies, and crew. The owner usually provides the cargo. It is not unusual for a ship manager to operate 50 to 100 ships. The ship manager will charge a fixed daily cost for the operation of a vessel. A typical analysis of the daily operating expenses of an ocean bulk carrier is shown in Table 1-7.

TABLE 1-7

Typical Daily Running Expenses for a Bulk Carrier in International Trade

Wages	.380
Victualling	.060
Stores	.126
Water	.002
Lubricants	.070
Maintenance	.022
Spares	.031
Taxes	.043
Crew & medical	.045
Hull insurance	.127
P & I insurance	.038
Administration	.051
Total	1.000

Sale and Purchase Brokers

A S & P broker is retained by an owner to find a purchaser for a ship. Since the 1950s, it has been common practice for some shipowners to seek profits from the timely sale and purchase of ships. There is considerable risk for the owner in depending on fortuitous timing in the sale and purchase of ships to earn profits but it has been done with some notable successes. The S & P brokers have worldwide contacts among the buyers and sellers and earn broker's commissions on the sales which they organize. In economic analysis, the broker's commission is included in the price of the vessel. In some cases the S & P broker plays a role in assisting shipyards and owners to reach agreements for new construction. Brokerage fees depend on the size of the vessel and are normally under 4 percent.

Cargo Brokers

Cargo brokers provide a service by assisting shippers to find an available vessel(s) adequate for the need to transport a cargo or a stream of cargo. Shippers also retain brokers to find minimum cost transport for their material. There is a concomitant need for shipowners who do not have a worldwide freight solicitation staff to use the services of cargo brokers. Typical fee for cargo brokers is 1.25 percent of the freight bill. The fee is paid by the shipowner regardless of who retains the broker.

Fuel Brokers

Main and auxiliary fuel bunkering is a major cost for an operating vessel. Since large costs are involved, it is desirable for the owner operating ships in a large area of the globe to seek the most advantageous fuel purchases possible in terms of quality and price. Quality is equally important as price since it affects the maintenance cost of the propulsion machinery. Fuel brokers bring consumers together with suppliers to produce long- and short-term contracts for fuel purchases.

Ship Financing Organizations

To purchase an existing ship or a newly constructed ship the buyer usually requires a loan. Such loans range from 60 to 80 percent of the purchase price of the vessel. They are secured by the vessel and sometimes by other vessels belonging to the owner. Loans are usually repaid in 10 years or less to the lending organization, which may be a large bank or a corporation. In some cases the lending corporation retains the title to the vessel and leases the vessel to the ship operator. Loan interest rates are determined by the economy and demand for money at the time of the loan. In 1993, a qualified borrower could obtain a loan for less than 10 percent interest.

Classification Societies

The concept of the ship classification society started in the late eighteenth century at Lloyd's Coffee House in London, where ship underwriters gathering to accept maritime risks would seek some guarantee of the fitness of the vessel concerned for the voyage in prospect. This led to an informal system of inspection of hulls and equipment and the assessment of the reputation of the master and owners. This information was gathered in handwritten "ship's lists" which were distributed to the underwriters. By 1834 these lists had grown into a formal technical organization called Lloyd's Register of Shipping, which rapidly grew into the large international operation it is today. The services grew to include the rules for the construction and classification of ships that provide minimum standards for the construction of seaworthy vessels; the registry of ships (which includes a list of all Lloyd's classified vessels and the current classification resulting from periodic survey); the inspection and classification of materials, machinery, and equipment used in shipbuilding; the inspection and approval of repairs to vessels following damage; and the approval and audit of quality assurance systems. These services are billed to clients such as shipbuilders, machinery manufacturers, and shipowners, and find their way into the capital costs and operating costs of ships.

Today most maritime countries have a classification society which provides the above services for national flag vessels and others as requested. These societies include in addition to the United Kingdom–based Lloyd's Register of Shipping: the American Bureau of Shipping (United States), Bureau Veritas (France), Det Norske Veritas (Norway), Germanischer Lloyd (Germany), Registro Italiano (Italy), Hellenic Register (Greece), and Nippon Kaiji Kyokai (Japan).

Technical Services

There are many service companies available to assist ship operators and related industries in the technical aspects of operations. These services include the following.

Salvage societies are technical organizations which estimate damage to vessels and cargo and maintain statistics on such damage.
Planned maintenance service is provided by companies that develop computer-based maintenance plans for a vessel or class of vessels.
Communications, both radio and satellite telephone, are furnished by large international companies.
Satellite Navigation is available from companies who have earth-orbiting satellites.

GOVERNMENT REGULATION

All maritime nations have laws and regulations to regulate safety, environmental impact, seaworthiness, personnel, health standards, security, crew qualification and certification, navigation, and numerous other matters. In the United States these laws and regulations are administered by the U.S. Coast Guard of the Department of Transportation. Other countries have similar organizations to administer their law. The cost of complying with a wide range of complex laws and regulations issued by numerous nations is probably expensive but is never identified as an item in the cost accounts since it is impossible to identify and segregate.

Maritime nations have a national ship registry which includes all the ships authorized to fly that nation's flag. These ships must conform to all laws, regulations, and crew citizenship requirements which the national government promulgates. After World War II, a few nations, for example, Panama and Liberia, established open registries, which welcomed any ship in the world to fly their flag for a modest fee and minimum regulation, including the important consideration that crew members could be hired from any country in the world. These countries became known as "flags of convenience" and many of the world's largest and finest bulk carriers flew such flags. In the late 1980s, the continued pressure of lower crew costs available from open registry ships caused many owners to abandon their traditional national registries for the open registries. In response, some European countries established a special national and open ship registry which specifically permitted the hiring of noncitizen crews. The Norwegian International Registry and the Isle of Man Registry are examples of this practice. Most of these special registries required that the senior officers be citizens of the flag country. This practice has led to the reduction of crew wages to the lowest international standard for qualified personnel and has left many countries which have not adopted the special registry with a noncompetitive high labor cost in their traditional flag vessels. A partial list of available flags of convenience with few if any restrictions on crew nationalities would include the following: Antigua, Bahamas, Cayman, Liberia, Malta, Panama, St. Vincent, Singapore, and Vanuatu.

SHIP DESIGN

In the English-speaking world, engineers who are trained to design ships are called naval architects. Naval architects have a wide range of engineering skills, including special training and competence in fluid mechanics and structural engineering. They have the overall responsibility for a ship's design. Marine engineers assist in the design of ships by taking responsibility for most of the systems internal to the ship's hull. Such systems include the propulsion plant, electrical power generation and distribution, refrigeration, air-conditioning, cargo-handling machinery, and other hull and deck machinery. Marine engineers are also employed in the maintenance, repair, operation, and management of ship machinery systems. When employed in the management team of a shipowner, marine engineers frequently play a key role in determining the technical characteristics of ships to be added to the owner's fleet. There are two general approaches to the business of ship design, i.e., the naval architect in private practice and the shipbuilders who employ naval architects and other engineers to design ships to meet the needs of a buyer.

A Naval Architecture Firm

While a naval architecture firm provides a variety of services to the maritime community, its principal role is to interpret the needs and desires of the prospective owner or purchaser of a new ship and produce a preliminary design, overall hull and arrangement drawings, and a specification which defines the vessel in sufficient detail to permit the prospective owner to obtain shipbuilder bids and ship financing. Since the modern ship is a large, complex structure, a typical naval architecture firm may employ, in addition to naval architects, engineers with skills in mechanical, marine, electrical, electronic, materials, and industrial fields. The design services of a naval architecture firm for a new ship design cost from 3 to 10 percent of the cost of the vessel depending on the type and size of the vessel. The number of independent firms of naval architecture in the United States decreased significantly during the 1980's following the downward trend in commercial shipbuilding.

Shipbuilder Design

All shipbuilders must undertake the detailed design of ships they have contracted to build. Detailed design includes the working drawings, production instructions, purchasing specifications, quality requirements, test memoranda, and other documents needed to build a ship. The detailed design requires different skills than preliminary design but involves the same range of engineering disciplines. Detailed design may account for between 5 and 15 percent of the ship direct labor cost depending on the type of vessel.

In addition to the necessary detailed design, some shipbuilders offer to undertake the preliminary design and enter into contracts to provide a package of all design work required for a ship. This shipbuilder activity has resulted in some innovative ship designs, including the large RO/RO ship and the double hull crude oil carrier.

SHIPBUILDING

Shipbuilding in America and many other places in the world is a highly cyclical industry. Wars and government subsidy policy have a great impact on shipbuilding activity. During World War I, America became a major shipbuilder, but activity in this field declined after the war. The Merchant Marine Act of 1936 started the recovery of American shipbuilding by providing construction and operating subsidies. This was soon followed by the commencement of World War II, which ushered in the world's largest mass production shipbuilding program in America. Thousands of ships were constructed in many newly organized shipyards, in some of which the production rate for the 10,807-dwt Liberty ships (EC-2) reached one launching each week. The Victory ship and the T-2 tanker were also built in large quantities during World War II. In the early 1980s, the government ended the construction subsidy program and commercial shipbuilding disappeared from the American industrial scene. Figure 1-15 shows the decline in American shipbuilding beginning in 1981.

Worldwide shipbuilding continued to do well during the late 1980s and early 1990s. Consistent with the growing demand for bulk transport, the construction of tankers and bulk carriers led the way, followed by freighters. Table 1-8 shows the total number of ships and the dwt by ship type for worldwide shipbuilding.

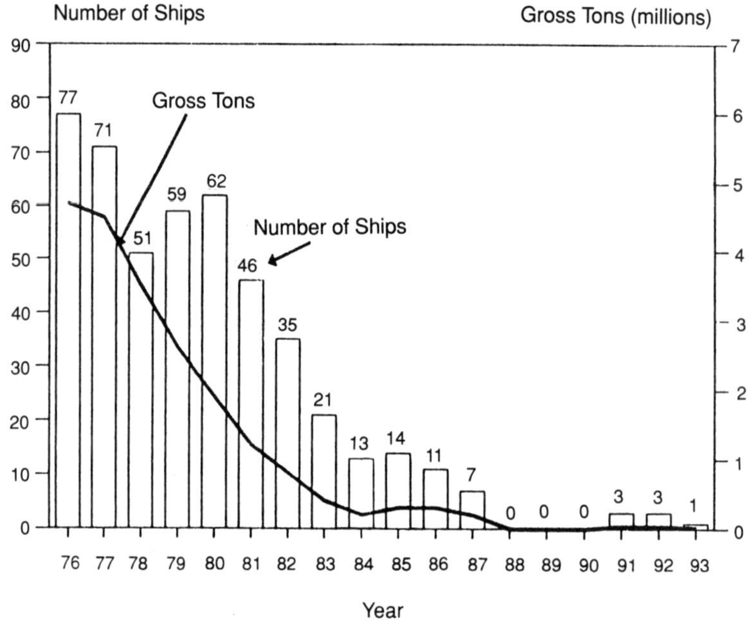

Figure 1-15. History of U.S. shipbuilders' commercial ship order book. Courtesy MarAd

TABLE 1-8

Worldwide Ship Deliveries, Vessels over 1,000 dwt

(Tonnage in Thousands)

Year	*No. Ships*	*Combination Ship dwt*	*Freighter dwt*	*Bulker dwt*	*Tanker dwt*
1986	542	15.2	3045.8	9425.2	4800.1
1987	436	67.7	2546.5	7718.3	5248.1
1988	692	130.2	4178.2	6229.8	13021.1
1989	372	34.4	1992.0	5732.3	6542.5
1990	478	59.0	2735.0	8799.0	7789.0
1991	601	34.0	3555.0	5994.0	11567.0

Shipbuilding Labor

Shipbuilding is a labor-intensive business. Therefore, if governments do not interfere, it should not be surprising to find that successful shipbuilding nations are those with the lowest-paid workers, assuming that the needed technology is available to these countries. During the decades from 1960 to 1990, Japan and then Korea became major world shipbuilders due primarily to favorable labor rates. In the 1990s, the United States has had favorable labor rates and therefore an opportunity to reenter commercial shipbuilding, which it had abandoned in the early 1980s. Table 1-9 shows the average compensation paid to shipyard workers in major shipbuilding countries. Some forecasters are confident that China will become the world's major shipbuilder at the beginning of the twenty-first century.

Figure 1-16 shows the average total employment in U.S. private shipyards during the 10 year period starting in 1982. Since very few merchant ships were constructed in U.S. shipyards during this time, the labor force was supported by naval shipbuilding and ship repairing. The projected reductions in naval shipbuilding for the late 1990s suggests that a return to merchant shipbuilding is essential to retain America's capability to build ships.

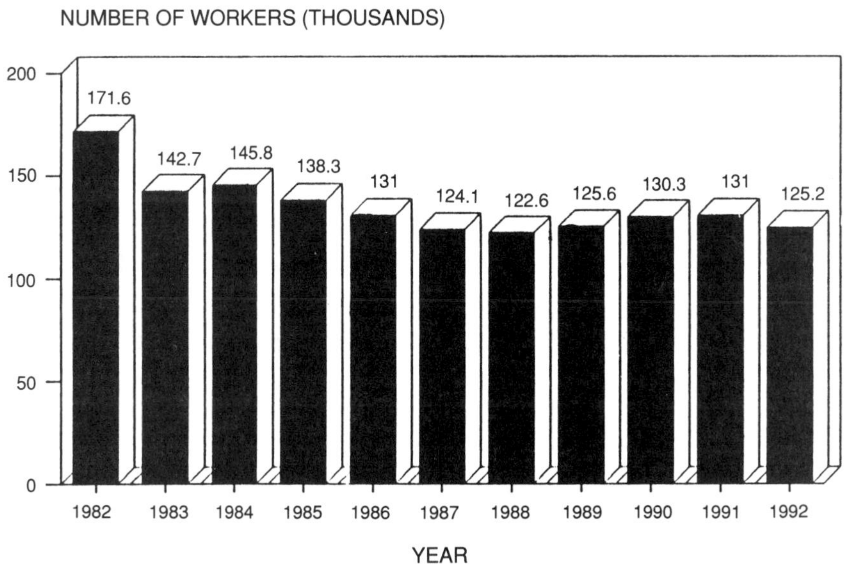

Figure 1-16. Average total employment in U.S. private shipyards. Courtesy MarAd

TABLE 1-9

Typical Hourly Compensation of Shipbuilding Workers

Country	*1988* *$/hour*	*1989* *$/hour*	*1990* *$/hour*
USA	14.33	14.77	15.50
Japan	14.83	14.67	15.80
Korea	4.40	6.35	10.00
U.K.	9.89	10.06	12.55
Germany	20.89	20.16	26.50
France	14.51	14.09	18.60

SHIP REPAIR AND MAINTENANCE

The international ship repair business is very large and diverse and includes major shipbuilding yards which accept repair work, yards which do only ship repair work, ship repair companies which do not have a yard but which dispatch repair workers to piers or anchorages where client ships are located, shipowner repair facilities located in a few key ports, and companies which provide riding crews who embark and travel with the ship while doing repair work. Most ship repairers are prepared to undertake the full range of ship repair activities. Others become specialists retained by a shipowner or a ship repair facility. Such contractors specialize in refrigeration, diesel, electrical, control, tank cleaning, in-water hull cleaning, sand blasting, and other activities.

Worldwide, there are approximately 200 shipyards with major ship repair facilities, i.e., a drydock capable of handling a 600-foot or larger ship. Most of these facilities are located in shipbuilding countries or along major trade routes.

Shipowner's Annual Expenses

Table 1-10 is an analysis of the typical annual ship maintenance and repair expenses which a vessel owner or manager may expect in ocean trade.

TABLE 1-10

Typical Shipowner Annual Repair Expense

Expense Category	*Percent*
Drydock and underwater cost	20.10
Machinery	23.30
Hull	37.60
Electrical	6.80
Exterior paint	5.80

TABLE 1-10 *(continued)*

Interior paint	3.80
Tank cleaning	1.40
Other	1.20
Total	100.00

In efforts to control ship repair costs, shipowners have recently employed sophisticated diagnostic techniques to reduce the risk of machinery failure and excessive wear. These methods include computer-based main engine diagnostic systems, vibration analysis, fuel analysis, lube oil analysis, and planned maintenance systems.

Shipowners make decisions concerning ship repair locations on the basis of the total repair costs which, for a vessel in a worldwide trade, must include the following considerations: currency exchange rates; overall price charged by shipyard, i.e., yard bid price; cost of owner supply items, spares and paint; ship diversion cost from last trading port; port, agency, and tug costs; tank-cleaning and gas-freeing costs before entering yard; special technical consultants to be paid for by owner.

Tables 1-11 and 1-12 are typical published catalogue prices for two ship repair tasks.

TABLE 1-11

Typical International Drydocking Published Prices, 1993

Gross Tonnage	*Dock and Undock Plus 2 days, $/GRT*	*Each additional day, $/GRT*
6,000	.705	.141
55,000	.705	.155
100,000	.733	.183
150,000	.790	.197
200,000	.847	.211
250,000	.917	.240

TABLE 1-12

Typical International Prices for Cleaning Boiler Firesides, 1993

Normal Boiler Evaporation	*Prices for fireside water wash and open drums for survey, $*
20,000	3,380
30,000	4,090
40,000	4,790
50,000	5,360
60,000	6,060

Ship Repair Contracts

Ship repair facilities and shipowners or managers enter into one of the following types of contract relationships for ship repair work: *time and material, negotiated price, sealed bid,* or *published price* contracts. The first two types of contracts provide some advantage to the ship repairer while the latter two types assure that the shipowner has reasonable control over the cost of the work completed. These contract types are discussed in Chapter 12.

Ship Repair Personnel

Ship repair is more labor intensive than shipbuilding. The personnel engaged in ship repair must be more skilled and flexible than those engaged in shipbuilding, where repetitive tasks with detailed instructions and controls are possible. The machinery ship repairperson is frequently required to determine the problem and its source before undertaking a repair. Most ship repair work does not require engineering input except in cases of structural damage or major machinery damage.

Ship Repair Price Index

There are several published ship repair indexes which reflect worldwide costs. *Motor Ship* publishes such an index. Table 1-13 contains a list of this index extended by linear regression to the year 1995. In this table, the index is increasing at an average rate of 2.8 percent per year.

TABLE 1-13
Ship Repair Price Index

Year	*Index*
1985	100
1986	95
1987	95
1988	95
1989	110
1990	129
1991	115
1992	118
1993	120
1994	125
1995	128

ENGINEERING ECONOMICS

The outcome of complex engineering decisions in the maritime industries is significantly influenced by questions concerning the costs and benefits, taxes, cash flows, revenues, profits, management goals for rates of return on investments, financing, interest rates, inflation, depreciation, and market research. The integration of these economic considerations into the engineering decision-making process—leading to a decision or act which is both technically sound and economically justified, if not optimal—is the subject of engineering economics. In applying the methods of engineering economics, the decision maker must decide if a deterministic or uncertain view of the future is most appropriate to his or her needs.

Most of the decision methods in this text are based on a single criterion or measure of merit used to rank the alternatives or acts available to the decision maker. The exception to the single criterion is the brief introduction to the multi-criteria decision method in Chapter 15. The principal economic measures of merit discussed in the following chapters and used for problem analysis throughout the text include the following.

Required freight rate is the total annual operating and capital cost of a vessel divided by the tons of cargo transported per year. The decision maker seeks to minimize this value.

Economic cost of transport is the same as required freight rate except the inventory cost of the cargo is added to the total operating cost. This value must be minimized.

Present worth is the discounted sum of a stream of annual cash flows generated during the life of a project or venture alternative. The decision maker seeks to maximize the present worth of benefits, or minimize the present worth of costs, or maximize the present worth on the net annual cash flow.

Equivalent uniform annual cash flow is a series of equal annual cash flows for a project or venture alternative which for a given discount rate has a value equivalent to the present worth. In this case, the equivalent uniform annual net cash flow must be maximized.

Benefit-cost ratio is the ratio of the present worth of the stream of benefit annual cash flows divided by the present worth of the cost annual cash flows. The decision maker seeks to maximize this ratio.

Internal rate of return is the interest rate which causes a stream of discounted net cash flows for a project alternative to sum to zero. The decision maker seeks to maximize this value.

To appreciate the meaning of cash flow, it is useful to think of the project, venture, or enterprise as a "black box" which exists during the economic life. Any cash added to the box is positive cash flow, cash removed from the box is negative cash flow, and cash moved about within the box is not cash flow. Revenues, initial capital investment, income, and other benefits flow into the box. Loan repayments, operating costs, taxes, etc., flow out of the box. Depreciation is a relocation of cash within the box.

REFERENCES AND BIBLIOGRAPHY

Ambrose, A. J. *Jane's Merchant Shipping Review.* London: Jane's Publishing Company, published annually.

Annual Report of the Maritime Administration. Washington: U.S. Department of Transportation; published annually.

Construction, Ship Repair, Manpower and Labor Issues, Contracts / Financial. Arlington, Va.: Shipbuilders Council of America, published annually.

Frankel, Ernst G. *Management and Operations of American Shipping.* Boston: Auburn House Publishing Company, 1980.

Hutchinson, J.L. "The Shipowner's View of Ship Repair Yards." *Transactions,* I. Mar. E. Vol. 92 (1980).

Marine Log Annual Yearbook and Maritime Review. New York: Simmons-Boardman Publishing Corp., published annually.

Maritime Reporter and Engineering News, World Yearbook. New York: Maritime Activity Reports, published annually.

"Merchant Shipbuilding Return." London: Lloyd's Register of Shipping.

Report on Survey of U.S. Shipbuilding and Repair Facilities. Washington: Maritime Administration, U.S. Department of Transportation, published annually.

Roseman, D., et al. "Technical and Financial Alternatives for Advanced Design Dry Bulk Carrier Concepts." *SNAME Symposia,* Sept 1982.

Seaborne Trade and Transport Reports. London: Drewry Shipping Consultants, published annually.

Walsh, Robert G. "Estimated Return-on-Investment of Oil Tankships." SNAME *Marine Technology*, Jan. 1978.

"U.S. Merchant Marine Data Sheet." Washington: U.S. Department of Transportation, published monthly.

QUESTIONS AND PROJECTS

1. What are the primary design variables to be considered when undertaking a preliminary study of a single new ship?
2. What does risk mean? Uncertainty? Are they the same?
3. Why do shipowners specify larger and larger ships for some ocean trades? What are the limitations on ship size?
4. How would you define an ocean shipping trade?
5. What type of vessels make up the major portion of the international merchant fleet?
6. What is the significant design and construction complication of a parcel tanker and why is it necessary?
7. What do we mean when tankers are classified as panamax, capesize, and handysize?

8. A rich lode of iron ore is discovered in Antarctica. Describe the characteristics of the vessel you would design for a long-term trade to transport this ore to North America.
9. Discuss the advantages and disadvantages of a RO/RO ship versus a container ship liner service.
10. What are the primary types of insurance required for an oceangoing vessel?
11. What is a ship management company? Why are they needed?
12. What is the role of a classification society?
13. Discuss the advantages and disadvantages for a country which maintains both a national and an open ship registry.
14. What is the difference between a ship design for bidding purposes and a detailed ship design for construction?
15. Suggest a scenario in which American-built ships would be competitive with most foreign built ships.
16. Define the term *required freight rate.* How is it different from economic cost of transport?
17. How did major American shipbuilders survive the decade of 1980–1990 without building merchant ships?
18. What is your understanding of engineering economics?
19. Why do small countries such as Liberia and Panama have the world's largest merchant fleets?
20. What considerations, other than low price, enter into the decision to select a ship repair facility?
21. What are the most common types of contracts between shipowners and ship repairers?
22. What are the major cost categories for a typical shipowner's annual maintenance and repair cost?
23. As a shipowner, what actions would you take to reduce and control your maintenance and repair costs?
24. What do we mean by cash flow?
25. What is present worth? Required freight rate?
26. How would you classify merchant ships based on their function?
27. List the auxiliary and support services required for normal ship operations.
28. Discuss the economic impact of specialized ship designs on auxiliary and support services.

Chapter 2

Principles of Engineering Economics

Let us first understand the facts and then we may seek the causes.
—Aristotle

DECISION PROCESS

To make a decision in the maritime environment or any other situation requires that alternatives be available from which to select the best choice. In the real world, the possible alternatives may be too numerous to deal with in practical terms. Therefore, we establish a model for our decision process. The model, usually based on our experience, includes a finite number of promising alternatives. For example, in selecting a propulsion prime mover for a tanker, we might consider the low-speed diesel and medium-speed geared diesel, while for an LNG carrier, experience might suggest that (as well as the diesel engine alternatives) the geared steam turbine and the geared gas turbine should be considered, each with several thermodynamic cycle options. As we expand our model to include other features of a ship, such as the speed, cargo capacity, hull parameters, and type of propeller, the problem of selecting among the feasible alternatives becomes increasingly complex.

To measure the desirability of one alternative relative to the others available, it is necessary to select a meaningful measure-of-merit, which must incorporate and sum all of the independent economic variables during the life of the ship, project, or venture. These independent variables include the best estimates of equity investment, cost of capital, operating costs, taxes, revenues, and the salvage value at the end of the economic life. The required freight rate is one such measure. In subsequent chapters, numerous measures-of-merit used in shipbuilding and ocean transportation will be discussed.

It is essential that alternatives considered in the decision process have the same economic capacity. It would be illogical to select alternatives for economic analysis which did not provide the same function or, in the case of ships, the same transport capacity. For example, it would be meaningless to compare a 100,000-dwt bulk carrier with a 50,000-dwt bulk carrier of the same speed. However, if the smaller bulker had a higher speed and faster turnaround time, permitting it to transport equal cargo tonnage per year, the comparison would be realistic for an equal lifetime. In complex transportation systems, it is not always obvious that the alternatives under consideration have the same economic capacity.

The life of the economic venture is another basic consideration in the decision process. The decision model must be designed so that all alternatives have the same economic life. For modern ships, this is usually about 25 years. A comparison between the economic benefit of a new ship with a 25-year life and an available ship of identical capacity with a remaining life of 10 years might result in the wrong decision. Notwithstanding, it is not uncommon to make such comparisons by using the annual economic benefit as a measure-of-merit. The correct comparison would require that the new ship be compared with two ships; i.e., after the 10-year life of the available ship, it would be necessary to purchase another with a 15-year remaining life. The possibility of predicting the cost and availability of the second ship would be problematic. However, the analyst must endeavor to ensure equal economic life of alternatives to achieve realistic decision models.

In any decision process, there exists a set of activities which lead to a choice. This is also true for engineering decisions in which economics are an important consideration. A general procedure for engineering decisions in the ocean transportation field would include the following:

1. Explicitly recognize the need for a decision. Perhaps a class of vessels in a commercial fleet are nearing the end of their economic life. It is necessary to decide on the type, size, speed, etc., of replacement vessels.
2. Search for data to define the environment in which the selected alternative must function. A successful decision process will require large quantities of data concerning the current and future cost and technology of design, construction, and operations.
3. Define the alternatives. A set of choices is compiled which experience suggests covers the economically and technically feasible possibilities. Use of computers makes it possible to consider a large number of alternatives during the conceptual phase.
4. Predict future states of the operating environment. Extensive study is required to make best possible prediction of rates of increase for cost elements such as labor, maintenance, insurance, etc. Will future revenues increase to provide continuing profitable operation? What new regulations will be imposed on design and operations? What

will happen to the cost of fuel? How and when will the planned trade change?

5. Organize data into decision model. Select a measure-of-merit and evaluate each alternative. If more than one future state scenario exists, the alternatives must be evaluated for each of the defined states. If possible, the probability of each future state scenario must be determined.
6. Evaluate the model. Use methods (to be discussed in later chapters) to calculate the expected measure-of-merit for each alternative and then to select the best alternatives including consideration of numerous future events which change value of measure-of-merit.
7. Apply management judgment. While a decision model is a useful device, it cannot incorporate all the nuances of a complex decision. Experienced management review is necessary to confirm the correct conclusion of the model or to require the process to be repeated with an expanded model. For example, political questions or competitors' behavior may become the dominant consideration.

In summary, a decision process will involve the development of a finite set of feasible alternatives or choices. If we are confident that the future can be represented by a single set of independent variables, then we employ these variables to calculate the measure-of-merit for each alternative and select the best alternative. If more than one set of independent variables is required to adequately describe a range of futures, a decision method which deals with uncertainty is required. For decisions under uncertainty, the model is expanded to include the evaluation of measure-of-merit of the alternatives under a collectively exhaustive and mutually exclusive set of future events.

TIME VALUE OF MONEY

The methods of engineering economics are principally related to the recognition of the time value of money. If we invest our personal funds or open a bank account, we expect to achieve a rate of return on our investment or interest on the bank deposit. This interest is paid or credited at the end of a time period: daily, monthly, quarterly, yearly, etc. The most convenient and common period for capital investments is one year. If we borrow money and pay interest, if we lend money and receive interest, then, implicitly, if we hold money without employing it, we lose or fail to gain whatever the available current interest rate would provide. Money has a time value.

A convention for analysis is that the interest is credited or paid (i.e., a transaction) at the last instant of the time period, e.g., one year. If the investment, loan, or bank account continues on for multiple years, the interest of the first year becomes a deposit or investment in subsequent years and earns its own interest. This is called compounding.

CASH FLOW

A venture, project, or company receives and distributes money. This is called cash flow. For analysis, it is useful to think of the venture as a black box. Funds added to the box are termed positive cash flow. Funds removed are called negative cash flow. The difference is net cash flow. Cash moved about within the box has no direct effect on cash flow. For example, depreciation charges of an asset are the company's recognition that the asset is being depleted; e.g., for a ship there is a reduction in its useful life. Depreciation has no direct influence on annual net cash flow but could affect the tax payments which are negative cash flow.

In the analysis of a venture over time, a graphical depiction of cash flow is often useful. Figure 2-1 illustrates the net cash flow for a five-year project. A horizontal line represents the time periods for the venture. Vertical lines at the end of each period indicate net cash flow. The zero time or beginning vertical line represents the initial investment, purchase, or construction cost which for this example is negative, i.e., a disbursement.

Much of the basic work of engineering economics is the calculation of a single sum which is equivalent to a time-oriented set of net cash flows. The single sum is usually calculated at time zero. The process is called discounting and the sum is called the net discounted cash flow.

Example 2-1

Two alternatives, *A* and *B*, are proposed for the selection of a device to reduce labor. Life is five years. The cost of alternative *A* is $10,000. Alter-

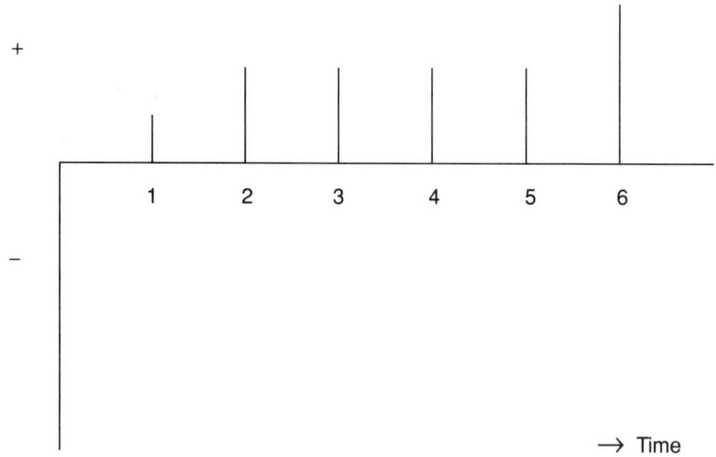

Figure 2-1. Net cash flow displayed on a time line

native *B* also costs $10,000. Alternative *A* will generate cost savings of $3,000 per year for the five-year life. Alternative *B* will save $1,500 for each of two years and $4,000 each year thereafter. There is a $2,000 salvage value at the end of life. There is no time value of the money. Which is the best choice?

Net Cash Flow

Year	*A*	*B*
0	–10,000	–10,000
1	3,000	1,500
2	3,000	1,500
3	3,000	4,000
4	3,000	4,000
5	5,000	6,000
Sum	7,000	7,000

These cash flows, shown in Figure 2-2, have equal sums with the unrealistic assumption of zero interest rate. Therefore, the alternatives are equally desirable. In a realistic case, with a time value of money, alternative *A* would be preferred. Why?

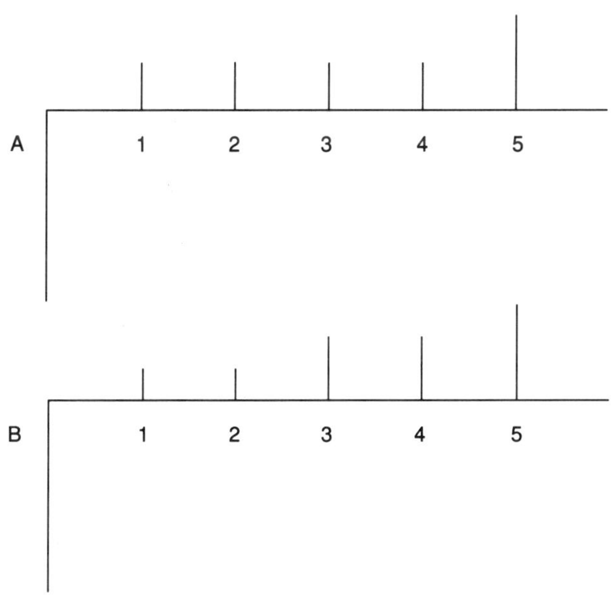

Figure 2-2. Net cash flow comparison

PRESENT AND FUTURE WORTH

Since money has a time value, we would expect that a *present sum* (i.e., at zero time) would increase over time to equal a *future sum* (i.e., after n periods) by accruing interest each period at some rate. Let P equal the present sum, F the future sum, i the interest rate, and n the number of periods. Then at the end of the first time period:

$$F = P + i\,P = (1+i)\,P$$

and at the end of the second period:

$$F = (1+i)\,P + i\,(1+i)\,P$$

$$= [(1+i) + i\,(1+i)]\,P$$

$$= (i^2 + 2i + 1)\,P$$

$$F = (1+i)^2\,P$$

For three periods (i.e., years), the *future sum* is:

$$F = (1+i)^3\,P$$

and for n periods, it is clear that:

$$F = (1+i)^n\,P$$

The factor $(1+i)^n$ is the *future worth factor* or the *compound-interest factor*. The notation for such factors used in this text will be:

$$\text{Future worth factor} = (F/P,\, i,\, n)$$

which is the operator applied to a present worth to calculate a future worth given i interest for n periods.

Example 2-2

Two sums of $10,000 each are deposited in different accounts. The first account F_a pays 5 percent interest compounded annually. The second account F_b pays 5½ percent annually. When the accounts are closed at the end of five years, what is the difference in their value or future worth?

$$F_a = \$10{,}000\ (F/A,\ .05,\ 5)$$

$$= \$10{,}000\ (1 + .05)^5 = \$12{,}762.82$$

$$F_b = \$10{,}000\ (F/A,\ .055,\ 5)$$

$$= \$10{,}000\,(1 + .055)^5 = \$13{,}069.60$$

$$F_b - F_a = \$306.78$$

Figure 2-3 is a plot of the future values of account F_a as the interest rate varies from 0 to 10 percent at five- and ten-year account lives.

Example 2-3

How many periods before a sum of money P will triple at 5 percent interest compounded annually?

$$F = 3\,P$$

$$P\,(F/P, i, n) = 3\,P$$

$$(1 + .05)^n = 3$$

$$n \ln 1.05 = \ln 3$$

$$n = \ln 3/\ln 1.05 = 22.5$$

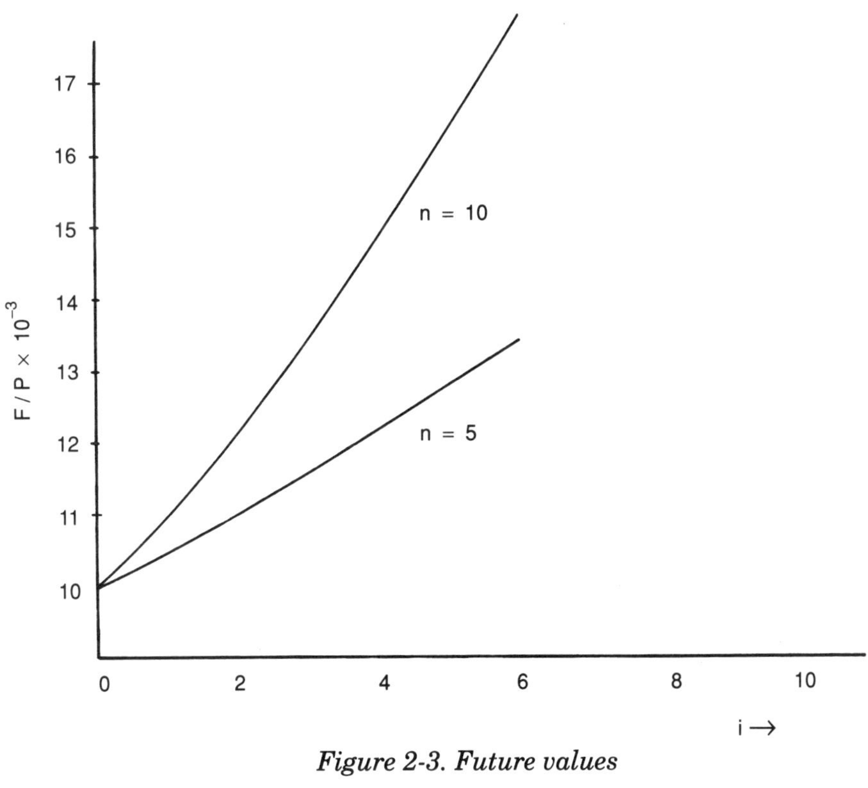

Figure 2-3. Future values

Note: A useful rule for quick estimates of time or interest to double a sum is the Rule of Seventy. A sum will double in 10 years at:

$$70/10 = 7 \text{ percent interest}$$

Or a sum will require:

$$70/5 = 14 \text{ years}$$

to double at 5 percent interest compounded annually.

It is frequently necessary to calculate an interest rate or *rate of return* when present worth, future worth, and number of periods are known.

Example 2-4

What rate of return is required to have a present worth of $12,000 grow to $24,000 in 10 years if the interest is compounded annually?

$$P = \$12{,}000$$

$$F = \$24{,}000$$

$$24{,}000 = 12{,}000\,(1+i)^{10}$$

$$2 = (1+i)^{10}$$

$$\ln 2 = 10 \ln (1+i)$$

$$.06931 = \ln (1+i)$$

$$1.0717 = 1+i$$

$$i = .0717 = 7.17 \text{ percent}$$

Using the equation for a future sum, it is possible to solve for present worth *P*, given some future worth *F*, an interest rate *i*, and the number of periods. For example:

$$F = (1+i)^n\ P$$

$$P = \frac{F}{(1+i)^n}$$

The factor $\frac{1}{(1+i)^n}$ is called the *present worth factor* and is identified as follows in this text:

$$\text{Present worth factor} = (P/F, i, n)$$

Therefore:

$$P = F\,(P/F, i, n)$$

and the present worth factor is an operator which converts a future sum to a present sum given an interest rate and a number of periods.

Example 2-5

In Example 2-1, the net cash flow at the end of the fifth year for alternative *A* was $5,000 and for alternative *B* was $6,000. If the interest rate is 7.5 percent, what is the difference between the present worth of these net cash flows in the fifth year only?

$$P_a = \frac{F_a}{(1+i)^n} = \frac{5{,}000}{(1+.075)^5} = 3482.79$$

$$P_b = \frac{6{,}000}{(1+.075)^5} = 4179.35$$

$$P_b - P_a = 696.56$$

The present worth is sometimes referred to as the discounted value of the future sum and the process of applying the present worth factor may be called discounting. In fact, we will be introduced later to a measure-of-merit called the *discounted average cost per unit throughput*.

The present worth of a series of annual net cash flows is a popular and useful measure-of-merit for a project or venture. Present worth will be used throughout this text to evaluate alternatives in the engineering decision process. A summation of discounted net cash flows can be represented as:

$$P = F_0 + \frac{F_1}{1+i} + \frac{F_2}{(1+i)^2} + \frac{F_3}{(1+i)^3} + \ldots + \frac{F_n}{(1+i)^n}$$

P is the present worth of *n* future net cash flow.

Example 2-6

If the time value of money is 5.5 percent, calculate the present value of the two net cash flow streams of Example 2-1.

Year	*Net Cash Flow* *a*	*Discounted Net Cash Flow* *a*	*Net Cash Flow* *b*	*Discounted Net Cash Flow* *b*
0	–10,000	–10,000	–10,000	–10,000
1	3,000	2,843	1,500	1,422
2	3,000	2,695	1,500	1,348

(Table continued on next page)

3	3,000	2,555	4,000	3,406
4	3,000	2,422	4,000	3,229
5	5,000	3,825	5,000	4,591
	$ 7,000	$ 4,340	$ 7,000	$ 3,996

Using the summation of the discounted net cash flows (i.e., present worth) at an appropriate interest rate reveals that discounting changes the conclusion so that alternative *A* is the best choice with a present worth $344 greater than alternative *B*. Since money has a time value (i.e., 5.5 percent interest), the alternative which returns the funds sooner is better even when the actual cash flow summations are equal at $7,000.

UNIFORM ANNUAL AMOUNT

A series of equal cash flows or amounts for *n* periods is called a *uniform series* or *uniform annual amount.* If *A* is a uniform amount, the present worth of a single future worth of *A* is determined as follows:

$$F = A\,(1+i)^n$$

and the sum of a series of *A*'s is:

$$F = A + A(1+i) + A(1+i)^2 + \ldots + A(1+i)^{n-1} + A(1+i)^n$$

Multiply both sides by (1 + i):

$$F(1+i) = A(1+i) + A(1+i)^2 + A(1+i)^3 + \ldots + A(1+i)^n + A(1+i)^{n+1}$$

then subtract the first equation from the second to yield:

$$i\,F = A\,\{(1+i)^n - 1\}$$

$$F = A\left\{\frac{(1+i)^n - 1}{i}\right\}$$

which is the future sum of a series of equal amounts *A* for *n* periods at an interest rate of *i*. The value:

$$\frac{(1+i)^n - 1}{i}$$

is called the *uniform series compound amount factor* which will be identified as (F/A, i, n).

Solving for *A* in the above relationship provides the *uniform series sinking fund factor* or (A/F, i, n):

$$A = F\left\{\frac{i}{(1+i)^{n-1}}\right\}$$

$$= F\,(A/F, i, n)$$

Example 2-7

A major refit of a passenger vessel will be necessary in 10 years at an estimated cost of $12 million. If the owner can earn 8 percent interest, how much money should be deposited at the end of each year, including the last year, to have the refit funds available when needed?

$$n = 10$$

$$F = \$12{,}000{,}000$$

$$i = 8 \text{ percent}$$

$$A = F\left[\frac{i}{(1+i)^n - 1}\right]$$

$$= 12{,}000{,}000\left[\frac{.08}{(1.08)^{10} - 1}\right] = \$828{,}353.$$

Example 2-8

A seaman planning for his retirement 20 years hence decides to deposit $5,000 each year in a bank where he is confident that 5 percent compounded interest will be earned. What will be the value of his personal fund when he retires?

$$n = 20$$

$$i = 5 \text{ percent}$$

$$A = 5{,}000$$

$$F = A\left\{\frac{(1+i)^n - 1}{i}\right\}$$

$$= 5{,}000 \left\{ \frac{(1.05)^{20} - 1}{.05} \right\}$$

$$= 5{,}000(33.065) = \$165{,}329.77$$

CAPITAL RECOVERY AND SERIES PRESENT WORTH

Given that a single future amount is equal to a present amount times the future worth factor, it is possible to develop a relationship for a uniform annual amount given a single future amount, that is:

$$F = P(1+i)^n$$

$$A = F\left[\frac{i}{(1+i)^n - 1}\right]$$

$$A = P(1+i)^n \left[\frac{i}{(1+i)^n - 1}\right]$$

$$A = P\left[\frac{i(1+i)^n}{(1+i)^n - 1}\right]$$

where the term in brackets on the right is called the *uniform series capital recovery factor* which will be identified in this text as (A/P, i, n).

Solving the capital recovery factor for *P*, the present worth of a uniform series of *A*'s provides:

$$P = A\left[\frac{(1+i)^n - 1}{i(1+i)^n}\right]$$

where the term in brackets on the right is called the *uniform series present worth* of (P/A, i, n).

Example 2-9

If $10 million is borrowed from internal company sources to construct a new container warehouse, how much must be set aside each year for 10 years, assuming that the funds could be invested externally for a 10 percent compounded return?

$$A = P\left[\frac{i(1+i)^n}{(1+i)^n - 1}\right]$$

$$A = 10{,}000{,}000 \left[\frac{.1\,(1.1)^{10}}{(1.1)^{10} - 1} \right]$$

$$A = 10{,}000{,}000\ (.1627) = \$1{,}627{,}553.$$

Example 2-10

What is the difference in present worth of two alternatives for a project as stated below if *i* equals 7 percent?

Year	*Project (a)*	*Project (b)*
0	–4,000	–10,000
1–10	1,000	2,000

$$P_a = -4{,}000 + 1{,}000 \left[\frac{(1.07)^{10} - 1}{.07\,(1.07)^{10}} \right]$$

$$P_a = \$3{,}023$$

$$P_b = -10{,}000 + 2{,}000 \left[\frac{(1.07)^{10} - 1}{.07\,(1.07)^{10}} \right]$$

$$P_b = \$4{,}047$$

$$P_b - P_a = \$1{,}023$$

ARITHMETIC GRADIENT

In ocean transport engineering economic analysis, cash flow factors such as maintenance cost, energy cost, revenue, etc., increase over time, sometimes in a linear or arithmetic fashion. Consequently, it is convenient to have relationships which permit the calculation of present *P* or future *F* worth—given an arithmetic gradient, an interest rate, and a number of years. On a time scale, a gradient for an annual maintenance cost starting at $1,000 and increasing at $100 for eight-year life is displayed in Figure 2-4.

The future value of the gradient-only portion of the cash flow vectors of Figure 2-4 can be expressed as follows:

$$F = G\,(1 + i)^{n-2} + 2G\,(1 + i)^{n-3} + \ldots + (n - 2)\ G\,(1 + i) + (n - 1)\,G$$

and numerically for the values of Figure 2-4:

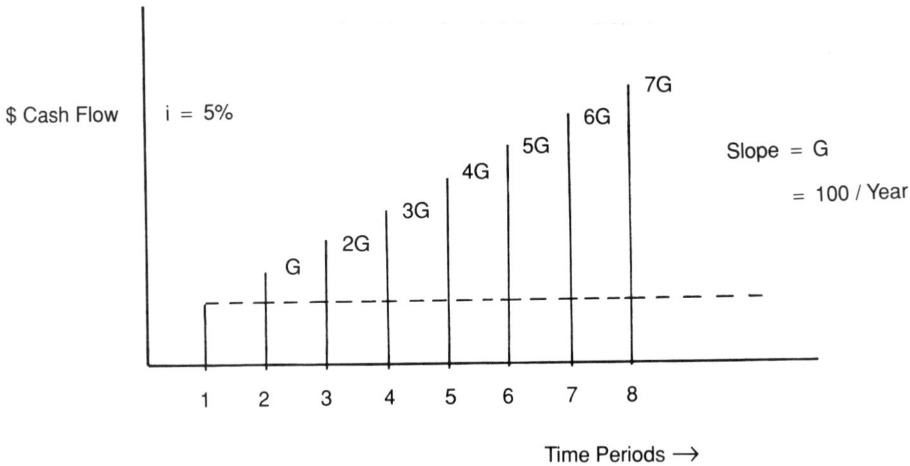

Figure 2-4. Arithmetic gradient

$$F = 100\,(1.05)^6 + 200\,(1+i)^5 + 300\,(1+i)^4 + 400\,(1+i)^3 + 500\,(1+i)^2 + 600\,(1+i) + 700$$

$$(1+i)\,F = G\,\{(1+i)^{n-1} + 2\,(1+i)^{n-2} + \ldots + (n-2)\,(1+i)^2 + (n-1)\,(1+i)\}$$

$$F = G\,\{(1+i)^{n-2} + \ldots + (n-3)\,(1+i)^2 + (n-2)\,(1+i) + n - 1\}$$

$$F + i\,F - F = G\,\{(1+i)^{n-1} + (1+i)^{n-2} + \ldots + (1+i)^2 + (1+i) + 1\} - n\,G$$

$$\therefore\; i\,F = G\left\{\frac{(1+i)^n - 1}{i}\right\} - n\,G$$

$$F = \frac{G}{i}\left\{\frac{(1+i)^n - 1}{i} - n\right\}$$

Multiplying this future worth of an arithmetic gradient by the single payment present worth factor provides:

$$P = \frac{G}{i}\left\{\frac{(1+i)^n - 1}{i} - n\right\}\left[\frac{1}{(1+i)^n}\right]$$

$$P = G\left\{\frac{(1+i)^n - i\ n - 1}{i^2\,(1+i)^n}\right\}$$

where the factor on the right is the arithmetic gradient present-worth factor of (P/G, i, n). Therefore,

$$P = G\,(P/G, i, n).$$

The arithmetic gradient uniform series factor (A/G, i, n) is obtained by multiplying the arithmetic gradient future-worth factor by the sinking fund factor to provide:

$$(A/G, i, n) = \left\{\frac{(1+i)^n - i\ n - 1}{i\,(1+i)^n - i}\right\}$$

$$A = G\,(A/G, i, n)$$

Example 2-11

The preventive maintenance cost for the main cargo pumps has been estimated at $5,000 during the first year, with increases of $1,000 each year for the next 10 years. At 5 percent interest rate, what annual equal payment could be used in a comparative analysis?

$$A_{10} = 5{,}000 + 1{,}000\,(A/G, i, n)$$

$$= 5{,}000 + 1{,}000\left\{\frac{(1+.05)^{10} - .05\,(10) - 1}{.05\,(1+.05)^{10} - i}\right\}$$

$$= 5{,}000 + 1{,}000\,(4.099)$$

$$= \$9{,}099$$

Example 2-12

A shipowner plans to increase the length of his vessel after five years of service. Since revenues are increasing, he plans to establish an account earning 7 percent interest in which he deposits an amount increasing at $1 million each year. The conversion will cost $20 million. How much equity will he have in the conversion? How much must he borrow?

$$F = G\left\{\frac{1}{i}\left[\frac{(1+i)^n - 1}{i} - n\right]\right\}$$

$$F = 1{,}000{,}000\left\{\frac{1}{.07}\left[\frac{(1.07)^5 - 1}{.07} - 5\right]\right\}$$

$$F = \$10{,}724{,}843$$

Since he will have accumulated $10,724,843, he must find financing for $9,275,157. Cash flow is shown in Figure 2-5.

Example 2-13

In evaluating a ship modification, we have selected present worth as our measure-of-merit. The modification will have an economic life of 12 years. The time value of money is 8.5 percent. A part of the modification is a new shaft-driven generator which has maintenance cost increasing at $100 per year above the first year cost of $950. What present-worth cost of maintenance should be used in this analysis? Cash flow is shown in Figure 2-6.

$$P = 950\ (P/A, .085, 12) + G(P/G, .085, 12)$$

$$P = 950\left\{\frac{(1+i)^n - 1}{i\,(1+i)^n}\right\} + 100\left\{\frac{(1+i)^n - i\,n - 1}{i^2\,(1+i)^n}\right\}$$

$$P = (950)\ (7.3447) + 100\ (33.3678)$$

$$P = 6{,}977.47 + 3{,}336.78$$

$$P = \$10{,}314.$$

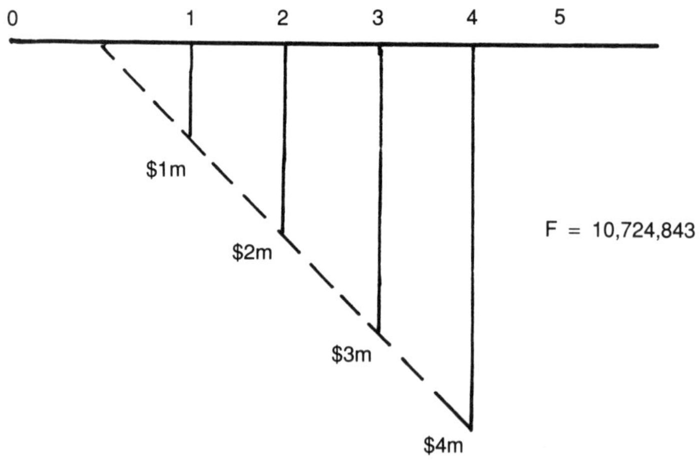

Figure 2-5. Cash flow

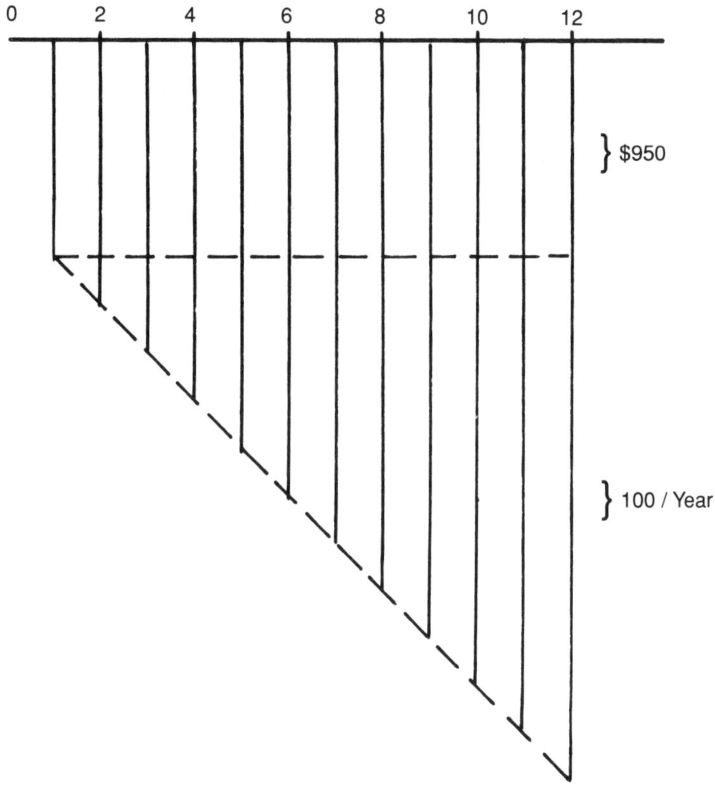

Figure 2-6. Maintenance cash flow

GEOMETRIC GRADIENT

When an end-of-period payment increases or decreases by a fixed percentage each year, it is called a geometric gradient series. In this series, the initial payment c increases at some fixed rate r_g at the end of each period, while an interest rate i is applied to the balance remaining each period. If r_g has a negative value, the series is decreasing. Therefore, a series starting with an initial payment of c will have a present worth of:

$$P = \frac{C}{1+i} + \frac{C\,(1+r_g)}{(1+i)^2} + \ldots + \frac{C\,(1+r_g)^{n-1}}{(1+i)^n}$$

where $i \neq r_g$.

Multiply by $\frac{1+r_g}{1+i}$ and subtract the result from the above to yield:

$$P - \left(\frac{1 + r_g}{1 + i}\right) P = \frac{C}{1 + i} - \frac{C}{1 + i}\left(\frac{1 + r_g}{1 + i}\right)^n$$

$$P = \frac{C}{1 + i}\left[\frac{1 - \left(\frac{1 + r_g}{1 + i}\right)^n}{1 - \left(\frac{1 + r_g}{1 + i}\right)}\right]$$

$$P = C\left[\frac{1 - (1 + r_g)^n (1 + i)^{-n}}{1 - g}\right]$$

and, if $i = r_g$, the result is:

$$P = C\frac{n}{(1 + i)}$$

Therefore, in the functional notation used previously, the geometric series present worth factor is:

$$(P/C, r_{g,}\, i, n) = \frac{1 - (1 + r_g)^n (1 + i)^{-n}}{i - r_g}$$

where $i \neq r_g$, and

$$(P/C, r_{g,}\, i, n) = \frac{n}{1 + i}$$

where $i = r_g$.

Example 2-14

The crew cost for a ferry venture is expected to increase at 2 percent each year during the 10-year life of the vessel. The time value of money is 8.5 percent annually. What crew cost must be incorporated into a present-worth evaluation of lifetime cost if the first year cost is \$1.26 million?

$$c = \$1{,}260{,}000$$

$$r_g = .02$$

$$i = .085$$

$$P = 1{,}260{,}000\left(\frac{1 - (1 + .02)^{10} (1 + .085)^{-10}}{.085 - .02}\right)$$

$$P = \$8{,}933{,}574$$

The solution cash flow is illustrated in Figure 2-7.

Example 2-15

If the interest rate i and the rate of labor cost increase r_g were equal in Example 2-14, the result would be:

$$P = 1{,}260{,}000\,\frac{(10)}{1 + .085} = \$11{,}612.903$$

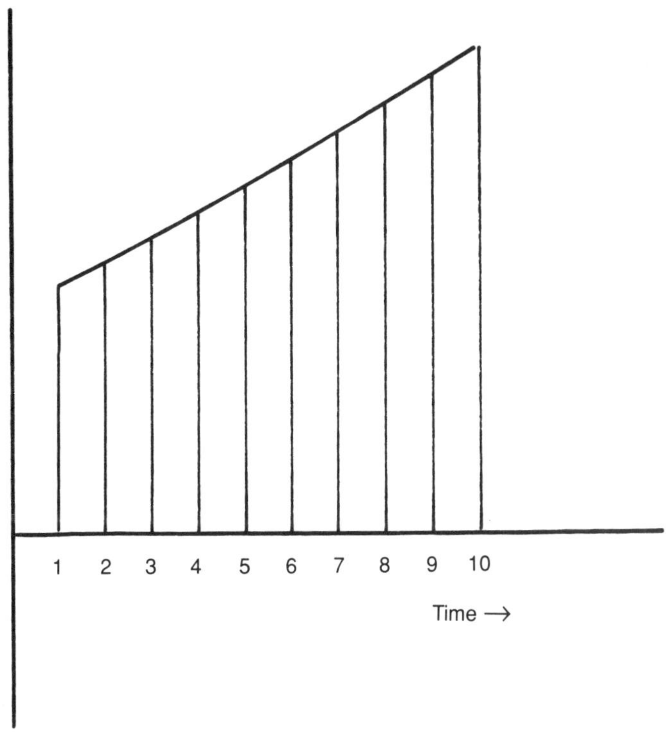

Figure 2-7. Geometric gradient cash flow

EQUIVALENCE

When a sum of money at a point in time is recognized by all interested parties to have the same value, including consideration of the interest, as a sum of money at a different time or a series of money sums over time, the sums are said to be equivalent. In making engineering decisions, any consistent set of equivalent sums may be used, although it is more common to use present worth or annual payments.

The knowledgeable engineer should find all of the following statements to be equally interesting from an economic viewpoint; i.e., he or she should be indifferent in making a selection among them because they are equivalent.

1. Receive \$10,000 today.
2. Receive \$12,763 in five years when $i = 5$ percent.
3. Receive \$6,381 in five years and \$8,144 in ten years when $i = 5$ percent.
4. Pay \$2,000 in two years and receive \$15,764 in five years when $i = 6$ percent.
5. Receive \$1,785 annually for six years when $i = 2$ percent.

The present worth of each of the above transactions is \$10,000 and they are equivalent.

TABLES

Algebraic equations have been used in all examples to calculate present, future, or series payments in all exercises up to this point. The availability of modern hand-held calculators with exponential and logarithmic functions facilitates such calculations. However, it is common practice in texts and handbooks to provide extensive tables of the factors discussed in this chapter. Appendix B contains such tables for interest rates and time periods which are normally of interest in the ocean transport industries.

In addition to the tables, the computer program used to generate the tables has been provided on the floppy disk accompanying this text. This should prove very useful to avoid table interpolation for commonly used interest rates, such as 6.33 percent or 9.75 percent.

Since the table values are factors, they can be multiplied together to provide a desired result. For example, the present worth of a uniform series which starts 10 years in the future and lasts five years can be calculated as follows:

$$P_a = (P/F, i, 10)\ (P/A, i, 5)\ A$$

Another approach to the same question, would be:

$$P_b = (P/F, i, 14)\ (F/A, i, 5)\ A$$

If the interest rate were 5 percent, the combined factor in both cases above would be 2.791, i.e.:

$$P_a = (.64461)\ (4.3294)\ A = 2.791\ A$$

$$P_b = (.50507)\ (5.5256)\ A = 2.791\ A$$

Example 2-16

Find the present worth, future worth, and the equivalent annual payment for the following stream of end-of-year payments given a money time value of $i = 4$ percent.

EOY	*Payment*		*(P / F, .04, n)(Payment)*
0	–10,000		–10,000.00
1	0		0
2	1,000	Arithmetic Series	924.55
3	1,100		977.89
4	1,200		1,025.76
5	1,300		1,068.50
6	1,500	Uniform Series	1,185.47
7	1,500		1,139.87
8	1,500		1,096.00
9	6,000	Future Worth	4,215.52
10	–9,000		–6,080.07
11	2,000	Geometric Series $r_g = .04$	1,299.16
12	2,200		1,374.11
13	2,420		1,453.38
14	2,662		1,537.23
15	2,928		1,625.81
$\sum_{i}^{n}$ (P/F, .04, n)			2,843.18

$$P = -10{,}000 + (P/G, .04, 4)\,(P/F, .04, 1)\,(100)$$

$$+ (P/A, .04, 4)\,(P/F, .04, 1)\,(1{,}000)$$

$$+ (P/A, .04, 3)\,(P/F, .04, 5)\,(1{,}500)$$

$$+ (P/F, .04, 9)\,(6{,}000) - (P/F, .04, 10)\,(9{,}000)$$

$$+ (P/C, .04, 5)\,(P/F, .04, 10)\,(2{,}000)$$

$$P = -10{,}000 + (5.267)\,(.9615)\,(100)$$

$$+ (3.63)\,(.9615)\,(1{,}000) + (2.775)\,(.8219)\,(1{,}500)$$

$$+ (.7026)\,(6{,}000) - (.6756)\,(9{,}000)$$

$$(5.2)\,(.6756)\,(2{,}000)$$

$$P = \$2{,}843.00$$

Of course, in this case, it would be no more difficult if we calculated (P/F, i, n) for each year.

$$F = 2{,}843\ (F/P, .04, 15) = \$5{,}120$$

$$A = 2{,}843\ (A/P, .04, 15) = \$255.58$$

CONTINUOUS COMPOUNDING

In some cases, an economic model is more realistic if the compounding is continuous. As the number of periods approaches infinity, it can be shown with the calculus that:

$$(1 + i) = e^{r}$$

where i = effective interest rate, the annual rate including the effect of compounding during the year, and r = nominal interest rate, the annual rate without the effect of compounding.

Then, for continuous compounding, the factors are:

$$(F/P, r, n) = e^{rn}$$

$$(P/F, r, n) = e^{-rn}$$

$$(A/F, r, n) = e^{r} - 1/e^{rn} - 1$$

$$(A/P, r, n) = \frac{e^{rn}(e^{r} - 1)}{e^{rn} - 1}$$

$$(F/A, r, n) = e^{rn} - 1/e^{r} - 1$$

$$(P/A, r, n) = \frac{e^{rn} - 1}{e^{rn}(e^{r} - 1)}$$

The effective and nominal interest rates are a consideration whenever it is necessary to compound more frequently or to determine the interest rate for a number of periods in one year. The effective interest rate per year i is:

$$i = \left(1 + \frac{r}{m}\right)^{m} - 1$$

where r is the nominal annual rate ignoring any effect of compounding during the year and m is the compounding periods per year.

REFERENCES AND BIBLIOGRAPHY

Collier, Courtland A., and William B. Ledbetter. *Engineering Economics and Cost Analysis,* 2d. ed. New York: Harper & Row, 1988. Extensive development and coverage of fundamentals and methods of comparing alternatives. Income tax treatment is good. Spreadsheet programs are proposed for economic analysis.

Fabrycky, W.J., and G.J. Thuesen. *Economic Decision Analysis,* 2d ed. Englewood Cliffs, N.J.: Prentice-Hall, 1980. Basic text with an introduction to accounting, decision making under risk, and linear programming.

Jelen, F.C. *Cost and Optimization Engineering,* 2d ed. New York: McGraw-Hill, 1983. An advanced text with minimum treatment of engineering economic fundamentals and emphasis on application of operation research methods to the problems of engineering economics. Prepared under sponsorship of American Association of Cost Engineers.

Newnan, Donald G. *Engineering Economic Analysis,* 3d ed. San Jose, Calif.: Engineering Press, 1988. Well organized and comprehensive approach to the subject. Depreciation and income tax are covered. Includes a floppy disk of microcomputer programs for elementary economic analysis.

QUESTIONS AND PROJECTS

1. Describe the steps in an engineering decision-making process.
2. What do we mean by feasible alternatives?
3. Is "do nothing" a feasible alternative? Why?
4. What transaction elements are included in net cash flow?
5. What does time value of money mean to you?
6. What is equivalence and why is it important in engineering economic analysis?
7. When is the customary time for transactions to occur in engineering economic analysis?
8. Describe the procedure to shift a single payment transaction one year forward or one year back on the time scale.
9. Draw the following cash flow diagrams from the borrower's viewpoint.
   a. Five annual equal payments to repay a borrowed sum.
   b. Five annual payments increasing by $100 each year to repay a borrowed sum.
   c. Five annual payments increasing 10 percent each year to repay a borrowed sum.
   d. One future payment at the end of five years to repay a borrowed sum.
   e. How should the above diagrams be modified to show the lender's viewpoint?
10. Discuss the difference between nominal and effective interest rates.

11. Make a table comparing the relationship for a single future amount when compounding occurs annually, semi-annually, quarterly, monthly, daily, and continuously. The annual interest rate is i and the number of years equals n.
12. A constant can be found which is useful for making a quick estimate of the time required to double a present sum at a given i or the interest required to double a present sum in a given time period n. What is this constant?
13. Discuss the importance and limitations of estimates in engineering economic analysis for ocean transport systems.
14. Prepare plots of the following factors as functions of n at $i = 0, .10$, and .20:

    a. (F/A, i, n)
    b. (P/A, i, n)
    c. (F/G, i, n)
    d. (A/G, i, n)
    e. (P/G, i, n)

15. When is it appropriate to ignore time value of money in engineering economic analysis?
16. Use the program "Table" to prepare a table of compound interest factors for $n = 5$ years to $n = 30$ years at an interest rate of 12.36 percent.
17. Use the program "Gtable" to prepare a table of geometric gradient compound interest factors for $n = 5$ years to $n = 30$ years at an interest rate of 12.75 percent and a minimum gradient of 3.5 percent and a maximum of 5.5 percent.

PROBLEMS

1. Prepare a cash flow diagram and table for the following transactions. A used shipboard trash incinerator is purchased for $25,000. The installation cost is $15,000. The maintenance costs for the first two years of a five-year life are zero but, during the last three years, the cost is expected to increase from $500 to $700 in equal steps. The annual operating cost is $1,000. There is a $3,000 salvage value.
2. A sum of money is borrowed to purchase new capital equipments for a marine terminal. The funds and interest will be returned in six years in uniform annual payments. Draw a cash flow diagram from lender's and borrower's viewpoints.
3. Five million dollars must be expended five years hence to upgrade cargo-handling equipment. If the cost of money is 8 percent, what value should be assigned today for this future expenditure when analyzing the venture?

4. You have been offered a payment of $100,000 in five years or $110,000 in seven years. At what time value of money would you be indifferent in the selection of the offers?
5. It is anticipated that it will be necessary to replace the economizer of a marine boiler in five years. How much must be invested in bonds that are due in five years and pay 12.5 percent interest to have the needed $48,000 available to pay for the replacement?
6. Ten thousand dollars is deposited with a commodity broker. The broker promises to pay a nominal interest of 6 percent compounded semi-annually. You request monthly compounding. What is the difference between the two methods if the money is left on deposit for two years?
7. If $3,000 is invested at 6 percent interest, how much will be available in 12 years?
8. What is the future amount of $100,000 after 7 years at 3 percent interest?
9. What is the present worth of a $100,000 payment 12 years from now at 4.83 percent interest?
10. What interest rate is required to double an initial investment in 14 years?
11. What is the present worth of the following stream of net cash flows if the time value of money is 3.72 percent?

Year	*Net Cash Flow*
0	–15,000
1	2,362
2	1,924
3	–1,638
4	5,926

12. What annual payment is needed to accumulate $40,000 in ten years if the time value of money is worth 7 percent?
13. What interest rate is required to accumulate $100,000 in five years with an annual payment of $12,000?
14. How long will it take to repay a $100,000 loan if $5,000 is paid annually and the interest rate is 9 percent?
15. How much must be deposited at the end of each quarter at 9 percent interest compounded quarterly to accumulate $87,000 in six years?
16. To pay for a major ship modification five years hence, a shipowner established an account at 8 percent interest in which equal annual payments would be made to accumulate the $500,000 cost of modification. How much must be deposited each year?
17. How much must be paid to a banker at the end of each year to repay a $50,000 loan in 15 years if the agreed interest is 12 percent?

18. What annual payment for 12 years will have a present worth of $25,500 at 5 percent interest?
19. What is the total annual cost for an emergency generator which costs $85,000, has no salvage value, and has an operating cost of $3,100 per year? The cost of money is 10 percent and the life is 20 years.
20. The net cash flow from a venture will be $40,000 per year. If the life of the venture is 25 years and money has a time value of 7.37 percent, what is the present worth and what is the future worth at the summation of the cash flows? Calculate the same values at 20, 15, 10, and 5 years.
21. If $5,000 increases to $7,500 in four years at compound interest, what is the rate of return (i.e., interest)?
22. Find the equivalent annual payment for five years to equal the following net cash flows if i = 7 percent.

Year	*Cash Flow*
0	–1,000
1	2,000
2	7,000
3	3,000
4	5,000

23. A $10,000 face value bond has a payment rate of 2 percent payable quarterly. The bond is due in 7 years. You wish to purchase this bond at a price which will yield you at least 8 percent per year. What is the maximum purchase price?
24. For an interest rate of 7⅔ percent and 20 years, calculate the following factors, using the program "Table."

    a. (F/P, .0766, 20)
    b. (P/F, .0766, 20)
    c. (A/F, .0766, 20)
    d. (A/P, .0766, 20)
    e. (F/A, .0766, 20)
    f. (P/A, .0766, 20)

25. What amount must be invested today at 9 percent interest to generate an income of $2,350 quarterly for the next five years?
26. A machine will have an annual operating cost of $2,000 for the next 15 years. If the time value of money is 12 percent, what is the present worth of cost?
27. It is proposed to add a heat recovery steam generator to a gas turbine propulsion plant at a cost of $600,000. It is estimated that the fuel savings will be $100,000 per year. The maintenance cost is estimated to be $12,000 per year. The salvage value at the end of 18 years is

estimated at $10,000. If the acceptable rate of return is 12 percent, will the savings justify the costs?

28. A shipbuilder has a ship module paint facility on which there is a $300,000 mortgage. The mortgage rate is 9 percent and is to be paid off in 15 equal end-of-year payments. After five years, a drop in interest rates provides an opportunity to refinance the balance at 6 percent to be paid in 20 equal end-of-year payments. What will be the new annual payments?
29. The machinery maintenance expense for a 206,000-dwt tanker with 12 years' remaining life is $900,000 per year. It is expected to continue to increase $50,000 each year. If the time value of money is 10 percent, what is the present value of this cost and the equivalent annual payment?
30. A gradient payment of $1,000 continues for 10 years. The cost of money is 4 percent. What is the equivalent uniform annual payment for the 10-year period?
31. Two net cash flow streams exist. Stream 1 is a uniform payment of $500 each month, while Stream 2 is increasing from zero at a gradient of $100 each month. If the time value of money is 24 percent, at what period will the monthly payments be equal?
32. At current load factors, a container vessel operator spends $40,000 per year in operating costs for refrigerated boxes. These costs are expected to increase at 4.2 percent each year for the 20-year life of the vessel. If interest is 8 percent, what is the present worth of this cost? What is the equivalent annual cost?
33. Use the computer program "GTable" to calculate the geometric gradient compound interest factors for five to nine years, eight percent interest, and a geometric gradient range of two to four percent.
34. If $12,000 is deposited in an account which pays 6 percent interest compounded continuously, how much will accumulate in three years?
35. Given an annual cash flow of $16,000 for 10 years with a time value of money of 12 percent compounded continuously, calculate the present worth and the future worth.
36. A main propulsion electronic diagnostic system for a new vessel has an installed cost of $80,000 and a life of 20 years. The owner is concerned that technology advances will make the package obsolete in 10 years. He considers buying an alternative package for $52,000 with a life of 20 years. He will replace this alternative in 10 years with a similar package if the anticipated technical advances do not occur. Operating and maintenance costs are equal and the cost of money is 12 percent. Which is the better choice?
37. To encourage an owner to install an electronic engine analysis system, the manufacturer offers special terms of payment. The equipment has an installed cost of $110,000 but the manufacturer has

quoted an installed cost of $119,200 with payment in three years. Comment on the value of this offer to the owner.

38. The diesel generators of a five-year-old container vessel are suitable for the addition of superchargers which have an installed cost of $400,000. Total fuel consumption will be reduced an average of 1.0 metric tons per day at sea and in port. The remaining life of the vessel is 15 years. The operator's MARR is 14 percent. Fuel cost is $200 per metric ton. Increased weight (i.e., reduction in cargo capacity) is negligible. The increased annual maintenance cost is $1,000. Should the equipment be installed?
39. A Middle Eastern ship operator is considering the purchase of an available 80,000-dwt bulk carrier for a price of $12 million. He has an offer from a coal company for a 12-year time charter which covers the vessel's estimated life. His MARR is 10 percent. The ship will be in trade 11.7 months per year. What is the minimum hire charge per deadweight ton per month if the total annual operating expense is $1.6 million?

Chapter 3

Deterministic Comparisons of Decision Alternatives

If a man will begin with certainties he shall end in doubts; but if he will be content to begin with doubts he will end in certainties.
—Francis Bacon

THE DECISION PROCESS

Engineering decisions are usually reduced to a finite set of alternatives for which an economic measure of merit has been selected and computed for each alternative. The economic measure of merit must incorporate into a single value a useful representation of capital costs, operating costs, revenues, taxes, economic life, and time value of money. If these inputs are assumed to be known with certainty, the decision process is referred to as deterministic. If some or all of the inputs to a decision process are provided as a range of values which may be described by a discrete or continuous probability distribution, the decision process, called a decision under uncertainty, is far more complex. Some aspects of decision under uncertainty will be considered in subsequent chapters.

The concepts of time value of money and equivalence permit the engineering analyst to compute a single value that encodes the information of numerous cash flows over a period of years which represent the economic life. The engineering problem for which the alternatives have been defined always has a time requirement even if the time is infinite. For example, a shipowner may receive a 20-year bulk freight contract. The alternatives in the analysis for the selection of a vessel(s) to fulfill that contract must be capable of providing the transport for 20 years. If one of the alternatives is to use a vessel for which there is a five-year remaining economic life, then this alternative must include a second vessel obtained at the end of the first five years of the contract. Each alternative considered must be capable of fulfilling all the requirements stated in the problem. Only one alternative may be selected; i.e., the alternatives are mutually exclusive. These are the characteristics of our decision model.

NET PRESENT WORTH ANALYSIS

As we have seen, given a stream of net cash flows over the economic life of a project or venture and an interest rate which represents the time value of money, it is possible to calculate a net present worth (NPW) of the cash flow.

$$NPW_k = \sum_{j=0}^{n} A_j (P/F, i, j)$$

where: NPW_k = net present worth, alternative k
A_j = net cash flow in year j
n = total economic life, years
(P/F, i, j) = future payment present worth factor, year j

Given a number of alternatives, calculating the NPW_k for each alternative and selecting the best will result in a rational decision. When feasible from an engineering viewpoint, it is important to include the "do nothing" alternative as a decision baseline.

Example 3-1

A ship manager is considering the installation of a new heavy lift crane on a breakbulk vessel which the company plans to operate for another 15 years. There are two approaches to the crane: (1) two operators are required and (2) one operator is required. The do nothing alternative will provide some revenue when doubled-up cranes can be used. The net cash flow streams are as follows. The time value of money is 6 percent.

j	A_{1j} *Do nothing, \$*	A_{2j} *Crane #1, \$*	A_{3j} *Crane #2, \$*
0	0	–165,000	–205,000
1	10,000	25,000	35,000
2	15,000	45,000	50,000
3	15,000	50,000	65,000
4-14	10,000	40,000	55,000
15	20,000	60,000	75,000
NPW_k	109,933	230,530	322,595

The more expensive crane is justified since it has the highest net present worth. There will be many opportunities in this text to employ NPW as a measure-of-merit. It is a classical measure of engineering economics.

In the above example, net cash flow (i.e., cost and revenue) was discounted to the zero year to provide the measure of merit. Therefore, the highest value is the current choice. If net costs only had been discounted, the obvious choice would be the alternative with the lowest net present worth of the cost.

In any net cash flow stream, the end of the final year includes a salvage value which is the final income when the equipment is sold or scrapped. For ships, this is frequently a significant sum of money depending on the ship-breaking activity and the price of steel.

ANNUAL NET CASH FLOW ANALYSIS

A useful and easily understood measure of merit is found by calculating the equivalent annual net cash flow for each alternative including the do nothing. As we have seen in Chapter 2, the equivalent annual net cash flow, like present worth analysis, requires a specific economic life and a time value of money. All capital and start-up costs, salvage costs, and other one-time costs are converted to equivalent annual costs. Annual costs that are the same for each year can be used directly. Annual costs which are not the same each year must be converted into equivalent annual uniform amounts. The results of this analysis for an identical problem in present worth are the same.

Example 3-2

A ship operator is given a contract to place a fast ferry in operation for 10 years. The contract permits him to sell the vessel to the government for 10 percent of the original price at the end of 10 years. He has three alternatives: (1) a hydrofoil which costs \$5.7 million, (2) an air cushion vehicle at a cost of \$6.3 million, and (3) a surface effect vehicle at a cost of \$4.14 million. The annual cash flow for the three vessels is as follows.

	Hydrofoil	*Air Cushion Vehicle*	*Surface Effect Vessel*
Operating cost	\$3,676,728	\$2,808,358	\$4,066,206
Revenue	\$7,040,000	\$7,040,000	\$7,040,000
Capital recovery	\$1,854,780	\$2,050,020	\$1,347,156
Net cash flow	\$1,508,492	\$2,181,622	\$1,626,638

Clearly, the three alternatives would be ranked in the following order based on deterministic and nonincreasing operating costs and revenue: (1) air cushion vehicle, (2) surface effect vehicle, and (3) hydrofoil.

The capital recovery was calculated by using the factor (A/P, .10, 10) to convert the initial cost of two vehicles to an annual payment.

A more sophisticated view of this problem would recognize that operating costs tend to increase each year. In some cases, one or more costs might decrease or remain constant. The analyst must rely on experience and judgment to predict the rate of increase of operating costs. Figure 3-1 shows a table of deterministically increasing operating costs for the surface effect vehicle alternative of this example. To use these values in an annual net cash flow comparison, it would be necessary to calculate the equivalent annual uniform cash flow as follows.

$$A_k = (A/P, i, n) \sum_{j=1}^{n} (P/A_j, i, n)$$

Or, if the costs are increasing at a known geometric rate, the equivalent annual uniform cash flow is:

$$A_k = (A/P, i, n)\ A_1\ [P/A, r_g, i, n]$$

Example 3-3

There are two alternatives for a temporary passenger terminal with a useful life of five years. Select the least costly on the basis of equivalent uniform annual cost.

	A	*B*
First cost	29,500	36,200
Annual maintenance	1,700	2,000
Salvage value	4,000	9,000
Interest rate is 12%		

$$A_A = 29{,}500\ (A/P, .12, 5) + 1{,}700$$

$$= -\ 4{,}000\ (A/F, .12, 5)$$

$$A_A = 29{,}500\ (.2774) + 1{,}700 - 4{,}000\ (.1574)$$

$$A_A = 9{,}253.7 \text{ Annual cost}$$

$$A_B = 36{,}200\ (A/P, .12, 5) + 2{,}000$$

$$= -\ 9{,}000\ (A/F, .12, 5)$$

$$A_B = \$10{,}625 \text{ Annual cost}$$

Therefore, the lowest-cost alternative is A.

Year	*Crew*	*Fuel*	*Lube oil*	*M & R*	*Insurance*	*Port*	*G & A*	*Total*
1	1,050,000	864,182	62,545	1,188,039	386,440	190,000	325,000	4,066,206
2	1,076,250	881,465	63,796	1,233,680	390,304	192,850	336,375	4,164,721
3	1,103,156	899,095	65,072	1,260,390	394,207	195,743	348,148	4,265,812
4	1,130,735	917,077	66,374	1,298,202	398,150	198,679	360,333	4,369,549
5	1,159,004	935,418	67,701	1,377,148	402,131	201,659	372,945	4,476,006
6	1,187,979	954,126	69,055	1,377,262	406,152	204,684	385,998	4,585,257
7	1,217,678	973,209	70,436	1,418,580	410,214	207,754	399,508	4,697,379
8	1,248,120	992,673	71,845	1,461,137	414,316	210,871	413,491	4,812,452
9	1,279,323	1,012,526	73,282	1,504,971	418,459	214,034	427,963	4,930,558
10	1,311,306	1,032,777	74,748	1,550,120	422,644	217,244	442,942	5,051,780
Totals	11,763,550	9,462,548	684,855	13,619,530	4,043,017	2,033,517	3,812,703	45,419,720

Figure 3-1. Operating costs for a surface effect ferry fleet of two vessels combined

RATE OF RETURN ANALYSIS

Rate of return (ROR) is a particularly useful method when the specific time value of money is not established or when there is a corporate policy which establishes a minimum attractive rate of return (MARR). ROR is the equivalent interest rate on a series of cash flows (receipts or payments) at any period in the series. It is calculated by setting the summation of the net cash flows for n periods equal to zero and solving for i, i.e.:

$$\sum_{j=1}^{n} A_j(1+i)^j = 0$$

The rate of return i on an investment is the interest rate earned on the unrecovered investment, at any point in time, such that the payment schedule makes the unrecovered investment equal to zero at the end of the economic life.

A computer algorithm is the most efficient method of calculating i for a series of unequal annual net cash flows. Trial and error can be used by estimating the value of i, then using the present worth, future worth, or equivalent annual payment compound interest factor to determine if the approach to zero is close enough to be accepted as a solution. If not, a revised estimate is made using the error as a guide to the revision.

Example 3-4

A $5,000 investment results in a $10,000 return in 12 years. What is the rate of return?

$$10{,}000\,(1+i)^{-12} - 5{,}000 = 0$$

$$(1+i)^{-12} - .5 = 0$$

$$(1+i)^{-12} = .5$$

$$i = \sqrt[12]{2} - 1$$

$$i = 5.95\%$$

Example 3-5

An alternative venture which has a five-year duration is presented by the following cash flow stream. What is the rate of return?

Year	*NCF*
0	−10,000
1	4,000
2	5,000
3	3,500
4	2,750
5	−200

$$-10{,}000+\frac{4{,}000}{(1+i)^1}+\frac{5{,}000}{(1+i)^2}+\frac{3{,}500}{(1+i)^3}+\frac{2{,}750}{(1+i)^4}-\frac{200}{(1+i)^5}=0$$

Determine i by trial and error.

0%	*20%*	*15%*
−10,000	−10,000	−10,000
4,000	3,333	3,478
5,000	3,472	3,780
3,500	2,025	2,301
2,750	1,326	1,572
−200	−80	100
5,050	75	1,031

$$i \approx 20\%$$

INCREMENTAL ANALYSIS

In making comparisons among alternatives when rate of return or internal rate of return is the criterion for comparison, it is not always the highest rate of return which is the best if a policy exists establishing a minimum attractive rate of return. In this case, the internal rate of return must be established for each of the mutually exclusive alternatives. Then the alternatives are ranked in order of increasing rate of return. Next, the incremental investment and the rate of return on this investment is calculated to determine if the incremental return meets the MARR criterion. If not, then the incremental investment necessary to achieve the higher total rate of return is not justified.

Example 3-6

Three alternative investments for a new portable grinder will provide the annual saving listed below.

Year	*Do Nothing*	*A*	*B*	*C*
0	0	–5,500	–8,500	–10,500
1–10	0	1,450	1,950	2,600

If the company policy has established a 15 percent minimum attractive rate of return, which is the best alternative among the four?

The traditional comparisons on the basis of present worth or equivalent annual cash flow yield the following results for i = 15 percent.

Alternative	*PW*	*EACF*
Do nothing	0	0
A	1,778	354
B	1,287	256
C	2,549	507

Since alternative *C* has the highest value for either present worth or equivalent annual amount, it is the best. However, let's verify that the incremental investments meet the company policy for MARR.

Compare the do nothing alternative to alternative *A*.

Alternative	*0 Year*	*1-10 Years*
Do nothing	0	0
A	–5,500	1,450
Increment DN to A	–5,500	1,450

$$-5{,}500 + 1{,}450\ (P/A, i, 10) = 0$$

$$i \approx 23\%$$

DN to A

which is greater than the MARR, permitting the elimination of the do nothing alternative and assigning alternative *A* as the current best.

Next, compare alternative *A* to alternative *B*.

A	–5,500	1,450
B	–8,500	1,950
Increment A to B	–3,000	500

$$-3{,}000 + 500\ (P/A, i, 10)$$

$$i_{A\ to\ B} \approx 11\%$$

which is less than the MARR, requiring alternative *A* to be retained as the best.

Now compare alternative *A* with alternative *C*.

A	− 5,500	1,450
C	−10,500	2,600
Increment A to C	−5,000	1,150

$$-5{,}000 + 1{,}150\,(P/A,\,i,\,10) = 0$$

$$i_{A\ to\ C} \approx 19\%$$

Therefore, alternative *C* is the best since it generates the maximum positive cash flow while meeting the company MARR criterion.

In any ordered list of alternatives, it is necessary to invest the incremental amounts while the yield exceeds the MARR. If there is no MARR policy, it is necessary to consider the question of preference of a very high return on a very small sum with a cash yield near zero or a modest return on a larger sum with a substantial cash yield.

REQUIRED FREIGHT RATE

A popular economic measure of merit for ocean transport is the required freight rate (RFR) or the amount which must be charged per ton of cargo transported annually to achieve a specified rate of return on an investment.

The required freight rate is calculated in dollars per ton for a particular year as follows:

$$RFR = \frac{SAC\,(A/P,\,i,\,n) + AOC}{ATC}$$

where SAC = cost of the vessel
A/P = capital recovery factor
i = rate of return
n = economic life
AOC = annual operating cost
ATC = annual tons of cargo carried

To calculate a required freight rate for use as a measure of merit for the life cycle of a ship venture, calculate the mean of the present worth of the RFR for each year *j*.

$$\text{PW of } \overline{RFR} = \frac{SAC + \sum_{j=1}^{n} (P/F,\ i,\ j)\,(AOC_j)}{\sum_{j=1}^{n} ATC_j}$$

or the equivalent uniform annual cost:

$$\text{EUAC of } \overline{\text{RFR}} = \frac{n\,(A/P, i, n)\,SAC + \sum AOC_j}{\sum_{j=1}^{n} ATC_j}$$

which, if the *AOC* and *ATC* are constant year to year, is simplified to:

$$\text{EUAC of } \overline{\text{RFR}} = \frac{(A/P, i, n)\,SAC + AOC}{ATC}$$

The annual operating costs include crew costs (salary, benefits, living), fuel, lube oil, stores and supplies, maintenance and repair, insurance, port/terminal fees, canal fees, stevedoring fees, general and administrative costs, and cargo brokerage fees.

The price of the ship includes the total delivered price for a new ship or the S & P brokerage price for an older ship plus required equipment such as containers, barges, etc.

The life of a ship is difficult to assign. However, vessels operating in salt water are frequently assigned 25 years or less of life.

The interest rate, tax rate, depreciation method, and rate of return are used to determine the capital recovery. It is common to calculate the RFR on a before-taxes basis as is done in these examples. However, after taxes, RFR could be determined.

Example 3-7

A design study for a tanker includes consideration of three speeds, i.e., 14, 15, and 16 knots. Using the data below, what is the best design speed employing RFR during the first year (before taxes) as a measure of merit?

Ship	*A*	*B*	*C*
Length, m	234	234	234
Beam, m	58.5	58.5	58.5
Depth, m	30.676	30.676	30.676
Sea speed, knots	14.0	15.0	16.0
Block	.835	.813	.802
Midship coefficient	.994	.994	.994
Prismatic	.840	.818	.807
Ship cost, $	75,670,140	76,108,260	76,479,060
R.T. per year	10.35	10.86	11.11
Day at sea/trip	23.81	22.22	21.51
Cargo moved, tonnes/year	2,088,268	2,191,139	2,240,994
Deadweight, MT	204,265	204,408	204,533
Annual operating cost, $	17,215,310	18,326,750	19,124,460
Ship cost × (A/P, .10, 20)	8,891,241	8,942,720	8,966,289

Each hull is slightly different to provide optimum resistance of each speed.

$$RFR_A = \frac{8{,}891{,}241 + 17{,}215{,}310}{2{,}088{,}268} = \$12.5\,/\,\text{tonne}$$

$$RFR_B = \frac{8{,}942{,}720 + 18{,}326{,}750}{2{,}191{,}139} = \$12.44\,/\,\text{tonne}$$

$$RFR_C = \frac{8{,}966{,}289 + 19{,}124{,}460}{2{,}240{,}994} = \$12.54\,/\,\text{tonne}$$

In this case, the tanker designed for 15-knot sea speed is close to optimum since both design speed increase and decrease result in a higher required freight rate.

Figure 3-2 shows an interesting study of the relationship among speed, length, and block coefficient versus the required freight rate for a bulk carrier. Such a study identifies the optimum design or helps to assess the trade-offs when the vessel design must be suboptimized.

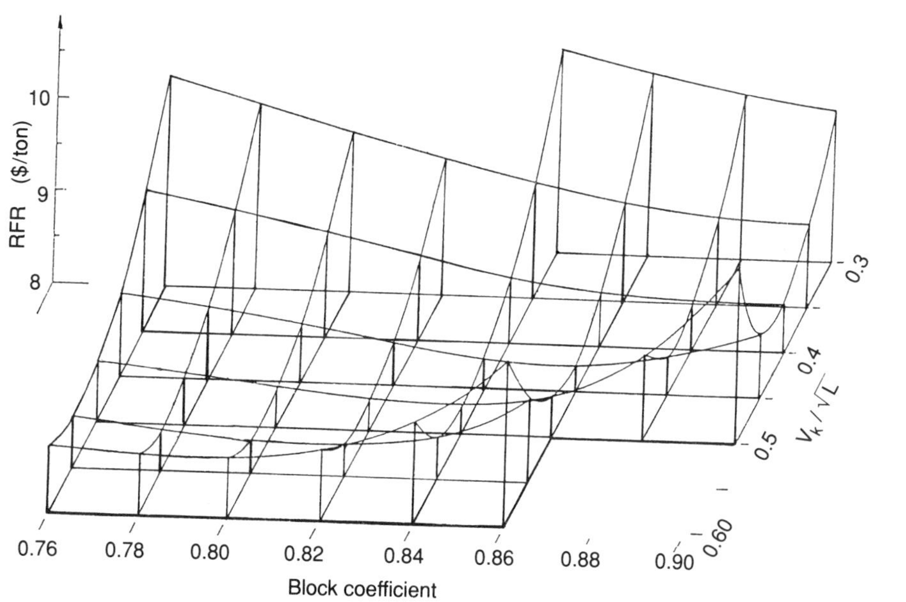

Figure 3-2. Required freight rate (RFR) as a function of block coefficient and design speed-length ratio; base fuel cost. Results are plotted for 6,000 n.m. round trip. Courtesy of Robert Sher, Harry Benford, and the University of Michigan

ECONOMIC COST OF TRANSPORT

In any transportation system, there is an inventory cost for the cargo while it is in transit. If this cost on a per-tonne basis is added to the RFR, a better representation of cost of transport or economic cost of transport (ECT) is available. Since increasing design speed increases fuel investment cost, it is desirable to include the inventory cost which decreases in a decision problem since the inventory cost contributes valuable information. The ECT for a single year of vessel operation is:

$$\text{ECT} = \text{RFR} + \frac{\text{Inv. cost}}{\text{ATC}}$$

Example 3-8

The cost of the crude oil transported in the previous example is \$17.50 per barrel. What is the economic cost of transport (ECT) if the cost of money is 10 percent and the specific gravity is .88?

$$42\ \frac{\text{gallons}}{\text{barrel}} \times 8.34\ \frac{\text{pounds}}{\text{gallon}} \times .88 \times .4536\ \frac{\text{Kg}}{\text{pound}} = 139.82\ \frac{\text{Kg}}{\text{barrel}}$$

$$\frac{\$17.5}{139} \times 1{,}000\ \frac{\text{Kg}}{\text{tonne}} = \$125/\text{tonne}$$

	A	*B*	*C*
Total cargo moved, tonne/year	2,088,268	2,191,139	2,240,994
Cargo days at sea	123.2	126.7	119.5
Interest at 10%/year	3.38	3.47	3.27
Cargo inventory cost/year	\$8,822,932	\$9,504,065	\$9,160,063

$$\text{ECT}_A = 12.50 + \frac{8{,}822{,}932}{2{,}088{,}268} = \$16.72$$

$$\text{ECT}_B = 12.44 + \frac{9{,}504{,}065}{2{,}191{,}139} = \$16.78$$

$$\text{ECT}_C = 12.54 + \frac{9{,}160{,}063}{2{,}240{,}994} = \$16.63$$

Therefore, with the cargo inventory cost considered, the best design is *C*, i.e., the fastest ship under consideration.

To calculate the economic cost of transport for a life cycle measure of merit, the following equation for present worth may be applied:

$$\text{PW of } \overline{\text{ECT}} = \text{PW of RFR} + \frac{\sum_{j=1}^{n} (P/F_j, i, j)\ (\text{inv. cost})}{\sum_{j=1}^{n} (ATC_j)}$$

AVERAGE ANNUAL BENEFIT INDEX

A useful measure of merit which can be applied to alternative projects, equipments, ventures, etc., is the average annual benefit index. This index is the ratio of the equivalent annual cash flow to the investment.

$$\text{AABI} = \frac{(A/P, i, n) \sum_{j=0}^{n} F_j\ (P/F_j, i, n)}{P_O}$$

Example 3-9

In the example of the three tankers designed for speeds of 14, 15, and 16 knots respectively, the total net cash flows discounted at 16.4 percent for the 20-year life are:

	Discounted NCF	*Vessel Cost*
Tanker A	\$96,650,540	\$75,670,140
Tanker B	\$97,557,860	\$76,108,260
Tanker C	\$97,954,930	\$76,479,060

$$\text{AABI}_A = \frac{.1723\ (96{,}650{,}540)}{75{,}670{,}140} = .2201$$

$$\text{AABI}_B = \frac{.1723\ (97{,}557{,}860)}{76{,}108{,}260} = .2209$$

$$\text{AABI}_C = \frac{.1723\ (97{,}954{,}930)}{76{,}479{,}060} = .2207$$

NET PRESENT VALUE INDEX

The net present value index is the ratio of discounted net present worth to investment.

$$\text{NPVI} = \frac{\sum_{j=1}^{n} F_j\ (P/F_j, i, n)}{\text{investment}}$$

Example 3-10

For the three tanker designs, the NPVI is as follows:

$$NPVI_A = \frac{96{,}650{,}540}{75{,}670{,}140} = 1.2773$$

$$NPVI_B = \frac{97{,}557{,}860}{76{,}108{,}260} = 1.2818$$

$$NPVI_C = \frac{97{,}954{,}930}{76{,}479{,}060} = 1.2808$$

Using this index, the three designs are ranked as: *B* or 15 knots, *C* or 16 knots, and *A* or 14 knots.

BREAK-EVEN COMPARISONS

Some engineers and managers find it useful to organize cost information for comparison of project alternatives into two categories: (1) fixed costs for design, construction, acquisition, and installation; and (2) variable costs which are associated with the maintenance, operating labor, energy, insurance, etc. These costs may be summed and compared to a revenue or cash flow benefit stream to find a break-even point on the time scale (or other time-related scale such as units produced, container boxes transported, round trips completed, ton-miles or tonne-miles per year transported, etc.). The break-even point occurs when the income or benefit cash stream equals the sum of the fixed and variable cash flow streams which are plotted versus time or the time related variable as shown in Figure 3-3. Since future costs and future revenues are reduced by discounting for time value of money and increased by inflation, it is desirable to include these features. However, this is not always done because considerations of time value and inflation introduce nonlinearity into an otherwise somewhat simple linear problem. For example, if the cost per ton of steel erected is a constant value V in dollars/ton and N is the number of tons erected, then NV is linearly dependent on production. C_F is the fixed cost which is independent of tons of steel. C_T, the total cost, equals $C_F + NV$. The profit or benefit from each ton of erected steel is P and the net sales contribution for each ton is S. Therefore, ignoring taxes:

$$P = NS - C_T = NS - (C_F + NV)$$

Since the break-even occurs at P = 0:

$$NS - C_F - NV = 0$$

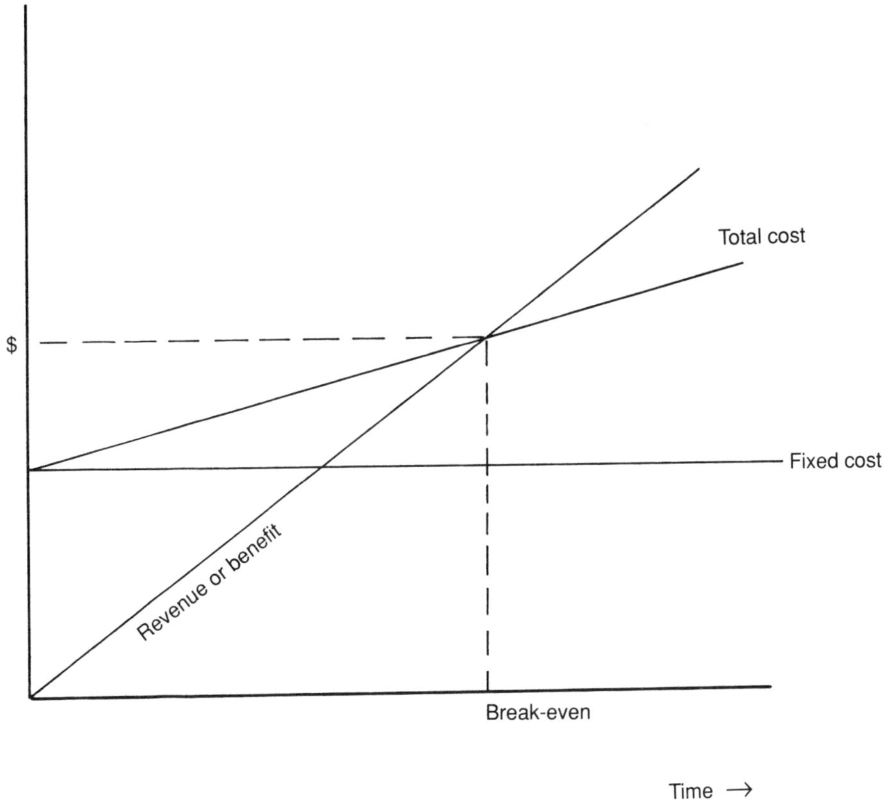

Figure 3-3. Break-even analysis

$$N = \frac{C_F}{S - V}$$

where N is the tons of erected steel per year necessary for break-even. The value S – V is called the contribution or the variable amount contributed to each unit of sales which, in this case, is the benefit of a ton of erected steel.

Example 3-11

A marine engineer must select a 175-HP motor for an anchor windlass. Motor *A*, with an efficiency of 91.7 percent, costs $3,933 and has a life of 20 years and no salvage value. Motor *B*, with an efficiency of 88.7 percent, costs \$3,200 and has a 20-year life and no salvage value. Operating and maintenance cost is \$220 per year for each motor. The cost of power is \$.042/kwhr. Money is worth 7 percent. What is the break-even point for selection of the motors?

Motor A	*Motor B*
P = \$3,933	P = \$3,200
Eff. = 91.7%	Eff. = 88.7%
n = 20	n = 20
i = .07	i = .07
Op = \$220	Op = \$220

Motor A

$$C_T = P\,(A/P, i, n) + Op + H\,\frac{(175\text{ Hp})\left(.746\,\frac{\text{kw}}{\text{np}}\right)}{\text{Eff.}}\left(.042\,\frac{1}{\text{kwhr}}\right)$$

$$C_T = 3{,}933\,(.0944) + 220 + \frac{5.483}{.917}\,H$$

$$C_T = 591.27 + 5.979\,H$$

Motor B

$$C_T = 3{,}200\,(.0944) + 220 + \frac{5.483}{.887}\,H$$

$$C_T = 522.08 + 6.182\,H$$

$$C_{T_A} = C_{T_B} \text{ at break–even time}$$

$$591.27 + 5.979\,H = 522.08 + 6.182\,H$$

$$H = \frac{69.19}{.203} = 340.8 \text{ hours}$$

SENSITIVITY ANALYSIS

In a deterministic approach to engineering economic analysis, it is assumed that the analyst knows all values of all independent variables and these variables are constants. Since the "what if" question is frequently asked during the decision process, the deterministic approach is not responsive unless the analysis has been completed for a range of the key input variables.

If the analysis is computer based, it is feasible to explore the impact of a number of independent variables on the solution. Table 3-1 is a list of the independent variables which may be considered in a comprehensive bulk carrier computer program used by the authors.

TABLE 3-1

Variables

Annual throughput
Density of cargo (API for oil or specific gravity for dry bulk or LNG)
One-way distance
Operating days per year
Shaft horsepower
Fuel rate
Loaded speed
Ballast speed
Loaded and ballast speeds
Deadweight
Cubic capacity
Time to load
Time to dock
Time to discharge
Amount of cargo lightered
Pump efficiency during lightering
Fixed lightering time
Variable lightering time per ton
Cost of lightering per ton
Ship price
Construction differential subsidy rate
Fixed terminal cost
Variable terminal cost
Depreciation period, ship
Depreciation period, terminal
Amortization period, ship
Amortization period, terminal
Average cost per crewperson
Fixed terminal size
Variable terminal size
Throughput cost per ton, terminal
Annual miscellaneous costs
Bunker cost at loading port
Bunker cost at discharge port
Engine maintenance and repair factor

Variables *(continued)*

Hull maintenance and repair factor
Value of cargo per unit
Discount rate, ship system
Discount rate, terminal system
Tax rate
Proportion of owner's equity, ship
Proportion of owner's equity, terminal
Salvage value, ship
Salvage value, terminal
Declining balance factor, ship
Declining balance factor, terminal
Interest rate on debt, ship
Interest rate on debt, terminal
Rate of insurance
Boil-off rate, ballast (LNG only)
Boil-off rate, loaded (LNG only)
Boil-off rate, both loaded and ballast (LNG only)
Escalation rate, bunkers
Escalation rate, crew costs
Escalation rate, miscellaneous costs
Escalation rate, insurance
Escalation rate, maintenance and repair
Escalation rate, terminal
Dwt factor, ship size/terminal size ratio
Escalation of value of cargo

Example 3-12

A 197,428-dwt vessel is being designed to make 12.77 trips per year to transport 2.5 million tons of crude oil per year. A computer was used to calculate the discounted average required freight rate for a change in the average cost of a crew member from $27,567 to $35,224 per year.

Cost per Crew Member, $	*Required Freight Rate, $/Ton*
27,567	7.258
29,098	7.268
30,630	7.280
32,161	7.293
33,693	7.305
35,224	7.316

PROJECT PROGRESS PAYMENT COSTS

In a major venture, such as the construction of a ship, the time between contract signing and delivery of the vessel is usually a period of years. A few progressive shipbuilders seek to control their costs by reducing construction time to 12 months. This requires the construction of an existing design and the prepurchase of the long-lead-time equipments, such as the propulsion engine.

A typical commercial vessel contract is about 27 months. During this period, the shipbuilder requires progress payments from the buyer. Such a schedule of payments might be as follows.

Event	*Months before Delivery*	*Fraction of Contract Price*
Contract signature	27	.10
1/3 of steel ordered	24	.10
2/3 of steel ordered	22	.10
Keel laying	20	.10
15% of steel erected	18	.10
30% of steel erected	16	.09
50% of steel erected	14	.09
70% of steel erected	12	.09
90% of steel erected	9	.08
Launching	6	.07
Dock trial	3	.02
Delivery	0	.05
End of warranty	(12)	.01
Total		100%

Therefore, in determining the investment at time zero on the cash flow time line from the owner's viewpoint, it is necessary to include the cost of progress payments advanced to the builder. In this case, if the cost of the ship is $75,670,000 and the cost of money is 10 percent, the zero time delivered cost of the vessel for purposes of analysis, assuming monthly compounding (i.e., i = .10/12 = .0083) is:

Month Before / after Delivery	*Payment, $*	*Payment (F / P, i, n), $*
27	7,567,000	9,466,317
24	7,567,000	9,231,740
22	7,567,000	9,080,400
20	7,567,000	8,936,627
18	7,567,000	8,785,287

16	6,810,300	7,777,362
14	6,810,300	7,647,966
12	6,810,300	7,525,381
9	6,053,600	6,525,780
6	5,296,900	5,567,041
3	1,513,400	1,551,235
0	3,783,500	3,783,500
(12)	756,700	684,980
Ship cost at time zero		86,653,616

It is obvious that a reduced time for design and construction will benefit the owner as well as the builder.

REFERENCES AND BIBLIOGRAPHY

Giffin, Walter C. *Introduction to Operations Engineering.* Homewood, Ill.: Richard D. Irwin, 1971. Includes a brief but nicely constructed chapter on the economics of decision making.

Grant, Eugene L., W. Grant Ireson, and Richard S. Leavenworth. *Principles of Engineering Economy.* New York: John Wiley & Sons, 1982. Text covers basic materials and some advanced topics. Sensitivity analysis and method of comparison among alternatives, including present worth, equivalent annual cash flow, and internal rate of return, are covered in detail.

Morris, William T. *The Analysis of Management Decisions.* Homewood, Ill.: Richard D. Irwin, 1966. A well-written description of the elements of deterministic decision making using principles of engineering economics.

Scher, Robert, and Harry Benford. *Some Aspects of Fuel Economy in Bulk Carrier Design and Operation.* Ann Arbor: University of Michigan, 1980.

QUESTIONS AND PROJECTS

1. By the convention used in engineering economic analysis, when do the following transactions occur in a period?

Present sum	P
Annual sum	A
Future sum	F

2. What are the risks in accepting a point of reference less than the entire firm for the comparison of the economic consequences of alternatives?

3. How do past costs (i.e., sunk costs) affect decisions concerning the future?
4. What is the meaning of rate of return on investment? How is this related to the MARR?
5. How does the engineering viewpoint of economic studies differ from the accounting viewpoint?
6. What do we mean by sensitivity? How is this concept employed by the decision maker in making a choice?
7. Discuss the problems encountered when the alternatives in a decision problem have different lengths of economic life.
8. In solving for the internal rate of return, what precautions are necessary to assure that the value of *i* is correct?
9. As $n \to \alpha$, what is the value of net present worth?
10. Do you see any advantage in using equivalent annual net cash flow as a measure of merit compared to net present worth? If so, what is it?
11. Required freight rate is usually calculated for the first year or a typical year of operation. Can you suggest a method for calculating an RFR which would represent the economic life of a vessel?
12. Employ the computer program TABLE to generate a table for n = 20 and i = 1% to 18%.
13. Solving the polynomial:

$$\sum_{m=0}^{n} \frac{X_i}{(1+i)^m} = 0$$

for the internal rate of return is a complicated problem for a large number of periods. Develop a computer algorithm to solve the problem.

PROBLEMS

1. Three alternatives are available to the decision maker who must select one. They are as follows:

	A	*B*	*C*
Initial cost	$500	$300	$400
Annual cost saving	$150	$100	$350
Equipment life, yrs.	6	6	3

For the three-year life, a second initial cost of $400 is required at the end of three years. On a net present worth basis and if i = 10 percent, which is the best choice?

2. Certain cargo tanks of a new vessel are scheduled for cleaning and painting at five-year intervals at a cost of $50,000. A new process is available which extends the interval to 10 years at a cost of $130,000. The anticipated life of the vessel is 20 years and money is worth 7 percent. Which is the lower-cost approach on present worth basis?
3. Three vessels of the same type have annual increasing maintenance costs for 10 years as follows (in thousands of dollars):

Vessel	*A*	*B*	*C*
First-year cost	500	480	510
Annual income	45	55	40

   What is the present worth comparison of the maintenance cost? Which vessel is lowest cost? The value of money is 7 percent.
4. Compare the costs of ship maintenance in problem 3 on an equivalent annual basis for 10 years at 7 percent cost of money.
5. A selection must be made between two alternatives for a new main feed pump. The pumps will last 20 years and have the following costs:

Pump	*A*	*B*
Installed cost	$80,500	$85,000
Salvage value	5,000	6,000
First-year operating & maintenance	2,000	1,700
Annual increase in O & M costs	400	350

   Money has a 10 percent time value. Which is the least expensive of the two alternatives?
6. A terminal productivity improvement modification has three alternatives with the following costs and benefits for the four-year life:

Year	*Do Nothing*	*Alternative A*	*Alternative B*
0	0	50,000	75,000
1	0	10,000	20,000
2-4	0	20,000	25,000

   If the corporate MARR is 17 percent, select the best alternative.
7. A recommended winch change will cost $10,000 and save $1,000 each year during the 20-year ship life. The MARR is 15 percent. The alternative is to do nothing. What is the correct decision?
8. During the first year of operation, a new tanker with 204,110 cargo deadweight has the following expected costs:

Crew	$ 628,000
Fuel	4,700,000
Maintenance	467,000
Insurance	1,690,000
Miscellaneous	300,000

The cost of the vessel at delivery is $96,449,300. The vessel completes 12.25 trips/year. The value of money is 12 percent. Calculate the required freight rate.

9. How many round trips must a 200,000-dwt tanker complete each year if the current freight rate is $10/ton, the operating costs are $7.9 million per year, the ship at delivery is $95 million, and the cost of money is 10 percent?
10. Three preliminary designs of tankers are in preparation. These designs have the following characteristics:

Ship	*A*	*B*	*C*
Length (ft)	1039.5	1157.1	1122.6
L/B	5.4	5.96	5.61
L/D	13.83	14.46	14.46
Block	.801	.850	.838
SHP	22,000	16,000	17,000
Average speed, knots	12.5	10.5	10.7
C dwt	204,400	240,800	237,100
Operating cost, $	7,750,000	7,100,000	7,210,000
Ship cost, $	96,800,000	108,900,000	106,700,000
Trips/year	12.30	10.30	10.5
Days port time/voyage	5.5	5.5	5.5

Compare the economic merit of these vessels on the basis of required freight rate if the life is 20 years and i = 12 percent.

11. For the ships in problem 10, the value of the cargo is $100/ton. Compare the three designs on the basis of economic cost of transport if the inventory cost of cargo is 10 percent per year.
12. You have an opportunity to purchase one of two available container vessels for a shuttle container trade. The vessels are both 15 years old. You plan to operate the vessels for 5 years, then sell for scrap. You have estimated the net cash flow for each of the 5 years for each vessel. Your management wishes to see the average annual benefit index. Calculate the index for the following data with time value of money equal to 9 percent:

	A	*B*
Vessel price	$5,800,000	$6,790,000
Year 1	1,000,000	2,000,000
Year 2	3,000,000	3,000,000
Year 3	3,500,000	4,000,000
Year 4	3,500,000	4,500,000
Year 5	3,500,000	4,500,000
Scrap	950,000	1,200,000

13. A new heavy lift boom is added to a break-bulk vessel trading on the west coast of Africa. The boom has an installed cost of $1.5 million. The increased revenue will be $400,000 per year. What will be the payback period at $i = 0$ and $i = .09$ if maintenance is $5,000 per year?

Chapter 4

Engineering Economic Analysis under Risk or Uncertainty

It is a truth very certain that when it is not in our power to determine what is true, we ought to follow what is most probable.
—*René Descartes,* Discourse on Method, *Part III*

INTRODUCTION

An objective of this text is to provide the marine engineer with the tools and methods needed to logically select the best from a comprehensive set of alternatives for which an economic measure-of-merit has been established. In the deterministic case, it is assumed that all present and future data required to calculate the measure-of-merit is known with certainty. This assumption, although obviously flawed, remains the most common approach to engineering economic analysis. However, there is a vast body of knowledge developed to assist the analyst in evaluating alternatives when some or all of the present and future values of the independent variables are less than certain. This chapter will review the fundamentals necessary for decision making under uncertainty or risk. These fundamentals will be applied in future chapters to problems of marine transportation and ship design.

Nondeterministic engineering economic decisions exist in two basic forms for the input variables. Each form requires a different approach. In the first form, i.e., a decision under uncertainty, there are no available probability measures for the independent variables and no logical method of making such measures. For this case, decision method researchers have suggested several principles which are useful in focusing the decision maker's attention on the essentials of the problems. In the second form, a decision under risk, the independent variables of the problem, such as escalation of labor costs, cargo load factors, fuel costs, insurance costs, etc., are known in terms of probability measures which take the form of probability distribution functions and describe the likelihood of a specific value of the variable, i.e., the variables are random variables. For example, in throwing a single perfect die, we do not know the outcome in advance. However, we do know that there is a 1/6 chance for each of the possible events, i.e., outcome values of 1, 2, 3, 4, 5 or 6 as shown in Figure 4-1.

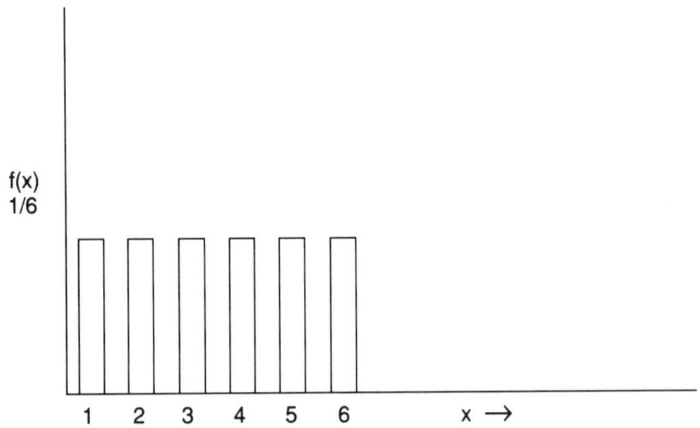

Figure 4-1. Discrete distribution for casting a single die

A MODEL FOR DISCRETE CHOICES UNDER UNCERTAINTY

To consider decisions under uncertainty, assume we are faced with a selection from among four alternatives. For each alternative, the measure-of-merit, for example, present worth of net cash flows, is calculated for each alternative under each of the five discrete conditions which describe the significant combination of future events. This information can be represented by a four-by-five matrix as shown in Example 4-1.

Example 4-1

A Payoff Matrix

	S_1	S_2	S_3	S_4	S_5
a_1	105	100	130	120	102
a_2	393	91	140	400	100
a_3	490	56	110	510	130
a_4	246	36	100	490	120

Inspection of the matrix shows that regardless of the future event or state of nature S_j, the outcome of the selection of alternative a_3 is superior to that of alternative a_4. Therefore, alternative a_3 dominates alternative a_4, which can be eliminated from further consideration. This is the Principle of Dominance.

Laplace suggested that in the absence of knowledge concerning the probabilities of future states, it is reasonable to assume that all events in a decision model are equally likely. Under the Laplace Principle, the analyst would select the alternative which:

$$P* = \max_{i=1}^{n} \left[\sum_{j=1}^{m} P_{ij} / m \right]$$

Evaluation of the matrix under this rule leads to the following vector:

a_1	112.0
a_2	224.6
a_3	259.0

where alternative a_3 is the best choice.

Decision rules like the Laplace Principle have obvious implications for the psychology of the decision maker. An optimistic decision maker may wish to employ the Maximax Principle which requires the selection of the alternative which is:

$$P* = \max_{i=1}^{n} \left(\max_{j=1}^{m} P_{ij} \right)$$

The maximum payoff vector is:

a_1	130
a_2	400
a_3	510

where alternative a_3 is the best choice.

The decision maker with a low tolerance for risk may wish to apply the Maximin Principle suggested by Wald. This very conservative principle requires the selection of the alternative which is:

$$P* = \max_{i=1}^{n} \left(\min_{j=1}^{m} P_{ij} \right)$$

The payoff vector for the Maximin Principle is:

a_1	100
a_2	91
a_3	56

leading to selection of alternative a_1.

Savage has proposed the philosophic argument that a decision maker should seek to minimize the regret which he will experience when the passage of time unfolds the event which actually occurs. To employ the Minimax Regret Principle, it is necessary to generate a regret matrix from the original payoff matrix. Regret is defined as:

$$R(P_{ij}) = \left[\max_{i=1}^{n} P_{ij} - P_{ij} \right]_{j=1}^{m}$$

and the regret matrix is:

	S_1	S_2	S_3	S_4	S_5
a_1	385	0	10	390	28
a_2	97	9	0	110	30
a_3	0	44	30	0	0

The selection criterion is:

$$P* = \min_{i=1}^{n} \left(\max_{j=1}^{m} R(P_{ij}) \right)$$

which leads to the following regret vector:

a_1	390
a_2	110
a_3	44

where alternative a_3 is the best choice under the Minimax Regret Principle.

Hurwicz developed a selection criterion which recognizes the problem boundary conditions of the Maximax and Maximin decision rules. The range of payoffs for each act are plotted as linear functions of the decision maker's optimism concerning the future events as defined for the decision matrix. The Hurwicz payoff function is defined as:

$$P_i = \left(\max_{j=1}^{m} P_{ij} \right) \alpha + \left(\min_{j=1}^{m} P_{ij} \right) (1 - \alpha)$$

where α is the optimism index arbitrarily assigned a scale of 0 to 1. When these functions are plotted, the decision maker has a tool to select an alternative based on his personal optimism.

Using these criteria, Figure 4-2 suggests that the decision maker would select a_1 for an optimism between 0 and .03, a_2 for optimism between .03 and .24, and a_3 for optimism index above .24. In short, great pessimism will lead the decision maker to alternative a_1 while an optimist will select a_3.

PROBABILITY AND STATISTICS FOR DECISIONS UNDER RISK

There are three useful interpretations of probability for the engineer engaged in economic analysis. (The mathematician may not accept all definitions as valid but, for the pragmatic engineer, all are useful.)

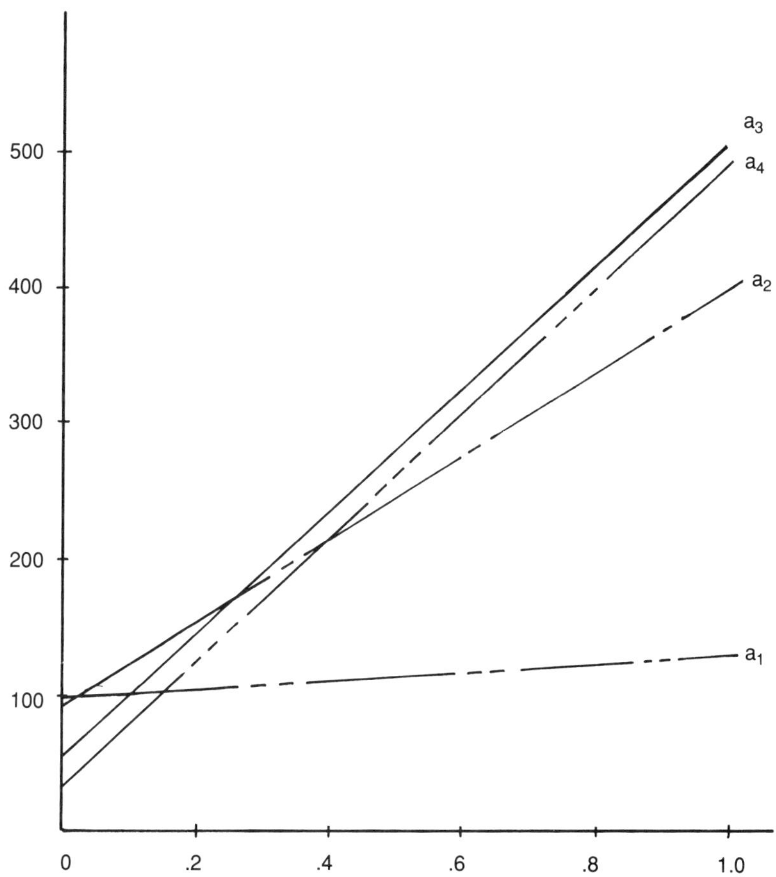

Figure 4-2. Hurwicz payoff functions

1. *Classical Probability.* If an event can occur in N different ways and n of these have the attribute E, then the probability of E is n divided by N, i.e.,

$$\text{Pr (E)} = \text{n/N}$$

For example, in throwing two dice, each die can result in numbers 1 to 6 on a throw. Therefore, there are $6 \times 6 = 36$ possible outcomes for each cast of the two dice. There are six ways in which the dice can sum to 7.

Die #1	*Die #2*
6	1
1	6
5	2
2	5
4	3
3	4

Therefore, the probability of a seven resulting from the single cast is:

$$\text{Pr (7)} = 6/36 = 1/6$$

2. *Objective Probability.* If an experiment is conducted N times and event E occurs n times, the limit of n/N as N tends to infinity is the probability of event E.

$$\text{Pr (E)} = \lim_{N \to \infty} \text{n/N}$$

Statistical methods of sampling permit us to obtain accurate estimates of such probability. Carefully planned and executed experiments are necessary to achieve required results.
3. *Subjective (or Personal) Probability.* Given a set of mutually exclusive and collectively exhaustive events, the probability of occurrence of any one of the events is a quantification of the degree of belief which a knowledgeable person has in the occurrence of that event expressed as a measure on a scale from 0 to 1 or:

$$0 \leq \text{Pr (E)} \leq 1$$

For engineering economic analysis, we will accept and use all of these interpretations of probability. Our objective will be to obtain a probability measure on a set of events such as the values which a random variable may assume and to use this measure to express our lack of knowledge concerning a specific value for an input variable to a problem. Good has described a rational procedure for assigning subjective probabilities to variable values. Schlaifer argues that "reasonable men base the probabilities which they assign to events in the real world on their experience with events in the real world, and when two reasonable men have had roughly the same experience with a certain kind of event, they assign it roughly the same probability." This is particularly useful in the complicated task of assigning probabilities when statistical information for some input variables such as annual growth of labor costs, insurance costs, maintenance costs, etc., is not available.

SET NOTATION

Set theory has been the formal tool for developing a calculus of probabilities. A brief review of set theory is needed to develop the principal axioms of probability.

Given a space *S*, *E* and *F* are defined as disjoint subsets of *S* provided they are both in *S* but are separate from one another, Figure 4-3.

The space of subsets *A* and/or *B* is called the union of *A* and *B*, i.e., $A \cup B$, Figure 4-4.

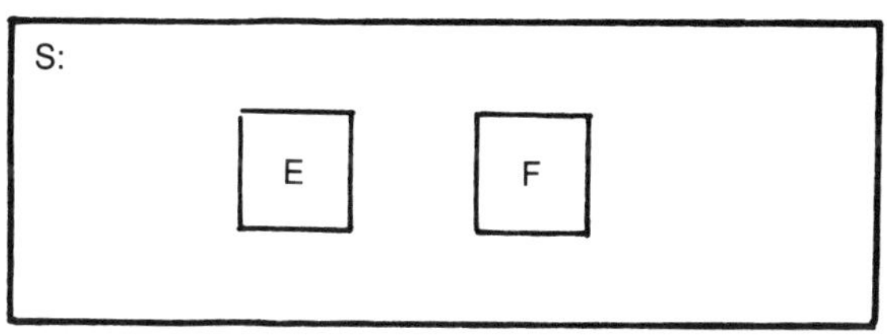

Figure 4-3. Sample space with disjoint sets

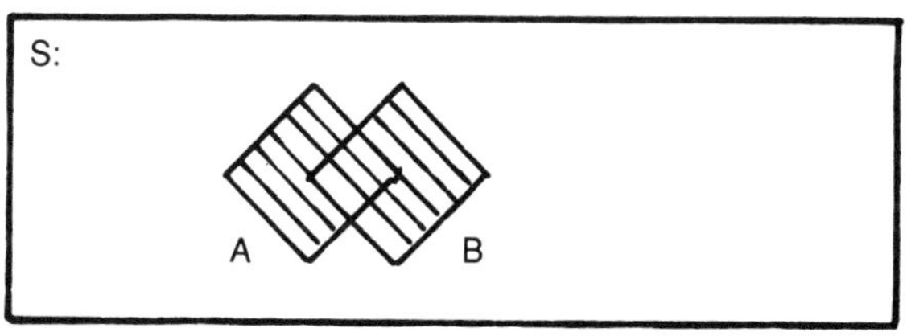

Figure 4-4. Union of sets A *and* B

The space which includes subsets *A* and *B* is the intersect of *A* and *B*, i.e., $A \cap B$, Figure 4-5.

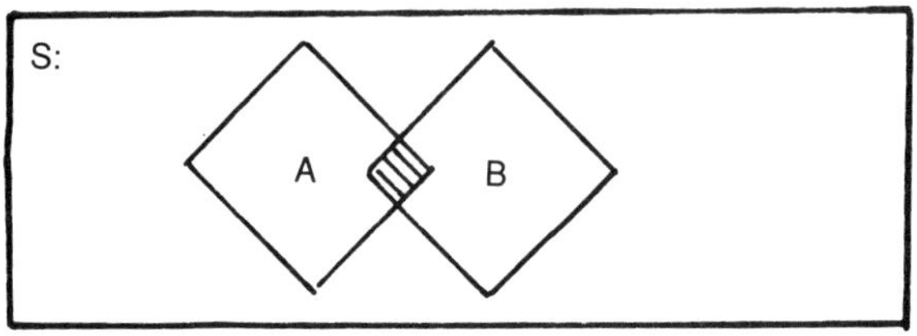

Figure 4-5. Intersect of sets A *and* B

FUNDAMENTAL THEOREMS OF PROBABILITY

E is a subset of the field *S* in Figure 4-3 which satisfies the following conditions:

Theorem 1. If *E* is any event in *S*, the probability of *E* is:

$$\Pr(E) > 0$$

$$\text{where: } \Pr(S) = 1$$

Theorem 2. If $E_1, E_2, E_3, \ldots E_n$ is any sequence of disjoint events, i.e., the events are mutually exclusive, then:

$$\Pr(E_1 \cup E_2 \cup E_3 \ldots \cup E_n) = \Pr(E_1) + \Pr(E_2) + \Pr(E_3) \ldots + \Pr(E_n)$$

Theorem 3. If $\bar{E}$ is the complement of *E*, i.e., $\bar{E}$ includes all events of *S* which are not part of *E*, then:

$$E \cup \bar{E} = S$$

$$\Pr(S) = 1$$

$$\Pr(\bar{E}) = 1 - \Pr(E)$$

Example 4-2

In our ship repair business, experience has shown that, in fully competitive situations, we have a 50 percent chance of submitting a successful bid, i.e., all events are equally likely. Next week, we will submit four bids. What is the probability that we will get at least one contract?

The sample space must include all the possible outcome combinations. In this case of winning or losing on each of four bids, there are 2^4 or 16 outcomes. Since each bid has the same chance of winning or losing, we can assign equal probability to each of the 16 events in the sample space or 1/16. Let E be the probability that we receive no contracts. There is only one event, #9, for receiving no contracts, which can be illustrated by arraying all the events in the space.

Event	*Bid #1*	*Bid #2*	*Bid #3*	*Bid #4*
1	W	L	L	L
2	L	W	L	L
3	L	L	W	L
4	L	L	L	W
5	L	W	W	W
6	W	L	W	W
7	W	W	L	W
8	W	W	W	L
9	L	L	L	L
10	W	W	W	W
11	L	L	W	W
12	W	W	L	L
13	L	W	L	W
14	W	L	W	W
15	W	L	W	L
16	L	W	L	W

The probability of receiving at least one contract is the complement of E or:

$$\Pr(\bar{E}) = 1 - \Pr(E) = 1 - 1/16 = 15/16$$

Theorem 4. If E_1 and E_2 are any two events in the sample space, then:

$$\Pr(E_1 \cup E_2) = \Pr(E_1) + \Pr(E_2) - \Pr(E_1 \cap E_2)$$

which can be extended to three events as follows:

$$\begin{aligned}\Pr(E_1 \cup E_2 \cup E_3) = {} & \Pr(E_1) + \Pr(E_2) + \Pr(E_3) \\ & - \Pr(E_1 \cap E_2) - \Pr(E_1 \cap E_3) \\ & - \Pr(E_2 \cap E_3) + \Pr(E_1 \cap E_2 \cap E_3)\end{aligned}$$

and the general case is found in Pfeiffer as well as most texts on probability. A Venn diagram shown in Figure 4-6 is used to illustrate the sets involved in the probability of the union of three events.

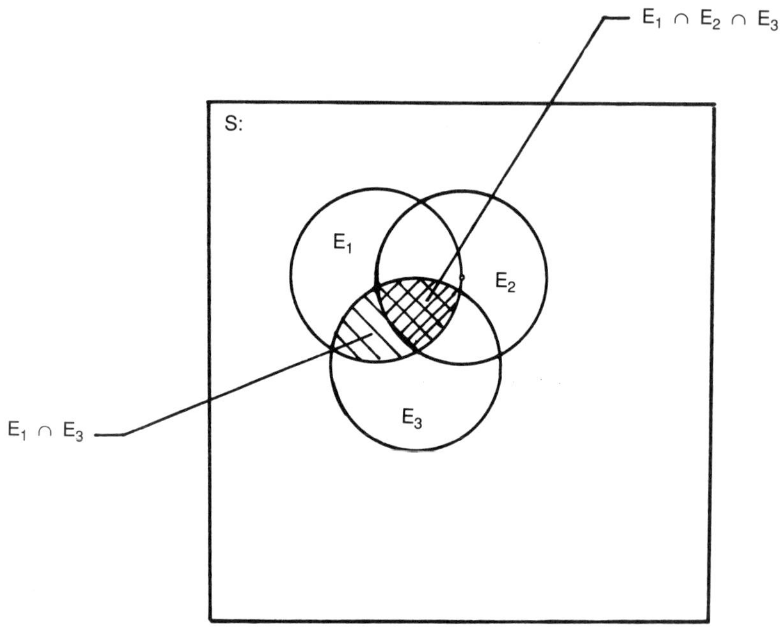

Figure 4-6. Venn diagram showing the union of three events

CONDITIONAL PROBABILITY AND BAYES THEOREM

If *A* and *B* are two events so that Pr (A) > 0 and Pr (B | A) is the probability of *B* given that *A* has occurred, then *A* is the revised sample space replacing *S* or:

$$\Pr(B \mid A) = \frac{\Pr(A \cap B)}{\Pr(A)}$$

and:

$$\Pr(A \cap B) = \Pr(A)\ \Pr(B \mid A)\,\}$$

Pr (B | A) is called the conditional probability that *B* occurs given that *A* has occurred.

If *A* and *B* are independent events, i.e., the occurrence of *B* is not conditional on the prior occurrence of *A*, then:

$$\Pr(A \cap B) = \Pr(A)\Pr(B)$$

Bayes Theorem, which facilitates the combining of initial or prior probability of the occurrence of some event with experimental data to obtain revised probability or posterior probability, follows from the definition of

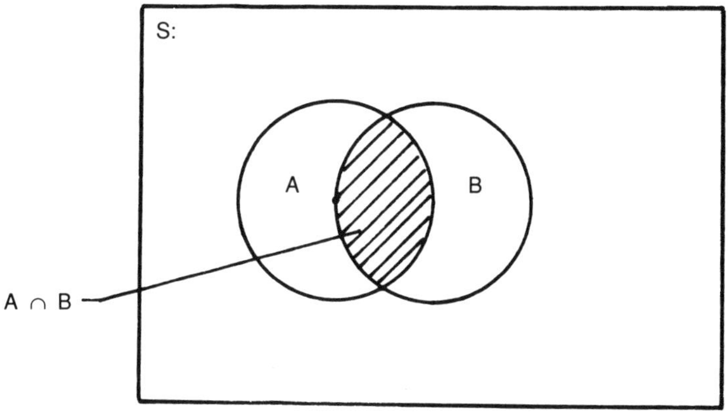

Figure 4-7. Sample space showing conditional probability of B *given* A *has occurred*

conditional probability. Given a prior probability $Pr(A_j)$, the conditional probability $Pr(B \mid A_j)$, and the reduced sample space, Bayes Theorem is stated as follows:

$$Pr(A_j \mid B) = Pr(A_j) \frac{Pr(B \mid A_j)}{\sum_{j=1}^{n} Pr(B \mid A_j)\, P(A_j)}$$

Bayes Theorem has many applications, one of which is Bayesian Analysis, which will be used in shipping venture analysis where market surveys or experimental data are used to revise prior information concerning passenger availability, cargo availability, etc.

Example 4-3

In a fast ferry transportation study, the management experience indicated a passenger availability per one-way trip as follows:

Prior Distribution

Passengers	*Pr (Passengers)*
120	.2
160	.7
180	.1

A market survey is conducted. The resulting sampling distribution is as follows:

Passengers	*Pr (Passengers)*
120	.2
160	.5
180	.3

Then, applying Bayes Theorem:

Event	*Passengers*	*Pr* (A_j) *Prior*	*Pr* $(B \mid A_j)$ *Conditional*	*Joint*	*Pr* $(A_j \mid B)$ *Posterior*
A_1	120	.2	.2	.04	.0952
A_2	160	.7	.5	.35	.8334
A_3	180	.1	.3	.03	.0714
		1.0		.42	1.0000

This indicates the probability of 160 passenger load is significantly increased from .7 to .83 by the sampling information.

COMBINATIONAL ANALYSIS

Counting the number of events or the sample points in a sample space, *S* is sometimes a complicated process. For a single case, consider a decision model defined by three independent variables, each of which has three discrete values. This will require $n_1 \times n_2 \times n_3$ or 27 events for sample points for the decision model.

If our problem has *n* distinct objects which we wish to arrange *r* at a time to define an event, a permutation of *n* objects taken *r* at a time is required.

$$nPr = \frac{n!}{(n-r)!}$$

In a permutation, the arrangement *a b c* is different from *b a c*. If we wish to make a selection without regard for order of the arrangement, a combination is required. The total combinations of *r* objects selected from a total of *n* taken *r* at a time is:

$$nCr = \frac{n!}{r!\,(n-r)!}$$

Example 4-4

A vessel is awaiting the delivery of 20 containers which will arrive randomly. Nine of the containers are empty, eight are refrigerated and three contain machinery. What is the probability that the first three arrivals are refrigerated containers and the probability that at least one of the first three contains machinery?

Probability that the first three are refrigerated containers:

$$\Pr = \frac{8C3}{20C3} = \frac{14}{285} \approx .0491$$

Probability that none are machinery:

$$\Pr = \frac{17C3}{20C3} = \frac{34}{57} \approx .5965$$

Probability that at least one is machinery:

$$\Pr = 1 - \frac{34}{57} = \frac{23}{57} \approx .4035$$

DISCRETE RANDOM VARIABLE DISTRIBUTIONS

In engineering economic analysis, it is frequently useful to build models which utilize discrete variables to provide a simplified representation of the problem. These variables may include representations of annual growth rates in costs, revenues, tax rates, ship speed, time in port, etc.

Given a sample space, it is possible to assign a number representing a variable to each point in the space. This is a function defined on the sample space. A random variable which has a finite number of values is called a discrete random variable (as compared to a continuous random variable, to be discussed later, which has an infinite number of values.)

If X is a discrete random variable which can assume values x_1, x_2, x_3, etc., these values may have probabilities:

$$P(X = x_j) = f(x_j) = 1, 2, 3, \ldots$$

Then the probability distribution is:

$$P(X = x) = f(x)$$

and from previous theorems:

$$f(x) \geq 0$$

$$\sum_{x=1}^{n} f(x) = 1$$

Example 4-5

A ship repair company plans to submit two bids for major work in a given week. The experience is that they have a 50 percent chance of success on each bid. What are the probability functions of success for these independent events?

$$\Pr(WW) = \tfrac{1}{2} \times \tfrac{1}{2} = \tfrac{1}{4}$$

$$\Pr(WL) = \tfrac{1}{2} \times \tfrac{1}{2} = 1/4$$

$$\Pr(LW) = \tfrac{1}{4}$$

$$\Pr(LL) = \tfrac{1}{4}$$

Let the random variable x be the number of wins. Then:

$$\Pr(x = 0) = \tfrac{1}{4}$$

$$\Pr(x = 1) = \Pr(WL \cup LW) = \tfrac{1}{4} + \tfrac{1}{4} = \tfrac{1}{2}$$

$$\Pr(x = 2) = \Pr(WW) = 1/4$$

or:

x	0	1	2
f (x)	1/4	1/2	1/4

This information can be plotted as a histogram. In the histogram, for discrete distribution, the width of each vertical bar is unity and the bar height is the probability. Therefore, for the discrete distribution, the area of the histogram, shown in Figure 4-8, is one or:

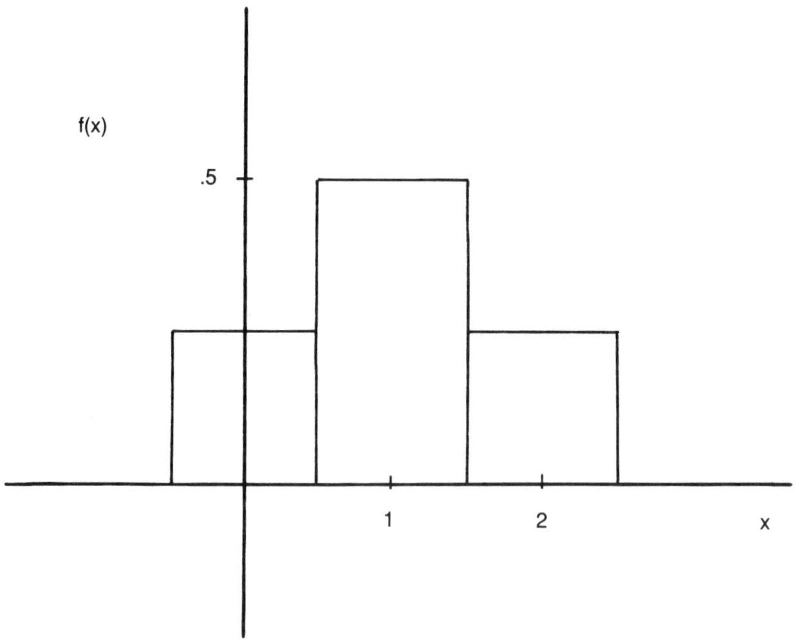

Figure 4-8. Histogram of discrete distribution

$$\sum f(x) = 1$$

A cumulative distribution function for the random variable x is:

$$P(X \leq x) = \sum F(x)$$

The cumulative distribution function $F(x)$ is obtained from the probability function $f(x)$ by summing:

$$F(x) = P(X \leq x) = \sum f(x)$$

The distribution function may be summed from the left or the right to form the cumulative function as shown in Figure 4-9 and Figure 4-10 where, respectively:

$$F(x) = \Pr(X \leq x)$$

$$1 - F(x) = \Pr(X \geq x)$$

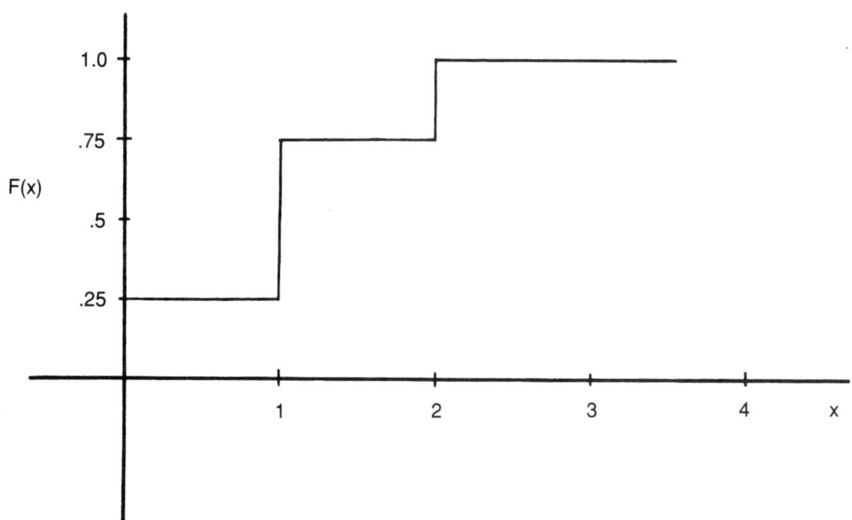

Figure 4-9. Left tail cumulative discrete distribution

DISCRETE DISTRIBUTIONS AND EXPECTATION

In all decision problems, the mathematical expectation of the random variables plays a key role. Given a discrete random variable X with possible values $x_1, x_2, x_3 \ldots xn$, the expectation of X is defined as follows:

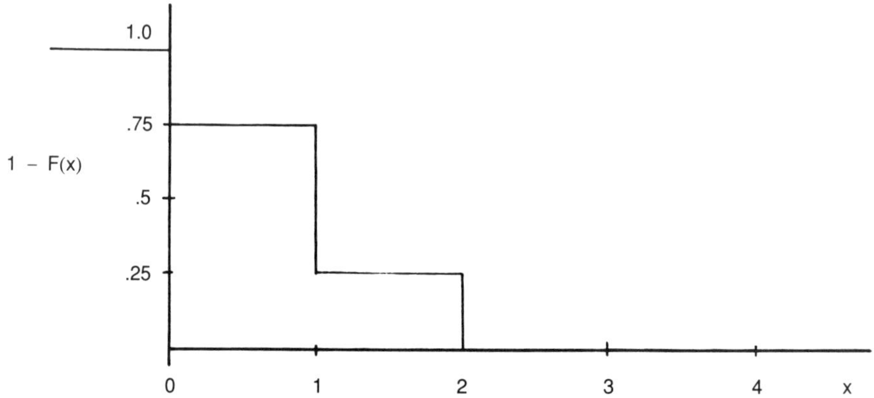

Figure 4-10. Right tail cumulative discrete distribution

$$E(X) = \sum_{j=1}^{n} x_j \Pr(X = x_j)$$

and if:

$$\Pr(X = x_j) = f(x_j)$$

$$E(X) = \sum_{j=1}^{n} x_j f(x_j)$$

which is equal to the mean μ.

Expectation of functions of discrete random variables are taken in the same manner, for example, given that Y = h(x), then *Y* is a random variable and

$$E(h(x)) = \sum_{j=1}^{n} h(x_j)\ f(x_j)$$

Example 4-6

Given that the daily fuel consumption varies with the square of the sea state as follows:

$$h(x) = c_1 x^2 + c_2$$

where:

$$c_1 = .010 \text{ and } c_2 = 27$$

On a voyage, the sea state likelihood is described by the probability distribution:

x_i	$f(x_i)$
1	.5
2	.3
3	.2

Then the expectation or mean of the daily fuel consumption is:

x_1	$f(x_j)$	$h(x_j)$	$h(x)$
1	.5	27.01	13.505
3	.3	27.09	8.127
5	.2	27.25	5.450
		E (h (x)) =	27.082

The mean is the first moment of the distribution about the origin. The rth moment of a random variable x about the mean μ is useful in describing a distribution:

$$\mu_r = \sum_{j=1}^{n} (x_j - \mu)^r \; f(x_i)$$

Higher moments describe the indicated properties:

μ_2	Variance
μ_3	Skewness, left or right
μ_4	Kurtosis or flatness

Relative measures of dispersion, skewness, and kurtosis have been defined for distributions in general and will be discussed under continuous distributions.

CONTINUOUS RANDOM VARIABLE DISTRIBUTION

A continuous random variable distribution function $F(x)$ is defined as follows:

$$F(x) = P(X \le x) = P(-\alpha < x \le x) = \int_{-\alpha}^{\alpha} f(\mu)\, dx$$

Example 4-7

Given that:

$$f(x) = \begin{cases} x & 0 < x < 10 \\ 0 & \text{elsewhere} \end{cases}$$

$$f(x) = \frac{1}{\mu_1 - \mu_2} \quad 0 < x < 10$$

$$f(x) = 0 \text{ elsewhere}$$

which is the uniform distribution:

$$\Pr\,(x_{1,} \leq x \leq x_2) = \int_{x_i}^{x_2} \frac{1}{10-0}\,dx$$

$$= \frac{x_2 - x_1}{\mu_1 - \mu_2} = \frac{x_2 - x_1}{10}$$

The probability is a function of the horizontal line as shown in Figure 4-11. The distribution function or cumulative probability is shown in Figure 4-12.

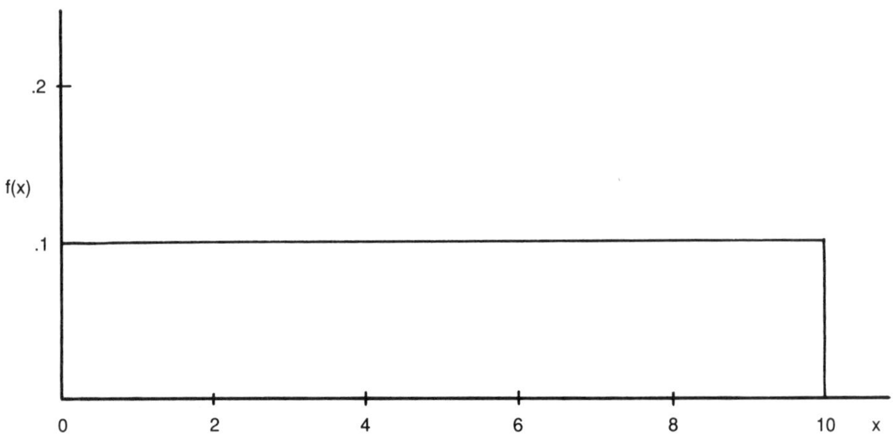

Figure 4-11. Uniform probability distribution

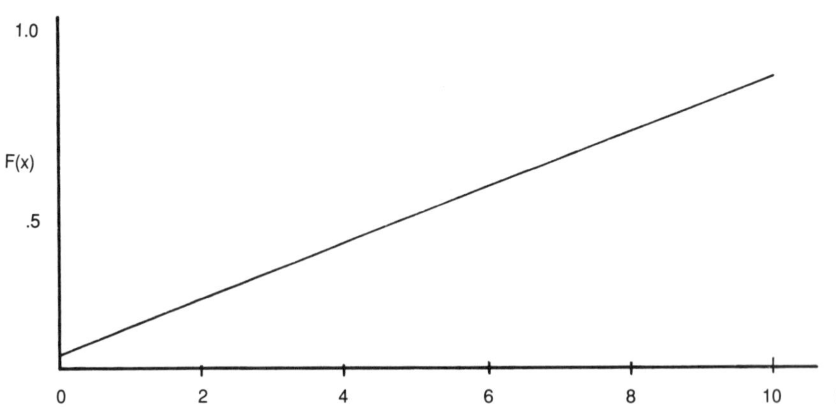

Figure 4-12. Cumulative uniform probability distribution

The normal probability distribution is the most useful model for economic decisions. The density function is:

$$f(x) = \frac{1}{\sigma\sqrt{2\pi}} \exp\left(-\frac{(x-\mu)^2}{2\sigma^2}\right)$$

where:

$$-\infty < \mu < \infty$$

$$\sigma > 0$$

The distribution function is:

$$F(x) = P(X \leq x) = \frac{1}{\sigma\sqrt{2\pi}} \int_{-\alpha}^{x'} e^{-(x-\mu)^2/2\sigma^2}\, dx$$

A very useful property of the normal distribution is that the mean is equal to the parameter μ and the standard deviation equals the second parameter σ. Figure 4-13 illustrates the normal density function. Figures 4-14 and 4-15 show the distribution function.

The moments of a continuous density function are defined:

$$\mu_k = E(x-\mu)^k = \int_{-\infty}^{\infty} (x-\mu)^k f(x)\, dx$$

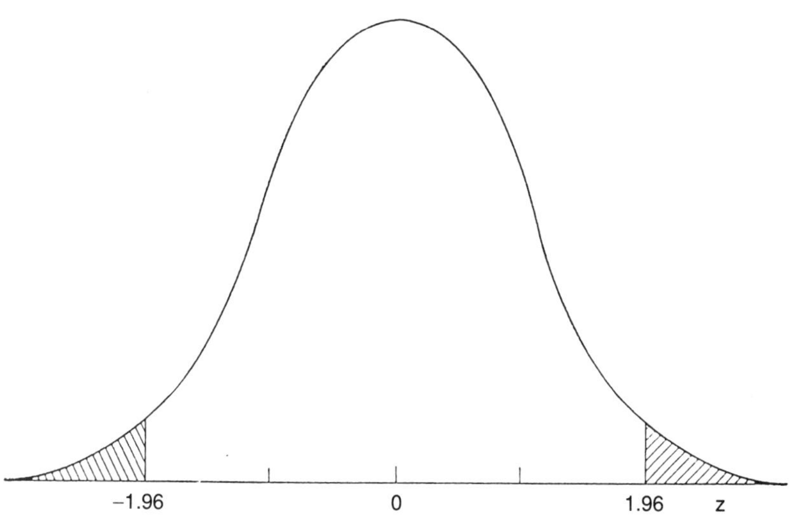

Figure 4-13. Normal density function

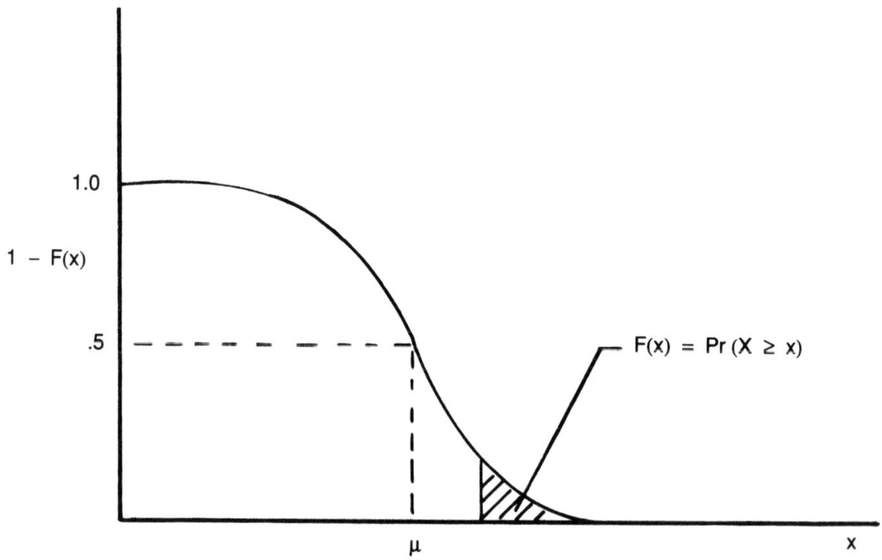

Figure 4-14. Normal right tail cumulative probability function

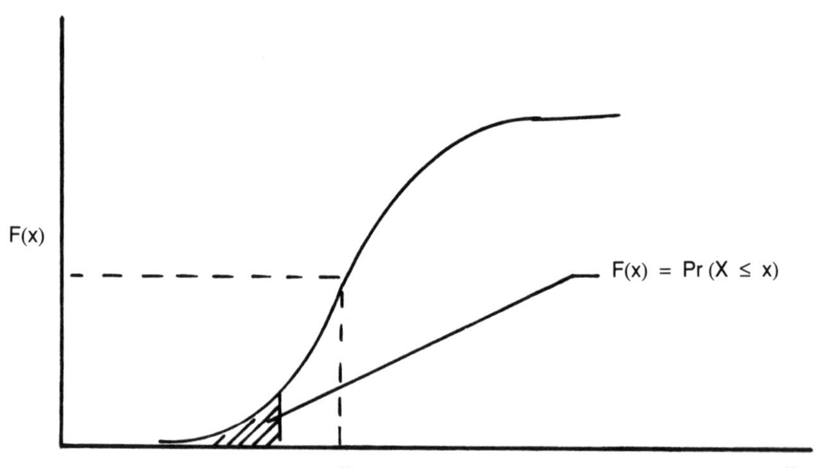

Figure 4-15. Normal left tail cumulative probability function

The mean μ is the first moment about zero. It is estimated from sample data as:

$$m = \sum x_i / n$$

The variance σ^2 is the second moment about the mean.

$$m_2 = \int_{-\infty}^{\infty} (x - \mu_1)^2 \; f(x) \, dx$$

It is estimated from a sample by:

$$m_2 = \sigma^2 = \sum (x_i - \mu)^2 / n$$

or for an unbiased variance:

$$m_2 = \sigma^2 = \sum (x_i - \mu)^2 / n - 1$$

The third moment about the mean is a measure of skewness and is estimated from sample data by:

$$m_3 = \sum (x_i - \mu)^3 / n$$

For the normal distribution $\mu_3 = 0$ and m_3 approaches 0 for large n. The quantity

$$\sqrt{\beta_1} = \frac{\mu_3}{(\mu_2)^{3/2}}$$

measures the skewness of the distribution relative to its degree of spread.

The fourth moment about the mean is estimated from data by:

$$m_4 = \sum (x_i - \mu)^4 / n$$

The fourth moment is a measure of peakedness or kurtosis. A common measure of relative kurtosis is:

$$\beta_2 = \frac{\mu_4}{\mu_2^2}$$

which, for the normal distribution, is 3 and for uniform distribution is 1.8. Another measure sometimes employed for relative kurtosis measure is:

$$\gamma_2 = \frac{\mu_4}{\mu_2^2} - 3$$

which, for the normal distribution, is equal to 0.

These higher moment measures of distribution shape are employed in the program RFRSIM discussed in chapter 7.

Example 4-8

The following daily fuel consumption of a small bulk carrier was observed on a recent voyage. Estimate the moments of the distribution of fuel consumption given in tons per day. Does fuel consumption approach a normal distribution for this 10-day voyage?

30.22
35.10
33.20
25.28
30.26
27.36
36.28
35.20
26.24
34.70

$$m_1 = \sum x_j / n = 31.38$$

$$m_2 = \sum (x_j - m_1)^2 / n - 1 = 16.72$$

$$\sqrt{m_2} = \hat{\sigma} = 4.09$$

$$m_3 = \sum (x_j - m_1)^3 / n = 70.18$$

$$m_4 = \sum (x_j - m_1)^4 / n = 348.41$$

$$\sqrt{\beta_1} = \frac{\mu_3}{(\mu_2)^{3/2}} = \frac{70.18}{16.72^{3/2}} = 1.03$$

$$\beta_1 = 1.013$$

$$\beta_2 = \frac{\mu_4}{\mu_2^2} = \frac{348.41}{16.72^2} = 1.25$$

Based on this small sample, the fuel consumption could not be considered normally distributed. However, if experience indicated that fuel consumption was indeed normally distributed, it would not be unreasonable to assume normality and use the mean and variance as estimates of the parameters of a normal distribution in the absence of better information.

STANDARD NORMAL DISTRIBUTION

If X is a normal random variable with mean μ and standard deviation σ, then we can define a standardized normal variable as:

$$z = \frac{x - \mu}{\sigma}$$

where:

$$E(z) = 0$$

$$Var(z) = 1$$

and z is a dimensionless quantity.

Standardized normal variables are needed to enter standard normal curve tables to find the area under the curve or the probability:

$$Pr(X \leq x)$$

or

$$Pr(Z \leq z)$$

in

$$F(z) = Pr(Z \leq z) = \frac{1}{\sqrt{2\pi}} \int_{-\infty}^{\infty} e^{-z^2/2}\, dz$$

Standard normal distribution functions are usually tabulated for the values of Z from 0 to 4. Since the curve is symmetrical, as shown in Figure 4-13, the probability for the negative values of Z may be easily obtained. The areas of the standard normal distribution at 1, 2, and 3 standard deviations are:

$$Pr(Z \leq 1) = .8413$$

$$Pr(Z \leq 2) = .9772$$

$$Pr(Z \leq 3) = .9987$$

A table of the standard normal distribution is included in Appendix B.

MONTE CARLO SIMULATION

In engineering economic problems where the independent variables are random variables, it is necessary to have a method of calculating the dependent variables or variables which encode the information contained in the distributions of the independent variables. Direct analytical methods as in Pfeiffer are exceedingly complicated and usually impossible except for simple functions. A numerical method based on the Taylor Series Expansion of $h(x_1, x_2, \ldots x_n)$ about

$$E(x_1), E(x_2) \ldots E(x_n)$$

is feasible and useful. However, with the high-speed calculations of the computer, Monte Carlo Simulation provides results irrespective of the complications with adequate accuracy.

The principle of Monte Carlo simulation involves the random drawing of individual values from the distributions of the independent variables to perform the required calculation and provide a sample value of the dependent variable. This process is repeated many times to provide a large sample of the dependent variable from which the dependent distribution moments are calculated.

The Central Limit Theorem suggests that, for a large sample, the dependent variable approaches a normal distribution irrespective of the distributions of the independent variables, that is:

$$\lim_{n \to \infty} \Pr\left(a \leq \frac{S_n - n_\mu}{\sigma\sqrt{n}} \leq b\right) = \frac{1}{\sqrt{2\pi}} \int_a^b e^{-u^2/2}\, du$$

To select a value from a discrete or continuous distribution for a Monte Carlo Simulation, it is necessary to relate a set of random numbers, say 000 to 999, to the values of the random variable according to the distribution function. For example, given the following discrete distribution, random numbers would be assigned as indicated:

Fuel Consumption	*Pr (f.c.)*	$\sum$ *Pr (f.c.)*	*Random Numbers Assignment*
22	.10	.10	000–100
24	.20	.30	101–300
26	.40	.70	301–700
28	.20	.90	701–900
30	.10	1.00	901–999

To draw a sample of $n = 10$ from this discrete distribution, it is necessary to select 10 numbers from a random number table starting at an arbitrary point or use a calculator to generate 10 random numbers.

Random Number	*Fuel Consumption*
549	26
023	22
982	30

Random Number	*Fuel Consumption*
118	24
968	30
810	28
561	26
573	26
245	24
665	28

From this sample, the mean and variance may be calculated as follows:

$$m_1 = {}^{264}\!/_{10} = 26.4$$

$$m_2 = {}^{62.4}\!/_{10} = 6.24$$

$$\text{unbiased } m_2 = {}^{62.4}\!/_{9} = 6.93$$

The actual mean of the daily fuel consumption is 26, so the simulation error is .4. This error can be reduced to an acceptable value by increasing n the sample size.

If the discrete distribution of the voyage length is:

Voyage Days	*Pr (vd)*	$\sum$ *Pr (vd)*	*Random Numbers Assignment*
15	.10	.10	000–100
16	.40	.50	101–500
17	.40	.90	501–900
18	.10	1.00	901–999

and the total fuel consumption (t.f.c.) = (voyage days)(fuel/day) then the total fuel consumption for the voyage on the basis of $n = 10$ is:

Random Number	*Tons Fuel Consumption / Day*	*Voyage Days*	*Total Fuel Consumption*
440	26	16	416
596	26	17	442
806	28	17	476
594	26	17	442
759	28	17	476
630	26	17	442
819	28	17	476
593	26	17	442
926	30	18	540
759	28	17	476
		$\sum$ (t.f.c.)	4628

and the resulting distribution may be estimated from the following sample statistics:

$$m_1 = 462.8$$

$$m_2 = 1057.76$$

$$m_3 = 60778.6$$

$$m_4 = 4118646.1$$

$$\sqrt{\beta_1} = 1.767$$

$$\beta_2 = 3.681$$

Simulation with continuous distribution or a mixture of discrete and continuous is accomplished in the same manner. Assume we wish to calculate a required freight rate where the acquisition cost, the annual operation costs, and the annual transport capacity are normal distributions as follows:

	μ	3σ
Acquisition cost, $\overline{AC}$	\$60,000,000	\$3,000,000
Annual operating costs, $\overline{AOC}$	\$ 1,700,000	\$ 150,000
Annual transport capacity, $\overline{ATC}$	450,000 tons	30,000 tons

$$RFR = \frac{(A/P, i, n)\ \overline{AC} \quad + \ \overline{AOC}}{\overline{ATC}}$$

If the vessel life is 20 years and the MARR is .10, the (A/P, .10, 20) = .11746. Then, the normal distributions of the annual capital cost for increasing Monte Carlo sample sizes is as follows:

n	100	200	300	400	500	600
E (RFR)	19.41	19.42	19.41	19.41	19.40	19.41
Var (RFR)	.27	.27	.26	.26	.25	.25
γ_1	.0392	.0715	.0269	.0964	.0596	.0425
γ_2	.0158	.0650	–.198	–.1555	–.1253	–.1073
$\sigma_{(RFR)}$	.523	.493	.509	.506	.499	.500

In this case, a measure of normality is that both γ_1 and γ_2 be close to zero.

If the input variable standard deviations were such that the dependent variables RFR for two alternatives, *A* and *B*, had the same expected values but significantly different standard deviations, the decision maker would select the alternative with the lower risk or lower standard deviation. This is illustrated in Figures 4-16a and 4-16b which show the output of the program RFR which provides the Monte Carlo Simulation. Figure 4-16c is a plot of Pr (RFR $\geq$ x) for the two alternatives and a vertical line to indicate a deterministic value (i.e., no variance) of RFR.

RFR	
Expected value	.194130E+02
Variance	.250107E+00
Standard deviation	.500107E+00
Coefficient of variation	.257615E–01
Gamma one, skewness	.425546E–01
Gamma two, kurtosis	–.107352E+00
Max. sample value	.206962E+02
Min. sample value	.178739E+02
Sample range	.282225E+01
Sample size	600

Figure 4-16a. Alternative A

RFR	
Expected value	.194303E+02
Variance	.631454–E01
Standard deviation	.251287E+00
Coefficient of variation	.129328E–01
Gamma one, skewness	.164730E+00
Gamma two, kurtosis	–.340933E+00
Max. sample value	.201552E+02
Min. sample value	.188346E+02
Sample range	.132055E+01
Sample size	600

Figure 4-16b. Alternative B

ECONOMIC ANALYSIS UNDER RISK DEFINED BY CONTINUOUS DISTRIBUTIONS

In Chapter 2, it was explained that a continuous equivalent to a year of discounting could be represented as:

$$e^{r} = 1 + i$$

where *r* is the equivalent continuous discount rate for a single year. If we wish to express more than one year, it is necessary to multiply the rate *r* by the number of years *n,* i.e.:

$$e^{rn} = (1 + i)^{n}$$

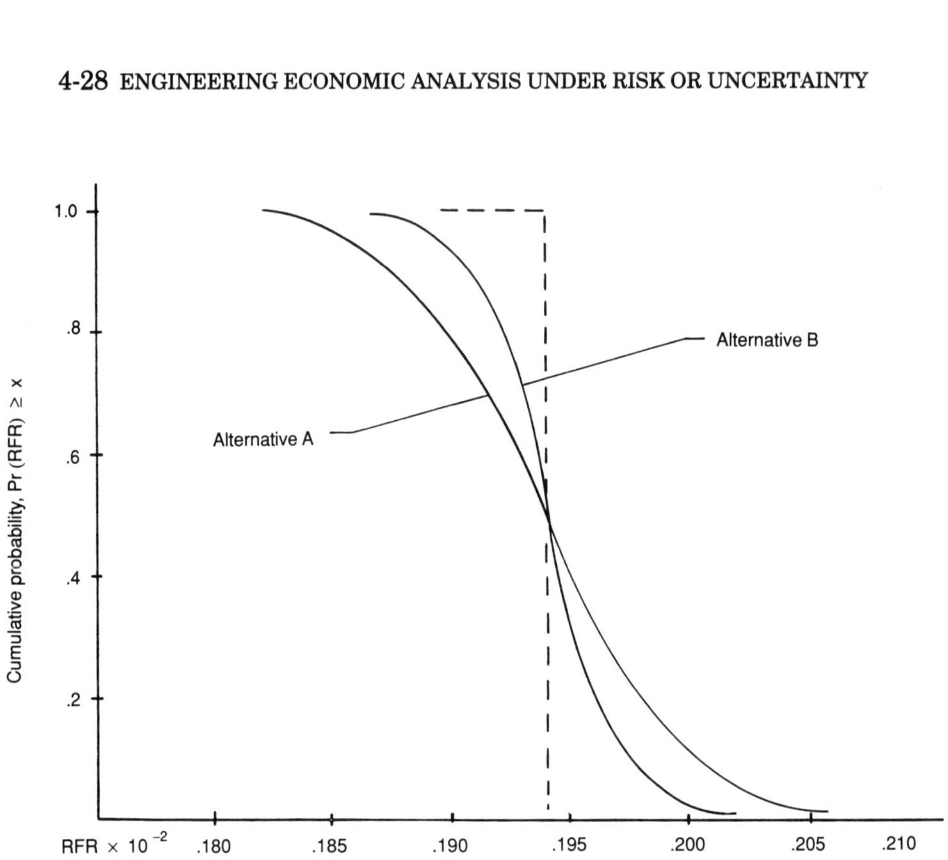

Figure 4-16c. Comparison of alternative RFR distributions

If we have a probability distribution *p(n)* for the occurrence time of a single future payment, then we can represent the expected value of the present worth *P* as follows:

$$E(P) = F \int_0^{\infty} F p(n)\ e^{-rn}\ dn$$

In the case of the normal distribution, *E(P)* becomes:

$$E(P) = F \int_0^{\infty} \frac{1}{\sigma\sqrt{2\pi}}\ e^{-(n-\mu)^2/2\sigma^2}\ e^{-rn}\, dn$$

and since:

$$E(P) = F\ E(e^{-rn})$$

then:

$$E(P) = F\ e^{-r\mu + (r^2\sigma^2/2)}$$

where P = Present worth of future sum
F = Future sum
n = Year of payment occurrence (random variable)
μ = Mean of the year of occurrence distribution
σ = Standard derivative of year of occurrence
r = Equivalent continuous rate

Buck and Tanchoco have provided tabulations of expressions for $E(e^{-rn})$ for commonly used distributions. The single cash flow occurrence time using the uniform distribution is:

$$E\,(e^{-rn}) = \frac{e^{-ra} - e^{-rb}}{r\,(b-a)}$$

where a = earliest time
b = latest time

$$E\,(n) = \frac{(a+b)}{2}$$

$$V\,(n) = \frac{(b-a)^2}{12}$$

and for the gamma distribution:

$$E\,(e^{-rn}) = (1 + rb)^{-a}$$

where E (n) = ab
V (n) = ab^2

Example 4-9

A single positive cash flow of $12,000 is expected to occur in five to seven years. The annual interest rate is 10 percent. What is the present worth of this single payment?

$$e^r = 1 + .1$$

$$r \ln e = \ln 1 \cdot 1$$

$$r = \frac{\ln 1.1}{\ln e} = .0953$$

Assume a uniform distribution, then:

$$E\,(P) = F\left[\frac{e^{-ra} - e^{-rb}}{r\,(b-a)}\right]$$

$$= \$12{,}000\left[\frac{e^{-.0953(5)} - e^{-.0953(7)}}{.0953\,(7-5)}\right]$$

$$= 12{,}000\,[.5653]$$

$$E\,(P) = \$6{,}783.8$$

The variances of these distributions can be used to express the measure of uncertainty on the present worth calculation. For the function e^{-rn}, the variance may be calculated as follows:

Uniform distribution:

$$V\,(e^{-rn}) = \frac{e^{-2ra} - e^{-2rb}}{2r\,(b-a)} - \frac{e^{-ra} - e^{-rb^2}}{r\,(b-a)}$$

Gamma distribution:

$$V\,(e^{-rn}) = (1 + 2rb)^{-a} - (1 + rb)^{-2a}$$

Normal distribution:

$$V\,(e^{-rn}) = e^{-2r\mu\, +\, 2\sigma^2 r^2} - e^{-2r\mu\, +\, \sigma^2 r^2}$$

For the case where the future single cash flow is deterministic and the occurrence time is probabilistic:

$$V\,(P) = V\,(e^{-rn})\;(E\,(F))^2$$

Example 4-10

Determine the variance of the present worth of the future single payment in Example 4-9. Determine the time limits for a 90 percent probability of the expected present worth. Assume a normal distribution for the year of occurrence.

$$V\,(e^{-rn}) = e^{-2r\mu\, +\, 2\sigma^2 r^2} - e^{2r\mu\, +\, \sigma^2 r^2}$$

$$r = .0953$$

$$\mu = 6$$

$$\sigma = 1$$

$$V\,(e^{-rn}) = e^{-2\,(.0953)\,(6)\, +\, 2\,(1)\,(.0091)} - e^{-2\,(.0953)\,(6)\, +\, (1)\,(.0091)}$$

$$V\,(e^{-rn}) = e^{-1.1254} - e^{-1.1345}$$

$$= .00294$$

$$\sigma = (V(e^{-rn})\mu^2)^{1/2}$$

$$= .3253$$

The limits on the occurrence time for 90 percent probability are:

$$\Pr(x_1 \le \mu < x_2) = .90$$

$$\mu \pm (1.645)(.3253) = 6 \pm .5352$$

$$x_2 = 6.5352$$

$$x_1 = 5.4648$$

In the case where the single future cash flow is a random variable as well as a time of occurrence, then it is necessary to convolve the distributions of F and n. If they are stochastically independent, which is usually the case, then:

$$E(P) = E(F)\,E(e^{-rn})$$

Example 4-11

Given an expected future payment with a range of \$10,000 to \$14,000 and uncertain time of occurrence which is a uniform distribution with a range of five to seven years.

In this case, the expected value of F is:

$$E(F) = \frac{10{,}000 + 14{,}000}{2} = \$12{,}000$$

and from Example 4-8, the $E(e^{-rn}) = .5653$. Therefore:

$$E(P) = 12{,}000(.5653)$$

$$E(P) = \$6{,}783.6$$

When both the future sum and the payment time of occurrence are random variables, the variance of the present worth, as shown by Buck and Tanchoco, can be determined by:

$$V(P) = V(F)\,V(e^{-rn}) + V(F)\,E(e^{-rn})^2 + V(e^{-rn})\,E(F)^2$$

Example 4-12

Find the variance of the present worth of Example 4-11 and estimate the upper and lower bounds on *P* for a probability of 95 percent.

$$V\,(F) = \frac{(b-a)^2}{12} = \frac{(14{,}000 - 10{,}000)^2}{12}$$

$$V\,(F) = 1.333 \times 10^6$$

Assuming a normal distribution for the time of occurrence, then:

$$r = .0953$$

$$\mu = 6$$

$$\sigma^2 = 1$$

$$E\,(e^{-r\,n}) = .567$$

$$E\,(F) = 12{,}000.$$

and:

$$V\,(e^{-r\,n}) = e^{-2r\mu + 2\sigma^2 r^2} - e^{-2r\mu + \sigma^2 r^2}$$

$$V(e^{-r\,n}) = .32456 - .3216 = .00295$$

$$V\,(P) = 1.333\,(10^6)\,(.00295) + 1.333\,(10^6)\,(.567) + .00295\,(12{,}000)$$

$$V\,(P) = 759{,}779.$$

$$\sigma = 871.6$$

Boundaries for 95 percent probability are:

$$\$6{,}783.6 \pm 1.96\,(871.6)$$

$$\$6{,}783.6 \pm 1{,}708.4.$$

REFERENCES AND BIBLIOGRAPHY

Buck, James R., and Jose M. Tanchoco. "Economic Risk Analysis." Chapter 9.5, *Handbook of Industrial Engineering*. Gavriel Salvendy, ed. New York: John Wiley & Sons, 1982.

Chance, William A. *Statistical Methods for Decision Making*. Homewood, Ill. : Richard D. Irwin, 1969. Introduction to useful statistical methods with direct applications to engineering economics. Includes some elementary program source codes which are of interest. A short chapter on Monte Carlo Simulation methods presents a good summary.

Good, I.J. *Estimation of Probabilities.* Cambridge, Mass.: M.I.T. University Press, 1965. Worthwhile reading for those in need of a logical approach to the estimation of probabilities for use in decision problems.

Griffin, Walter, C. *Introduction to Operations Engineering.* Homewood, Ill.: Richard D. Irwin, 1971. Methods of forecasting, economics of decision making, and simulation are discussed.

Hahn, G.J., and S.S. Shapiro. *Statistical Models in Engineering.* New York: John Wiley & Sons, 1967. Basic probability, continuous and discrete distributions, and Monte Carlo Simulation applied to engineering problems.

Morris, William T. *The Analysis of Management Decisions.* Homewood, Ill.: Richard D. Irwin, 1964. Includes interesting discussion of prediction of important economic variables such as sales, maintenance costs, and economic life. A chapter on decision under uncertainty provides further insights to topics of this text.

Pfeiffer, Paul E. *Concepts of Probability Theory.* New York: Dover Publications, 1978. Rigorous treatment of probability for engineers.

Pierre-Simon, Marquis de Laplace. *Théorie Analytique des Probabilités.*

Savage, L.J. *The Foundations of Statistics.* New York: Dover Publications, 1972.

Schlaifer, Robert. *Probability and Statistics for Business Decisions.* New York: McGraw-Hill Book Company, 1959. Introductory text on application of probability to business problems. Does not require formal mathematical background.

Spiegel, M.R. *Theory and Problems of Statistics.* New York: Schaum Publishing Co.; McGraw-Hill Book Company, 1961. A self-teaching outline of basic statistical methods.

———. *Theory and Problems of Probability and Statistics.* New York: Schaum Publishing Co.; McGraw-Hill Book Company, 1975. A self-teaching outline and reference of basic probability including set concepts, random variables, expectation, distribution, sampling theory, and estimation.

QUESTIONS AND PROJECTS

1. In a decision under uncertainty, i.e., no probability measures for the independent variable, which principle would you apply? Why? What rules would you formulate for selection of an uncertainty decision principle?
2. In a decision context, what is regret? How is it calculated? What is the underlying reasoning of regret?

3. What is the advantage, if any, of the Hurwicz decision principle? Would you expect difficulties in applying this principle?
4. What are the three useful definitions of probability? Discuss the differences and comment on the relative value to a decision analyst.
5. Write an expression for Bayes Theorem where the prior, conditional, and joint distributions are continuous distributions.
6. The probability function for the Poisson Distribution is:

$$P(x) = \frac{\lambda^{x} e^{-\lambda}}{x!}$$

   where the parameter λ equals both the expected value and the variance. If the average port days per voyage for a given service is eight, calculate the probability that the port days on a given voyage is five.
7. What is the difference between a discrete and a continuous random variable? What are the disadvantage and advantage of using discrete variables to approximate continuous variables?
8. What is the total area under a histogram or the curve of a continuous density function?
9. What is the difference between a left-tail and right-tail cumulative probability distribution?
10. What are the first four moments of a distribution function?
11. What is the area under the normal density function for $\pm\sigma$, $\pm 2\sigma$ and $\pm 3\sigma$?
12. Given the moments of a large sample, how can it be shown to be a normal distribution? uniform distribution?
13. In a Monte Carlo Simulation, why is it reasonable to assume that the dependent variable is normally distributed for large samples?
14. Assuming that the annual growth in labor costs is normally distributed, make a personal estimate of the mean and standard deviations. Compare your estimate with estimates of your associates.
15. Undertake research in reference text to identify five continuous and five discrete probability distributions used by engineers. What are the typical applications of each distribution?
16. In the application of a payoff decision matrix, what adjustment is needed to apply the decision principles if the matrix has cost elements?
17. Describe the meaning of the following symbols and the properties of the individual elements involved.

$$\bigcap_{i=1}^{4} A_i \qquad \bigcup_{i=1}^{4} A_i$$

18. Given an event A in a field S, draw a Venn diagram of the partition of A so that:

$$A = \bigcup_{i=1}^{4} A_i$$

19. What is a density function? What is a distribution function?

PROBLEMS

1. Given the following cost matrix, apply the decision principles:

17	17	9	13
13	13	13	13
4	25	9	9
13	21	9	9
9	11	11	9

2. To the matrix in Problem (1), add 2 to each value and multiply the result by 2 to form a new matrix. Apply the decision principles again and comment on the results.
3. List the elementary outcomes that are possible when a fair coin is flipped three times. If an event is defined by all outcomes where a head occurs in the third flip, what is the probability of that event? What is the probability of three heads?
4. Among 10 random digits, what is the probability that zero will appear three times and three times or less? (Combinations are called for.)
5. Among 100 new gaskets, 10 percent or 10 are defective. If ten are used, what is the probability that none are defective? (Combinations should be used.)
6. The distribution function F(x) is a step function with steps of 1/4 at $x = 0$, 3/8 at $x = 1$, 1/4 at $x = 2$, and 1/8 at $x = 3$. Draw a histogram and the mass distribution function. Determine $Pr\ (1 \le x \le 2)$, $Pr\ (x > 1.5)$.
7. A uniform distribution between 5 and 15 is defined as f(x). Plot f(x) and F(x). Determine:

$$Pr\ (5 \le x \le 15)$$

$$Pr\ (6 \le x \le 7)$$

$$Pr\ (x \ge 7)$$

8. Given the following table:

x	$Pr(x)$
−1	.3
0	.1
3	.2
5	.3
7	.1

Calculate: $E(x)$

$E(x - 2.5)$

$E((x - 2.5)^2)$

9. Given a normal distribution $\mu = 16$, $\sigma = 2$, calculate:

$Pr(x < 13)$

$Pr(x \leq 15)$

$Pr(4 \leq x \leq 15)$

Use a standardized normal distribution table.

10. Draw a histogram, five cells, of your personal estimate of the cost of a ton of Bunker "C" on East Coast USA.
11. What is the present worth of three future payments which are characterized as follows:

    a. $5,000 to be received in six to eight years.
    b. Between $12,000 and $16,000 to be received in seven to 10 years.
    c. A sum between $13,000 and $17,000 to be paid by contract in exactly 11 years.

    The interest rate is 8 percent. Experience suggests that the normal distribution is a reasonable model for the future payment *F* and the time to the transaction *t*. Determine the boundary values of the present worth which would give us 92 percent chance of predicting the correct number.

Chapter 5

Depreciation and Taxes

There is nothing permanent except change.
—Heraclitus

Once a planned company acquisition has been identified as an asset, it is necessary to make provision for its depreciation as the economic life is expended over time. Depreciation is essential to permit proper measurement of book income and taxable income. Four considerations are required; (1) the cost of the asset, (2) the economic life, (3) the rate at which the asset will be devalued over time, and (4) the salvage value at the end of the economic life. In short, it is necessary to find an equitable and rational means of allocating the initial valuation to operations over the period in which the asset makes economic contributions to the company.

The cost of the asset, such as a ship or a port facility, includes the purchase price and other one-time expenses necessary to place the asset in service.

The economic life, which for seagoing ships is about 20 or more years, is a complex determination. Physical deterioration from wear, corrosion, damage, etc., is an important factor which can be determined from engineering analysis and experience. Technical change and consequential economic obsolescence are more difficult to predict. Engineers with an understanding of the history of the pertinent technologies may overcome the difficulty. It is a major task for management to take full economic advantage of technological progress by recognizing asset obsolescence problems and developing depreciation policies which deal with such problems. For ships, it is usually possible to forestall unacceptable early obsolescence by additional investment to convert or refit the vessel. These activities include change of propulsion machinery to improve fuel performance, replacement of cargo-handling gear to reduce port time and stevedoring costs, or an increase in the length of a vessel to increase speed and/or cargo capacity.

Finally, tax laws usually define the boundaries for the economic life of assets for purposes of establishing depreciation periods.

The rate at which the book value of an asset is decreased over time depends on the method of depreciation selected. This is the major topic of this chapter. The methods include the traditional straight-line or equal annual devaluation and several accelerated methods which provide high devaluation initially with gradual reduction in the annual devaluation over time.

Salvage value is an important consideration in shipping since it sometimes represents a significant sum. Salvage value for a ship is almost totally dependent on the world price of scrap steel, which changes with supply and demand forces.

If a vessel can be sold for continuing use by a ship operator, the seller's salvage value is the sales price.

DEPRECIATION POLICY

The underlying objective of a depreciation policy is to match as closely as possible the period revenues and expenses in the most meaningful manner. This suggests that it may be desirable to use an accelerated method so that the shipowner can take advantage of lower maintenance costs during early life to offset the higher depreciation. Each enterprise must develop a rational policy which fits its particular needs and meets the approval of the tax authority and accounting standards.

The U.S. Internal Revenue Service guidelines and rules for depreciation specify a 20-year life for wharfs, docks, and water transportation equipments other than ships. Shipbuilding and boatbuilding equipments have a 12-year life by IRS rules, and floating dry docks have a 16-year life.

STRAIGHT-LINE DEPRECIATION

Under a straight-line depreciation policy, annual devaluation is uniform each year. This may be the most commonly used method. The annual devaluation fraction of the initial cost is;

$$D_f = \frac{1}{n}$$

where *n* is the depreciation period in years. The accumulated depreciation in the *m*th year is

$$\sum_{n=1}^{m} D_m = C_d \frac{m}{n}$$

and the asset book value at the end of *m* years is

$$B_m = C_a - C_d \frac{m}{n}$$

where C_a = initial value of asset

$$C_d = C_a - C_s$$

C_s = salvage value

Example 5-1

A RO/RO vessel trailer has an initial value of $12,000, a salvage value of $1,599 and an economic life of six years. What is the annual straight-line depreciation? What is the book value at the end of three years? Prepare a depreciation schedule.

$$C_d = C_a - C_s = \text{v}12{,}000 - 1{,}500 = \$10{,}500$$

$$D_m = C_{d/n} = \$1{,}750$$

$$B_3 = C_a - C_d \frac{m}{n} = 12{,}000 - 10{,}500 \left(\frac{3}{6}\right) = \$6{,}750$$

The depreciation schedule is:

Year	*Depreciation, $*	*Book Value, $*
0	0	12,000
1	1,750	10,250
2	1,750	8,500
3	1,750	6,750
4	1,750	5,000
5	1,750	3,250
6	1,750	1,500

It is sometimes useful to convert a depreciation series to a present worth for comparison among depreciation methods. Straight-line depreciation can be converted to a present worth as follows:

$$P_{Sl} = D_m\ (P/A,\ i,\ n)$$

DECLINING-BALANCE DEPRECIATION

Declining balance is an accelerated method of depreciation. Each year, the depreciation is a fixed fraction of the book value for that year. If F_{db} is a declining-balance fixed fraction, the book value at the end of the first year is

$$B_1 = C_a - C_a F_{db}$$

where the first-year depreciation is

$$D_1 = C_a F_{db}$$

The depreciation and book value for any year *m* are

$$D_m = C_a (1 - F_{db})^{m-1} F_{db}$$

$$B_m = C_a (1 - F_{db})^m$$

At the end of the economic life, i.e., *n* years, the book value should equal the salvage value. This will be true if the fixed fraction annual reduction in book value is

$$F_{db} = 1 - \left[\frac{C_s}{C_a}\right]^{\frac{1}{n}}$$

Example 5-2

A new RO/RO ship-loading tractor has an expected economic life of eight years. The purchase price is \$25,000 and the salvage value is \$2,000. Calculate a depreciation schedule for this asset.

$$F = 1 - \left[\frac{2{,}000}{2{,}500}\right]^{1/8} = .270733$$

Year	*Depreciation, \$*	*Book Value, \$*
0	0	25,000
1	6,227	18,773
2	5,083	13,691
3	3,706	9,985
4	2,766	7,219
5	1,954	5,265
6	1,421	3,840
7	1,040	2,801
8	800	2,000
	23,000	
Salvage	2,000	
Total	25,000	

Tax law requires the book value to be equal to the salvage value in the final year of the economic life.

Tax law also allows a more accelerated rate called double-declining-balance depreciation. In this case the constant annual fraction reduction in the book value may be any value from 1.1/n up to 2/n. For a life of eight years and a fraction of 2/n, the fraction would be 2/8 or .25 annual devaluation of the book value. Use of this fraction will result in a book value less than the salvage value, which is not allowed. Therefore, it is necessary to switch to a straight-line depreciation so that the book value equals the salvage value at the end of the economic life. The switch may occur at any convenient point between one-half and two-thirds economic life.

SUM-OF-THE-YEARS-DIGITS DEPRECIATION

Another form of accelerated depreciation, i.e., relative to straight-line depreciation, is the sum-of-the-years-digits depreciation. The fraction of the quantity to be depreciated each year under this method is the ratio of the years remaining to the sum-of-the-years digits for the economic life. For an eight-year economic life, the denominator would be 1+2+3+...+8 or 36. The fraction depreciated in the second year would be 6/36 or .1666. For the *m*th year in a total life of *n* years, the fraction is

$$F_{Sd} = \frac{n - m + 1}{.5n(n + 1)}$$

Example 5-3

Determine a sum-of-the-years-digits depreciation schedule for the problem of Example 5-2.

Year	*Fraction*	*Depreciation, $*	*Book Value, $*
0	0	0	25,000
1	.222	5,106	19,894
2	.194	4,462	15,432
3	.156	3,841	11,591
4	.139	3,197	8,394
5	.111	2,553	5,841
6	.083	1,909	3,932
7	.056	1,288	2,644
8	.028	644	2,000
	1.000	23,000	
		Salvage 2,000	
		Total 25,000	

ACRS DEPRECIATION

In engineering economic studies, the analyst seldom has information on the exact dates on which transactions will occur. Therefore, most engineering analysts and this text adopt the convention that all cash transactions occur on the last day of the year and all new assets are placed in service on the first day of the year. Assets are removed from service on the last day of the year and the positive cash flow representing the salvage value occurs at the same instant.

Except for unusual situations, the accelerated cost recovery system (ACRS) must be used by U.S. taxpayers. Furthermore, the accounting practices specified by the Internal Revenue Service must be followed in determining before-taxes net cash flow for the actual results of a company. The U.S. Tax Reform Act of 1986 provided for modified accelerated cost recovery system (MACRS) depreciation. The two features which provide the rapid depreciation are (1) salvage value is assumed to be zero and (2) a schedule of property lives which are less than the useful or economic life are approved for use by classes of property. An accelerated depreciation method is also permitted. Under this schedule, a cargo vessel might be classed as a 10-year property and a small fast ferry might be classed as a five-year property.

The method is based on the fixed 200 percent declining balance for the 10-year property and the 150 percent declining balance for the five-year property, both with a switch to straight-line depreciation at a point which maximizes the total depreciation deduction. Generally, a ½-year convention is used. This convention assumes that a new asset is placed in service at the midpoint of the taxpayer's year. However, for all equipment placed in service during the year, if equipment representing more than 40 percent of the total cost of the asset is placed in service during the last three months of the tax year, the midquarter convention must be used. If the midquarter convention is used, the asset is considered placed in service at the midpoint of the quarter that the asset is placed in service. For example, an asset placed in service by a calendar year taxpayer between October 1 and December 31 is considered placed in service on November 15.

Effective in 1994, the alternative minimum tax will be available which will allow a 150 percent declining balance over a class life of 18 years.

Example 5-4

Calculate the schedule for a 10-year ACRS depreciation of a $70 million vessel which has a $5 million salvage value with and without the ½ -year convention.

	ACRS *Engineering Analysis*		*ACRS* *½-Year Convention*	
Year	*Depreciation, $*	*Book Value, $*	*Depreciation, $*	*Book Value, $*
0		70,000,000		70,000,000
1	14,000,000	56,000,000	7,000,000	50,400,000
2	11,200,000	44,800,000	12,600,000	40,320,000
3	8,960,000	35,840,000	10,080,000	32,255,000
4	7,168,000	21,504,000	8,065,000	25,800,000
5	4,300,800	17,208,200	6,455,000	20,641,000
6	3,441,640	13,766,560	5,159,000	16,055,000
7	3,441,640	10,324,920	4,586,000	11,469,000
8	3,441,640	6,883,280	4,586,000	6,883,000
9	3,441,640	3,441,640	4,586,000	6,883,000
10	3,441,640	0	4,586,000	2,297,000
11			2,297,000	0
	70,000,000		70,000,000	

EXCEPTIONS FROM MANDATORY MACRS DEPRECIATION

As noted above, U.S. taxpayers are required to use MACRS depreciation with the following exceptions:

1. Taxpayers may elect to use either a 150-percent declining balance method or a straight-line method, in each case without regard to salvage value, over the recovery period, e.g., 10 years. In addition, taxpayers may use the straight-line method over the asset's class life, which for most vessels is 18 years.
2. For U.S. taxpayer property other than ships used predominantly outside the United States, i.e., not physically present in the United States for at least 50 percent of the year, depreciation must be calculated using the straight-line method over the property's class life as defined by MACRS. Salvage value may be ignored. U.S flag vessels are eligible to be depreciated using MACRS depreciation even if the vessel never enters U.S. waters. U.S.-owned foreign flag vessels must apply straight-line depreciation over the vessel's class life. Salvage is ignored.
3. If a vessel is leased to a foreign corporation which is not subject to tax in the United States, the vessel must be depreciated on a straight-line basis over the greater of its class life, 18 years, or 125 percent of the lease term. Leases less than three years are excluded from this rule. To determine the length of a lease, all renewal or extension options are taken into account.

TAXES ON SHIPPING COMPANIES

The following potential taxing sources must be included in any economic analysis of a shipping venture. (Also see the Bibliography for more information.)

1. Direct U.S. taxes on operations of U.S. corporations.
2. Gross transportation tax on income of foreign vessel owners operating to and from the United States. This tax is currently 4 percent of a foreign owner's U.S. source transportation income. There is no withholding mechanism. The U.S. exempts, by treaty, many countries from this tax when the foreign government does not tax American owners.
3. Worldwide withholding taxes.
4. Reciprocal exemptions are the result of treaties with foreign governments which specify that the parties will not tax each other's shipping.
5. Subpart F, Shipping Rules, requires a U.S. parent corporation of a foreign shipping company to pay taxes on the income of its foreign subsidiary regardless of whether such income is returned to the parent as dividends.

COMPARISON OF METHODS

After taxes, the advantage or disadvantage of the various depreciation methods must be determined for the specific project and tax liabilities. It is obvious that accelerated depreciation will lead to lower taxes in the early years and higher taxes in the later years. If a high risk of technical obsolescence exists, rapid depreciation is indicated. This is not usually the case in marine transportation. In the case of accelerated depreciation, some companies open special accrual accounts to offset the higher taxes in later years.

An approach to comparison of different methods of depreciation is to calculate the present worth of the annual depreciation amounts at the company MARR. This should be done on the before and after taxes basis.

Example 5-5

Assume a project with a $25,000 investment and a $2,000 salvage value at the end of eight years. If the company MARR is 12 percent, compare the three methods to determine which has the greatest present worth before taxes.

For straight-line depreciation, the present worth is

$$PW_{Sl} = \frac{1}{n}(P/A, i, n)(C_a - C_S) + C_S(P/F, i, n)$$

$$PW_{Sl} = .125(4.9676)(23{,}000) + 2{,}000(.40388)$$

$$PW_{Sl} = 14{,}282 + 807 = \$15{,}089$$

For declining-balance depreciation, the present worth is

$$PW_{db} = \sum_{m=1}^{8} C_d(1 - F_{db})^m F_{db}(P/F, i, n) + C_S(P/F, i, n)$$

$$PW_{db} = 5{,}561 + 4{,}051 + 2{,}638 + 1{,}758 + 1{,}108 + 720 + 472 + 323 + 808$$

$$PW_{db} = \$17{,}438$$

For sum-of-the-years-digits depreciation, the present worth is

$$PW_{Syd} = \sum_{m=1}^{8} C_d \left[\frac{n - m + 1}{.5n(n+1)}\right](P/F, i, n) + C_S(P/F, i, n)$$

$$PW_{Syd} = 4{,}559 + 3{,}556 + 2{,}734 + 2{,}035 + 1{,}447 + 967 + 584 + 260 + 808$$

$$PW_{Syd} = \$16{,}950$$

If other considerations are acceptable, the decision in this case would be to employ the declining-balance method since it provides the highest present worth.

For a comparison of methods on the after taxes basis, the lifetime cash flow for a shipping venture is useful. Three container vessels are built and employed in a North Atlantic trade. The following data apply to the venture:

Vessel life	15 years
Cost of vessel, each	$70,000,000
Salvage value, each	$4,000,000
Tax rate	33
Depreciation period	15 years
MARR	.135

Tables 5-1 through 5-4 show the cash flow projections for this venture with (1) no depreciation, (2) straight-line depreciation, (3) sum-of-the-years-digits depreciation, and (5) declining balance depreciation. The present worth of the lifetime cash flows including the salvage value under the various depreciation methods is depicted as follows in table 5-5.

TABLE 5-1

Shipping Venture Net Cash Flow With No Depreciation

Cash Flow Projections (Thousands of U.S. Dollars)

Year	*Net Revenue*	*Operating Costs*	*Loan Payment*	*Cash Flow Before Taxes*	*Interest Paid*	*Depreciation Cost*	*Taxable Income*	*Taxes Due*	*Cash Flow After Taxes*
0	0	0	0	–57,000	0	0	0	0	–57,000
1	203,229	94,913	29,032	79,284	22,680	0	85,636	28,260	51,024
2	210,601	97,545	29,032	84,024	21,822	0	91,234	30,107	53,917
3	218,243	100,253	29,032	88,959	20,849	0	97,142	32,057	56,902
4	226,167	103,039	29,032	94,096	19,744	0	103,384	34,117	59,979
5	234,383	105,906	29,032	99,445	18,491	0	109,987	36,296	63,149
6	242,902	108,856	29,032	105,013	17,067	0	116,978	38,603	66,410
7	251,734	111,892	29,032	110,809	15,452	0	124,390	41,049	69,761
8	260,892	115,016	29,032	116,843	13,619	0	132,256	43,645	73,198
9	270,387	118,231	29,032	123,123	11,538	0	140,617	46,404	76,719
10	280,232	121,540	29,032	129,660	9,176	0	149,516	49,340	80,319
11	290,441	124,946	29,032	136,463	6,496	0	159,000	52,470	83,993
12	301,027	128,451	29,032	143,544	3,453	0	169,123	55,811	87,733
13	312,003	132,058	0	179,945	0	0	179,945	59,382	120,563
14	323,385	135,771	0	187,614	0	0	187,614	61,913	125,701
15	335,187	139,593	0	200,594	0	0	195,594	64,546	136,048
Totals	3,960,813	1,738,009	348,388	1,822,416	180,388	0	2,042,416	673,997	1,148,419
	PW of NCF after taxes		13.5%						433,976

1. No depreciation taken.
2. Project taxed at 33%.
3. Amount of equity invested in project shown as cash flow in year zero.
4. Investment tax credit is 0.00%.

TABLE 5-2

Shipping Venture Net Cash Flow With Straight-Line Depreciation

Cash Flow Projections (Thousands of U.S. Dollars)

Year	*Net Revenue*	*Operating Costs*	*Loan Payment*	*Cash Flow Before Taxes*	*Interest Paid*	*Depreciation Cost*	*Taxable Income*	*Taxes Due*	*Cash Flow After Taxes*
0	0	0	0	–57,000	0	0	0	0	–57,000
1	203,229	94,913	29,032	79,284	22,680	13,200	72,436	23,904	55,380
2	210,601	97,545	29,032	84,024	21,822	13,200	78,034	25,751	58,273
3	218,243	100,253	29,032	88,959	20,849	13,200	83,942	27,701	61,258
4	226,167	103,039	29,032	94,096	19,744	13,200	90,184	29,761	64,335
5	234,383	105,906	29,032	99,445	18,491	13,200	96,787	31,940	67,505
6	242,902	108,856	29,032	105,013	17,067	13,200	103,778	34,247	70,766
7	251,734	111,892	29,032	110,809	15,452	13,200	111,190	36,693	74,117
8	260,892	115,016	29,032	116,843	13,619	13,200	119,056	39,289	77,554
9	270,387	118,231	29,032	123,123	11,538	13,200	127,417	42,048	81,075
10	280,232	121,540	29,032	129,660	9,176	13,200	136,316	44,984	84,675
11	290,441	124,946	29,032	136,463	6,496	13,200	145,800	48,114	88,349
12	301,027	128,451	29,032	143,544	3,453	13,200	155,923	51,455	92,089
13	312,003	132,058	0	179,945	0	13,200	166,745	55,026	124,919
14	323,385	135,771	0	187,614	0	13,200	174,414	57,557	130,057
15	335,187	139,593	0	204,594	0	13,200	182,394	60,190	144,404
Totals	3,960,813	1,738,009	348,388	1,826,416	180,388	198,000	1,844,416	608,657	1,217,758
	PW of NCF after taxes		13.5%						462,013

1. Straight-line depreciation with asset value reduced by salvage value, if any.
2. Project taxed at 33%.
3. Amount of equity invested in project shown as cash flow in year zero.
4. Investment tax credit is 0.00%.

TABLE 5-3

Shipping Venture Net Cash Flow With Sum-Of-The-Years-Digits Depreciation

Cash Flow Projections (Thousands of U.S. Dollars)

Year	*Net Revenue*	*Operating Costs*	*Loan Payment*	*Cash Flow Before Taxes*	*Interest Paid*	*Depreciation Cost*	*Taxable Income*	*Taxes Due*	*Cash Flow After Taxes*
0	0	0	0	–57,000	0	0	0	0	–57,000
1	203,229	94,913	29,032	79,284	22,680	24,750	60,886	20,092	59,192
2	210,229	97,545	29,032	84,024	21,822	23,100	68,134	22,484	61,540
3	218,243	100,253	29,032	88,959	20,849	21,450	75,692	24,978	63,980
4	226,167	103,039	29,032	94,096	19,744	19,800	83,584	27,583	66,513
5	234,383	105,906	29,032	99,445	18,491	18,150	91,837	30,306	69,139
6	242,902	108,856	29,032	105,013	17,067	16,500	100,478	33,158	71,855
7	251,734	111,892	29,032	110,809	15,452	14,850	109,540	36,148	74,661
8	260,892	115,016	29,032	116,843	13,619	13,200	119,056	39,289	77,554
9	270,387	118,231	29,032	123,123	11,538	11,550	129,067	42,592	80,531
10	280,232	121,540	29,032	129,660	9,176	9,900	139,616	46,073	83,586
11	290,441	124,946	29,032	136,463	6,496	8,250	150,750	49,747	86,716
12	301,027	128,451	29,032	143,544	3,453	6,600	162,523	53,633	89,911
13	312,003	132,058	0	179,945	0	4,950	174,995	57,748	122,197
14	323,385	135,771	0	187,614	0	3,300	184,314	60,824	126,790
15	335,187	139,593	0	204,594	0	1,650	193,944	64,002	140,593
Totals	3,960,813	1,738,009	348,388	1,826,416	180,388	198,000	1,844,416	608,657	1,217,759
	PW of NCF after taxes		13.5%						469,669

1. Straight-line depreciation with asset value reduced by salvage value, if any.
2. Project taxed at 33%.
3. Amount of equity invested in project shown as cash flow in year zero.
4. Investment tax credit is 0.00%.

TABLE 5-4

Shipping Venture Net Cash Flow With Declining-Balance Depreciation

Cash Flow Projections (Thousands of U.S. Dollars)

Year	*Net Revenue*	*Operating Costs*	*Loan Payment*	*Cash Flow Before Taxes*	*Interest Paid*	*Depreciation Cost*	*Taxable Income*	*Taxes Due*	*Cash Flow After Taxes*
0	0	0	0	–57,000	0	0	0	0	–57,000
1	203,229	90,927	29,032	83,270	22,680	37,800	51,822	17,101	66,169
2	210,601	93,479	29,032	88,089	21,822	30,996	64,303	21,220	66,869
3	218,243	96,105	29,032	93,106	20,849	25,417	75,872	25,038	68,068
4	226,167	98,809	29,032	98,326	19,744	20,842	86,772	28,635	69,691
5	234,383	101,591	29,032	103,760	18,491	17,090	97,211	32,080	71,680
6	242,902	104,455	29,032	109,414	17,067	14,014	107,365	35,430	73,984
7	251,734	107,403	29,032	115,298	15,452	11,491	117,387	38,738	76,561
8	260,892	110,438	29,032	121,422	13,619	9,423	127,412	42,046	79,376
9	270,387	113,561	29,032	127,793	11,538	7,727	137,561	45,395	82,398
10	280,232	116,777	29,032	134,423	9,176	6,336	147,943	48,821	85,602
11	290,441	120,087	29,032	141,322	6,496	5,196	158,663	52,359	88,963
12	301,027	123,495	29,032	148,500	3,453	4,260	169,819	56,040	92,460
13	312,003	127,003	0	185,000	0	3,493	181,507	59,897	125,103
14	323,385	130,615	0	192,770	0	2,865	189,906	62,669	130,101
15	335,187	134,334	0	205,854	0	13,050	187,804	61,975	143,878
Totals	3,960,813	1,669,078	348,388	1,891,347	180,388	210,000	1,901,347	627,444	1,263,902
	PW of NCF after taxes		13.5%						492,348

1. Declining-balance depreciation applied to full asset value.
2. Project taxed at 33%.
3. Amount of equity invested in project shown as cash flow in year zero.
4. Investment tax credit is 0.00%.

TABLE 5-5

Comparison of Depreciation Methods

Depreciation Method	*Present Worth After-Taxes Net Cash Flow, $*
No depreciation	433,976,000
Straight-line	462,013,000
Sum-of-the-years-digits	469,669,000
10-year ACRS (no ½ yr. conv.)	471,765,000
Declining balance	492,348,000

Figure 5-1 is a graphical illustration of the comparison among methods of depreciation for a $5 million asset with a $400,000 salvage value at the end of an eight-year economic life.

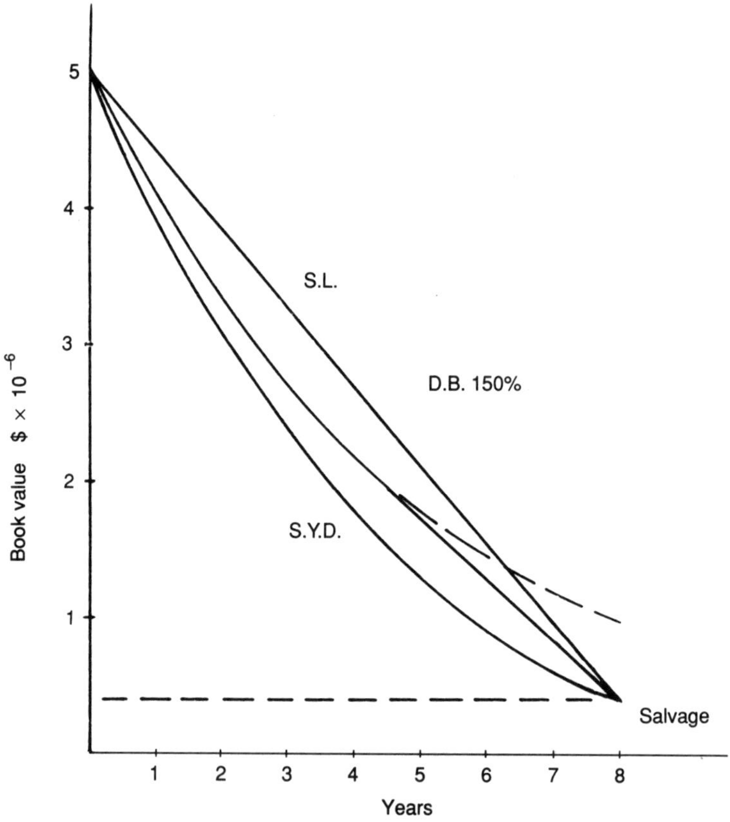

Figure 5-1. Comparison of depreciation methods showing book value of a five million dollar asset with eight-year life

REFERENCES AND BIBLIOGRAPHY

Garrison. "U.S. Income Taxation of Foreign Transportation Companies After the Tax Reform Act of 1986." *Fairplay International Shipping Weekly,* December 3, 10, 1987.

Grant, E. L., and P. T. Norton. *Depreciation.* New York: The Roland Press Company, 1955.

Granwell, A.W., and K. Klein. *The U.S. Federal Taxation of International Shipping Income.* Washington: Cadwalader, Wickersham and Taft, 1992.

Griffin, Walter C. *Introduction to Operations Engineering.* Homewood, Ill.: Richard D. Irwin, 1971.

Johnson. "Foreign Shipping Income Taxation after the Tax Reform Act of 1986." *International Tax Journal* Vol. 14, No. 53 (1988).

Salvendy, Gavriel. "Engineering Economy." Chapter 9, *Handbook of Industrial Engineering.* New York: John Wiley and Sons, 1982.

U.S. Internal Revenue Code of 1986, Sections 863, 883, 887, and 951-960.

QUESTIONS AND PROJECTS

1. What factors should a shipowner consider when selecting a method of depreciation for a new vessel? Are the factors different for a cargo-handling facility ashore?
2. Under what circumstances would it be proper for a shipping company to maintain two sets of accounts which include different depreciation methods for tax authorities and the shareowners?
3. How would you approach the determination of the salvage value and the economic life of a new vessel? Is there a difference in the approach for tankers and cargo vessels?
4. Why does the straight-line depreciation of ships continue to be a popular depreciation method?
5. Would it be practical to depreciate a ship on the basis of the cargo transported? What would be the advantages and disadvantages of such a method?
6. Derive an algebraic relationship for calculating the fraction of the initial book value which must be depreciated each year under the declining-balance method.
7. As an engineer, you have several valuable patents which have been assigned to your employer and are the source of company income. Should such patents be depreciated? Why? Other intangible properties include designs, models, copyrights, licenses, etc. Should these properties be depreciated also?
8. Why does the IRS require the use of a midquarter convention?
9. What is the ½ year convention? Is it an advantage to the ship owner?

10. Why is it necessary to study tax codes and international treaties before estimating the tax rate to be applied to a shipping venture study?

PROBLEMS

1. A shore-based cargo crane is purchased for $4 million. The life is 12 years and the salvage value is $100,000. Develop a straight-line depreciation schedule including the book value.
2. A shipyard painting facility is installed in a shipyard at a cost of $2.5 million. The life is 10 years with no salvage value. Prepare a sum-of-the-years-digits depreciation schedule including the annual book value.
3. A new heavy-lift crane is installed on a breakbulk cargo vessel at a cost of $800,000. The vessel has a remaining economic life of 10 years. It is estimated that the salvage value of the crane will be $75,000. Prepare a declining-balance depreciation schedule for 150 percent of the straight-line depreciation. Plot the result and indicate the year at which a switch to straight-line depreciation would be most favorable.
4. A new pipe fabrication facility has been designed for a shipyard. The following data has been estimated:

Asset installed cost	$12,000,000
Life	20 years
Salvage value	$500,000
Annual maintenance cost	$15,000
Straight-line depreciation	
Interest rate	9%

   What is the present worth of the capital cost and the maintenance cost of the facility?
5. A small chemical tanker purchased at a cost of $20 million will soon be placed in service. The owners plan to sell the vessel in 10 years at a price with the following estimated probability distribution:

P(price)	*Price*
.05	6,000,000
.20	7,500,000
.30	9,000,000
.25	9,750,000
.15	9,900,000
.05	10,500,000

The owners wish to recover cash as quickly as possible for other investments. What depreciation method would you recommend? Demonstrate the merit of your recommendation.

6. A stevedore pays a 40 percent tax on annual income. The company has traditionally used straight-line depreciation for capital assets. A major new piece of equipment has been purchased and placed in service at a cost of $300,000. The life and salvage value have the following estimates, which are presented as probability distributions due to the inherent uncertainty.

P (life)	*Life*	*P (salvage)*	*Salvage*
.03	8.0	.10	3,000
.40	9.3	.40	4,500
.30	10.0	.30	5,500
.20	11.0	.20	6,250
.07	12.0		

The minimum attractive rate of return (MARR) is 15 percent. What is the present worth of the gain for selecting the sum-of-the-years-digits depreciation instead of the straight-line depreciation?

7. Calculate a schedule of factors to use in developing a 10-year ACRS depreciation plan. The factors are to be applied to the initial asset cost to determine the annual depreciation.
8. An asset has the following possible depreciation schedules. The salvage value is $100. Identify the depreciation methods used to prepare each schedule.

Year	*A*	*B*	*C*	*D*
1	600	633	380	800
2	420	506	380	480
3	293	380	380	288
4	293	253	380	173
5	293	128	380	159
	1,900	1,900	1,900	1,900

9. Use the program RFRD to calculate the required freight rate for a shipping venture under declining-balance, sum-of-the-years-digits, and straight-line depreciation methods. Use the basic input data provided in the documentation for the program of a data set which you have developed. Discuss the impact of changing depreciation methods on the required freight rate.
10. Run the program VENTURE using the input set provided with the documentation or an input which you have developed. What depre-

ciation method does the program employ? Comment on the adequacy of a program which is limited to the method which you uncover.

11. A bulk cargo ship is purchased from a shipyard for $20 million by a U.S. shipping company. The shipping company registers the ship with the U.S. Coast Guard for operation in the U.S. domestic and foreign trade, but operates the ship exclusively between the United Kingdom and Sweden. Calculate the shipowner's regular tax rate and the AMT depreciation deductions assuming the ½-year convention.
12. Same as problem 11, but assume the U.S. shipping company registers the ship under the flag of Malta.
13. In 1994, XYZ Leasing Company, a U.S. corporation, purchases a VLCC from ABC Oil Company for $26 milion. After $5 million of capital improvements are made, XYZ leases the vessel to Oilé Petroleum Company, a French corporation, for 20 years. Calculate the U.S. tax depreciation schedule assuming a ½-year convention and a U.S. flag. Recalculate the tax depreciation for this vessel under the same arrangements except it flies a Liberian flag.

Chapter 6

Fast Ferry Transport System Decisions under Uncertainty or Risk

The theory of probabilities is at bottom nothing but common sense reduced to calculus; it enables us to appreciate with exactness that which accurate minds feel with a sort of instinct for which ofttimes they are unable to account.

—*Pierre Laplace*, Théorie Analytique des Probabilités

The development of technology of high-speed marine vehicles has made it possible for ferry services in river and coastal areas to compete effectively with the automobiles and trains used by commuters. The high-speed marine vehicles are listed below (see Burg in the References) in the general order of increasing capital cost per passenger capacity.

		Typical Speed Range, Knots
Catamaran	Cat	22-34
Surface effect vehicle	SEV	23-33
Hydrofoil	Hyd	30-45
Small waterplane twin hull	SWATH	20-25
Air cushion vehicle	ACV	30-40

A major advantage of high-speed ferry service is the lack of traffic congestion on waterways of major metropolitan areas.

In the analysis of choice among the competing vehicles for a fast ferry service, several of the key independent variables must be estimated. These variables include passenger load factor and the rates of increase during the vehicle life for labor, fuel, maintenance, insurance, cost, and fares. Therefore, the analysis of the selection of a fast ferry vehicle type is an instructive example of the application of the uncertainty and risk methods of Chapter 4 to a comprehensive case.

THE PROBLEM

Bridgeport on Long Island Sound to Manhattan is a 43-nautical-mile trip. There is a large commuter population which uses trains and automobiles for this trip (see Hunt in the References).

Preliminary market surveys indicate that a 200-passenger vessel making a total of 2,000 round trips per year will attract passengers from the usual land travel patterns.

Hydrofoil, air cushion, surface effect, catamaran, and SWATH vehicles are available in the 200-passenger capacity. All five will be included in the study. The characteristics of each of these vessels are included in Table 6-1.

TABLE 6-1

Vehicle-Dependent Deterministic Costs and Factors Used in Economic Evaluation of Fast Ferry Venture, Bridgeport to Manhattan

Cost or Factor	*Hydrofoil*	*ACV*	*SEV*	*Catamaran*	*SWATH*
Speed, knots	35	40	33	26	25
Crew size	6	4	5	4	6
Fuel, gal/hour	157	91	150	98.2	176
Oil, gal/hour	3	2	3	2	3
Maintenance hours /operating hour	2.1	2.33	6.44	.44	1.61
Fare, one way, $	10	10	10	8	8
Vessel cost, $*	5,700	6,300	4,140	1,990	2,200
Owner equity, $*	1,140	1,260	828	398	440
Salvage, $*	1,140	1,260	828	398	440
Hull insurance*	131.1	145	95.2	45.8	50.6

*Thousands of dollars

The anticipated life of these vehicles in the planned service is 10 years. At that time, they will be sold and replaced with newer equipments. A salvage value will represent the estimated sale price.

Deterministic input variables, costs, and factors are tabulated in Table 6-2. Since the project has a 10-year life, it is reasonable to expect some growth in costs. The deterministic or fixed-growth rates are included in Table 6-3. Estimates of deterministic growth rates are frequently made by fitting a linear regression line to historical data for similar costs in the same marketplace.

For purposes of this problem, three key variables have been identified as uncertain. They are number of passengers per one-way trip, the factor for annual fuel cost increase, and the factor for annual fare increase. Other factors, such as annual labor cost increase, might also be selected as uncertain variables. Each such selection increases the size and complexity of the decision model. Furthermore, the number of discrete values of the uncertain variables will increase the model size. Judgment is necessary to avoid an unmanageable model size.

TABLE 6-2

Project-Dependent Deterministic Costs and Factors Used in Economic Evaluation of Fast Ferry Venture, Bridgeport to Manhattan

Project life	10 years
Passenger capacity	200 persons
Maintenance mat'l & parts	$45/operating hour
Load/unload time	20 minutes
Operating days	250 days/year
Trips per day	8 trips/vessel
One way distance	43 nautical miles
Average crew cost	$210/day
Fuel cost	$.98/gallon
Lube oil cost	$4.4/gallon
Shipyard pay rate	$18/hour
Shipyard overhead rate	.55 of direct labor
Annual drydocking expense	$8,500/year
P & I insurance	$98,000/year
Port & terminal costs	$95,000/vessel/year
G & A expenses	$325,000/year
Concession income	$1.00/passenger
Interest rate	10 percent
Tax rate	34 percent

TABLE 6-3

Project-Dependent Deterministic Annual Adjustments to Costs and Factors Used in Economic Evaluation of Fast Ferry Venture, Bridgeport to Manhattan

Annual increase crew cost	.025
Annual increase maintenance cost	.030
Annual increase insurance premiums	.010
Annual increase port costs	.015
Annual increase G & A costs	.035
Annual increase concession income	.030

Computers are essential when dealing with most practical problems. For this example, experience has suggested that the following discrete values of the uncertain independent variables will be considered as an exhaustive set:

Passengers per One-Way Trip	*Annual Fuel Cost Increase Factor*	*Annual Fare Increase Factor*
100		
120	.02	.02
160	.06	.05

These three sets of uncertain future events must be combined to form the problem set. For modeling purposes, this discrete set of combined events represents all the possibilities for the future as shown in Table 6-4.

TABLE 6-4

Fast Ferry Problem Events

Event	*Passengers Carried*	*Annual Fare Increase*	*Annual Fuel Increase*
1	100	.02	.02
2	100	.02	.06
3	100	.05	.02
4	100	.05	.06
5	120	.02	.02
6	120	.02	.06
7	120	.05	.02
8	120	.05	.06
9	160	.02	.02
10	160	.02	.06
11	160	.05	.02
12	160	.05	.06

THE ANALYSIS

The measure-of-merit for the selection of the best vehicle for the situation presented is the discounted net cash flow after taxes.

This model requires 5 × 12, i.e., 60, discounted net cash flow calculations to include the five alternatives and 12 future states. A computer program, Fast Ferry Computer Analysis (© TransTech Marine Company), was used to calculate the cash flows. Table 6-5 illustrates the output of this program for the air cushion vehicle under Event No. 11. The tabulation of all 60 measures of merit forms the payoff matrix presented in Table 6-6. Management decision makers must select the best alternative from among the five in this matrix. Such selections will be dependent on the experience, judgment, and risk aversion of a given decision maker or makers. The logical decision principles in the face of uncertainty presented in Chapter 4 will be of assistance.

TABLE 6-5

Air Cushion Vehicle Cash Flow Projections under Event 11

Year	*Net Revenue*	*Operating Costs*	*Loan Payment*	*Cash Flow before Taxes*	*Interest Paid*	*Depreciation Cost*	*Taxable Income*	*Taxes Due*	*Cash Flow after Taxes*
0	0	0	0	−2,520,000	0	0	0	0	−2,520,000
1	7,040,000	2,808,358	2,016,000	2,215,643	1,008,000	2,520,000	703,643	239,238	1,976,404
2	7,379,200	2,872,690	1,915,200	2,591,310	907,200	2,016,000	1,583,310	538,325	2,052,984
3	7,734,975	2,938,669	1,814,400	2,981,907	806,400	1,612,800	2,377,107	808,216	2,173,690
4	8,108,144	3,006,339	1,713,600	3,388,205	705,600	1,290,240	3,105,965	1,056,028	2,332,177
5	8,499,564	3,075,748	1,612,800	3,811,016	604,800	1,032,192	3,786,824	1,287,520	2,523,496
6	8,910,135	3,146,943	1,512,000	4,251,192	504,000	825,754	4,433,439	1,507,369	2,743,823
7	9,340,802	3,219,973	1,411,200	4,709,629	403,200	825,754	4,891,876	1,663,238	3,046,392
8	9,792,558	3,294,889	1,310,400	5,187,269	302,400	825,754	5,369,516	1,825,635	3,361,634
9	10,266,440	3,371,743	1,209,600	5,685,100	201,600	825,754	5,867,347	1,994,898	3,690,202
10	10,763,550	3,450,590	1,108,800	6,204,161	100,800	825,754	6,386,407	2,171,379	4,032,782
	PW of NCF after taxes Disc. 10%								13,556,480

Notes:

1. Declining balance with switch to straight-line method of depreciation applied to full asset value.
2. Project taxed at 34%.
3. Amount of equity invested in project shown as cash flow in year zero.
4. Investment tax credit is 0%.

TABLE 6-6

Payoff Matrix

		Events											
		1	*2*	*3*	*4*	*5*	*6*	*7*	*8*	*9*	*10*	*11*	*12*
	1	–6,794	–7,816	–4,131	–4,897	–1,588	–2,320	–1,137	433	7,077	6,402	10,518	9,843
	2	–2,693	–3,078	–373	–740	1,962	1,606	4,583	4,227	10,126	9,775	13,566	13,216
Acts	*3*	–6,415	–7,447	–3,583	–4,338	–1,031	–1,770	1,680	969	7,333	6,652	10,774	10,094
	4	–1,628	–2,443	260	–325	2,093	1,539	4,158	3,604	8,419	7,865	11,172	10,618
	5	–12,167	–13,560	–9,560	–11,102	–7,376	–8,917	–4,344	–5,789	1,387	361	4,139	3,122

Note: Matrix outcome values are present worth of project lifetime net cash flow after taxes divided by 1,000.

THE DECISION UNDER UNCERTAINTY

The decision analysis program DECIANA, included in this text, will be of assistance in applying the Laplace, Maximax, Maximin, Minimax Regret, and Hurwicz principles to the matrix (see Morris in the References). Table 6-7 is the DECIANA output for the first four of these principles. The best alternative under each of these principles is:

Principle	*Best Vehicle Choice*
Laplace	ACV
Maximax	ACV
Maximin	CAT
Minimax regret	ACV

TABLE 6-7

Decision Analysis under Uncertainty

Act	*Laplace Principle Payoff*	*Maximax Principle Payoff*	*Maximin Principle Payoff*	*Savage Minimax Regret*
1	.65580E+06	.10518E+08	–.78157E+07	.53731E+07
2	.43481E+07*	.13566E+08*	–.30784E+07	.10653E+07*
3	.10645E+07	.10774E+08	–.74468E+07	.50043E+07
4	.37777E+07	.11172E+08	–.24425E+07*	.25979E+07
5	–.53296E+07	.41389E+07	–.13708E+08	.11266E+08

* Best act under given principle.

This suggests that the air cushion vehicle is a robust choice for the Bridgeport-Manhattan service for the problem stipulated costs and operating environment. Only the most pessimistic decision principle indicates another alternative.

The Hurwicz principle output shown in Table 6-8 indicates a similar result. On an arbitrary decision maker optimism scale of 0 to 1, the decision maker would select the ACV over the catamaran for all optimism levels above .2.

TABLE 6-8

Hurwicz Decision Principle

Optimism Index	*Best Act*	*Payoff*
.00	4	–.24425E+07
.10	4	–.10811E+07
.20	4	.28030E+06
.30	2	.19151E+07

(continued on next page)

TABLE 6-8 *(continued)*

.40	2	.35796E+07
.50	2	.52440E+07
.60	2	.69085E+07
.70	2	.85730E+07
.80	2	.10238E+08
.90	2	.11902E+08
1.00	2	.13566E+08

DECISION UNDER RISK AND BAYESIAN ANALYSIS

Let's return to the three independent variables which were judged to have significant uncertainty for the fast ferry problem. On the basis of experience, judgment, and some historical data (see Good in the References), the analyst is able to assign probability measures to the passengers per one-way trip, annual fuel cost increase, and annual fare increase. These values are as follows:

Average Daily Passengers per One-Way Trip	P_r *(Passengers)*	*E (Passengers)*
100	.30	30
120	.50	60
160	.20	32
	1.00	122

Annual fuel cost increase	P_r *(Fuel Incr.)*	*E (Fuel Incr.)*
.02	.80	.016
.06	.20	.012
	1.00	.028

Annual Fare Increase	P_r *(Fare Incr.)*	*E (Fare Incr.)*
.02	.70	.014
.05	.30	.015
	1.00	.029

Since these probabilities are independent, i.e., disjoint, they may be convolved into a single discrete probability by the probability theorem (see Hahn and Shapiro in the References) which states:

$$P_r\ (E_j) = P_r\ (\text{Pass})\ P_r\ (\text{Fuel})\ P_r\ (\text{Fare})$$

for each of the 12 events. The resulting probability distribution is shown in Table 6-9.

TABLE 6-9
Convolution of Independent Probabilities

Event	*Passengers Carried*	*Annual Fare Increase*	*Annual Fuel Increase*	*Probability*
1	100	.02	.02	.168
2	100	.02	.06	.042
3	100	.05	.02	.072
4	100	.05	.06	.018
5*	120	.02	.02	.280
6	120	.02	.06	.070
7	120	.05	.02	.120
8	120	.05	.06	.030
9	160	.02	.02	.112
10	160	.02	.06	.028
11†	160	.05	.02	.048
12	160	.05	.06	.012
				1.000

* Most-likely event
† Event with highest payoff

With the discrete probability distribution for the 12 events available, it is possible to calculate the expected values of the payoff, the regret, i.e., opportunity loss, and the cost of irrationality for each of the alternative marine vehicles (see Schlaifer in the References). The output of the program DECIANA provides this information as shown in Table 6-10. The highest expected value of discounted net cash flow after taxes is $2,934,400 for the air cushion vehicle. Therefore, our knowledge of the probability distribution for future events did not change the choice of the best vehicle.

TABLE 6-10
Probabilistic Decision Analysis

Act	*Expected Payoff*	*Expected Opportunity Loss*	*Cost of Irrationality*
1	–.71799E+06	.39478E+07	.36523E+07
2	.29344E+07	.29543E+06	.21875E+00
3	–.27307E+06	.35029E+07	.32074E+07
4	.27646E+07	.46520E+06	.16977E+06
5	–.64908E+07	.97206E+07	.94251E+07

Note: Value of decision under certainty is .32298E+07.
Cost of uncertainty is .29543E+06.

The expected opportunity loss, tabulated for each alternative, is the expectation of the rows of the regret matrix under the probability distribution. Therefore, the lowest expected opportunity loss of $295,430 is associated with the best alternative, the ACV. The lowest expected opportunity loss may be termed the cost of uncertainty (see Schlaifer), since this loss is a result of the variance of the probability distribution. A high cost of uncertainty suggests that more research is warranted to reduce this value. The cost of uncertainty is the difference between the expected value of decisions which would be made if it were known in advance which events would actually occur and the expectation of payoffs under all events:

$$E(P_i) = \sum_{j=i}^{m} P_{ij} \cdot P_r(P_{ij})$$

$$E(R_i) = \sum_{j=i}^{m} R_{ij} \cdot P_r(R_{ij})$$

$$E(CP) = \sum_{j=i}^{m} \max_{j=i}^{n} P_{ij} \cdot P_r(P_{ij})$$

$$\text{Cost of uncertainty} = E(CP) - E(P_i)^*$$

where P_i = NPW of alternative i
P_{ij} = NPW of alternative i under event j
$P_r(P_{ij})$ = Probability of NPW alternative i event j
R_i = Regret NPW of alternative i
R_{ij} = Regret NPW of alternative i under event j
$E(P_i)^*$ = Highest expected payoff NPW, alternative i
CP = Payoff under certainty

BAYESIAN DECISION ANALYSIS

Since the cost of uncertainty in the fast ferry decision has been calculated to be $295,430, it may be reasonable to spend some fraction of this amount to improve information concerning one of the significant random variables. A market research firm is retained for a fee of $25,000 to prepare questionnaires and undertake a sampling study (see Deming in the References) of the Bridgeport commuters to determine the average daily one-way passengers which could be expected. The result of this study is a sampling distribution of the average daily passengers during the project life. The sampling distribution is as follows:

Passengers per One-Way Trip	*P_r (Passengers)*	*E (Passengers)*
100	.10	10
120	.70	84
160	.20	32
	1.00	126

Based on this information, the probability for the 12 events is recalculated using the sample distribution for one-way passenger loads. The recalculated distribution is shown in Table 6-11. Using this sampling distribution, Bayes Theorem (see Hahn and Shapiro) can be applied to revise the original distribution to incorporate the sampling information (see Schlaifer):

$$P_r(A_i|B) = P_r(A_i)\frac{P_r(B|A_i)\ P_r(A_i)}{\sum_{i=i}^{12} P_r(B|A_i)\ P_r(A_i)}$$

where $P_r(A_i)$ = prior probability for the events i = 1 to 12, and
$P_r(B/A_i)$ = sampling study probability of event i.

TABLE 6-11
Sample Distribution from Passenger Load Study Convolved with Fuel and Fare Increases

Event	*Probability (E_j)*
1	.056
2	.014
3	.024
4	.006
5	.392
6	.098
7	.168
8	.042
9	.112
10	.028
11	.048
12	.012
	1.000

The computer program BAYES may be used to calculate the expected values of the net cash flow for the payoffs, regrets, and irrationality costs of the five alternatives. This information is presented in Table 6-12. The decision analysis is presented under both the prior $P_r(A_i)$, and the posterior, $P_r(A_i/B)$ distributions of the probability of the twelve events of the analysis. As a result of the market study, the cost of uncertainty has been reduced from $295,430 to $156,410. The most favorable choice is still the ACV; however, the expected net present worth has decreased from $2,934,400 to $2,801,700 on the basis of the additional information.

TABLE 6-12

Bayesian Analysis Based on Commuter Market Study

Discrete Probability Distributions

Event	*Prior*	*Conditional*	*Joint*	*Posterior*
1	.16800	.05600	.00941	.05680
2	.04200	.01400	.00059	.00355
3	.07200	.02400	.00173	.01043
4	.01800	.00600	.00011	.00065
5	.28000	.39200	.10976	.66261
6	.07000	.09800	.00686	.04141
7	.12000	.16800	.02016	.12170
8	.03000	.04200	.00126	.00761
9	.11200	.11200	.01254	.07573
10	.02800	.02800	.00078	.00473
11	.04800	.04800	.00230	.01391
12	.01200	.01200	.00014	.00087

Payoff Under Prior Distribution

Act	*Expected Payoff*	*Expected Opportunity Loss*	*Cost of Irrationality*
1	–.71799E+06	.39478E+07	.36523E+07
2	.29344E+07	.29543E+06	.21875E+00
3	–.27307E+06	.35029E+07	.32074E+07
4	.27646E+07	.46520E+06	.16977E+06
5	–.64908E+07	.97206E+07	.94251E+07

Note: Value of decision under certainty is .32298E+07.
Cost of uncertainty is .29543E+06.

Payoff Under Posterior Distribution

Act	*Expected Payoff*	*Expected Opportunity Loss*	*Cost of Irrationality*
1	–.74568E+06	.37038E+07	.35474E+07
2	.28017E+07	.15641E+06	.00000E+00
3	–.23186E+06	.31900E+07	.30336E+07
4	.27250E+07	.23316E+06	.76746E+05
5	–.65090E+07	.94671E+07	.93107E+07

Note: Value of decision under certainty is .29581E+07.
Cost of uncertainty is .15641E+06.

A comment on the Bayes Theorem application is appropriate. Before the market field study, event 5 (which predicted 120 average passengers per daily one-way trip, an annual fare increase of 2 percent, and an annual fuel cost increase of 2 percent) had a probability of .2800. The market study sampling distribution applied by Bayes Theorem increased the probability significantly to .6626. The value of good market research is obvious. Poor market research can cause great problems.

Comment on the Problem Analysis

This illustrative case of a complex decision under uncertainty or risk has generally led to the conclusion that the 40-knot air cushion vehicle is the best choice among the alternatives selected for consideration. The value of these decision models is diminished to the extent that the analyst fails to include all the viable alternatives and possible future events. Another shortcoming of the procedure is the discrete nature of the future events. Obviously, the independent variables are not discrete. Discrete independent variables have been selected to make the decision model manageable for the analyst. Failure to use judgment based on experience in selecting the range of discrete variables will produce an unsatisfactory decision model.

REFERENCES AND BIBLIOGRAPHY

Burg, Donald E. *A Practical Waterborne Rapid Transport System*. Washington, D.C.: American Institute of Aeronautics and Astronautics, 1986. An excellent summary of cost impact of vessel type, vessel speed, and docking times on a rapid transport system. Includes SWATH, hydrofoil, ACV, catamaran, and SES.

Deming, William Edwards. *Some Theory of Sampling*. New York: Dover Publications, 1966. An outstanding text on field sampling to develop probability measures on population, business, process, etc., data. A review of this text will convince the engineering analyst of the need for competent assistance in sampling.

Good, I.J. *Estimation of Probabilities*. Cambridge, Mass.: MIT Press, 1965. A useful review of the methods for developing rational subjective probability estimates.

Hahn, G.J., and S.S. Shapiro. *Statistical Models in Engineering*. New York: John Wiley & Sons, Inc.; 1967. Reference for understanding of probability and statistical theory necessary for solution of many practical engineering problems.

Hunt, Everett C. "Fast Ferry Transport System Decisions under Uncertainty, A Comprehensive Case." Washington, D.C.: I.A.M.V.C., June 1989. Data from this paper were used for the example of this chapter.

Morris, W.T. *The Analysis of Management Decisions.* Homewood, Ill.: R.D. Irwin, 1964. Text includes summary of decision methods under uncertainty and risk.

Schlaifer, R. *Probability and Statistics for Business Decisions.* New York: McGraw-Hill Company, 1959. Practical descriptions of applications of Bayesian methods to solution of problems in business and engineering economics.

QUESTIONS AND PROJECTS

1. What is the meaning and usefulness of the cost of uncertainty for a measure of merit?
2. How does one calculate the value of a decision under certainty?
3. Discuss the possibilities of a different result in the Fast Ferry model if the number of discrete random variable values is increased from three to six without changing the range of the variable.
4. Suggest another alternative which might have been included in the decision model.
5. Suggest other non-numerical factors which might have been considered in the Fast Ferry alternative selection.
6. Can anything be learned by studying the expected values of the regret matrix of the decision model?
7. Is the cost of irrationality a useful decision measure? What does it mean?
8. Discuss other measures of merit which might be used in the Fast Ferry decision model.
9. Suggest how this decision model might be applied to other shipping services such as bulk, container, tanker, etc.
10. The Fast Ferry model is based on a declining balance with a switch to the straight-line method of depreciation. Suggest possible changes in the decision outcome for other depreciation methods.

PROBLEMS

1. In the Fast Ferry model, assume that the model was designed with three discrete values for fuel cost escalation, three for fare escalation, and four values for one-way passenger loads. How many calculations of net present value would be necessary to complete the matrix?
2. A light manufacturing company has a three-year contract to manufacture 50,000 units per year of stainless steel bolts. The selling price is \$10 per unit. The cost is \$8 per unit. The contract has one more year to run with an opportunity for renewal for three more years. The company plans to reduce the cost of the product \$1.50 per unit.

An investment of $100,000 is required to achieve the cost reduction. The MARR is 10 percent. Analyze this decision.

3. In Problem 2, the probability of contract renewal is .60. Analyze the decision with this new information.
4. Use the program DECIANA to analyze the following decision matrix of net cash flow present worth:

	S_1	S_2	S_3	S_4	S_5
a_1	105	100	130	120	102
a_2	393	91	140	400	100
a_3	490	56	110	510	130
a_4	246	60	100	490	120

5. Given the following payoff matrix, find the best alternative under Minimax regret criteria.

310	296	286	274	262	250
280	288	294	303	305	260
270	281	293	295	308	265
310	306	302	298	297	295
315	308	301	297	294	290
290	297	295	299	302	305

6. Use the program DECIANA to analyze the decision matrix of problem 5.
7. If the events of problem 5 have the following probability distribution, find the best alternative. The program DECIANA may be used.

Event	*Pr (Event)*
1	.005
2	.150
3	.500
4	.200
5	.140
6	.005
	1.000

8. Further study of the project in problem 7 yields a sample distribution, i.e., conditional distribution, Pr (B Event_i) (see next page). With this new information, does the best alternative change? The program BAYES may be used.

Event	*Pr (B Event)*
1	.20
2	.20
3	.40
4	.10
5	.05
6	.05
	1.00

Chapter 7

Required Freight Rate Simulation

Lest men suspect your tale untrue, keep probability in view.
—John Gay

Perhaps the most commonly used measure of merit in ship design and ship operating economic evaluations is the required freight rate (RFR). This measure has an immediately apparent corollary: it is directly related to the total cost of vessel operations. When required freight rate is employed in marine transport problems, the numerous operating and capital costs are usually treated as deterministic despite the knowledge, based on experience, which suggests that there is considerable uncertainty in most of the numerous independent variables involved. If the analyst has the information to calculate and/or estimate the continuous probability distributions for all the independent variables, then the computer program RFRSIM, included in this text, can be employed to calculate the distribution of the required freight rate as well as some other intermediate dependent variables (such as voyages per year, annual fuel cost, annual operating cost, etc.).

RFRSIM INPUT DATA

A complete description of the input file format for RFRSIM is included in the program documentations of Appendix A. The RFRSIM input includes 23 independent random variables, which are represented by probability distributions input in one of several ways. In every case, three values of the independent random variable are required: a minimum value, a mode value, and a maximum value.

A fourth number, entered after the maximum value, indicates the area of the tails of the distribution. If the value .135 is entered, the program interprets this to mean that the maximum and minimum values are at the

+3 and −3 standard deviations. If an integer between 1 and 49 is entered, this is interpreted as the area in each of the tails beyond the minimum and maximum values. If the value 15.8655 is input, the program assigns the high and low values at +1 and −1 standard deviation.

In contractual agreements, it is not uncommon to fix an upper and/or lower value of an independent variable. In that case, the analyst will wish to set limits on independent variables which reflect such agreements. This is accomplished by truncating the independent variable probability distribution. For such a variable, the program requires two additional input values for a total of six: minimum value, mode value, maximum value, probability value, low truncation value, and high truncation value.

Additional details needed to construct an input data file for RFRSIM are presented in the documentation of Appendix A.

It should be noted that several of the input random variables, i.e., unit capacity of the vessel, owner equity, and total annual loan payment of interest and principal, will commonly be expressed as constants, i.e., fixed values. However, to provide the maximum program application flexibility, they have been assigned the role of random variables When it is desirable to express these values as fixed values, this can be practically achieved by specifying very small standard deviations. The program input value, design speed, is indeed a constant and the input format allows only a constant value to be used.

RFRSIM

RFRSIM employs simulation to calculate a probability distribution for a required freight rate and the economic cost of transport. This is accomplished in the following simulation steps:

1. Round-trip voyages per year are simulated from distributions of distance, vessel speed, operating days, and port time.
2. Annual cargo throughput is simulated from voyages per year, vessel capacity, and load factor.
3. Fuel costs and lube oil costs are simulated, given distributions of daily fuel and lube consumption at design speed.
4. Annual operating costs are simulated from distributions for each cost component.
5. Finally, the required freight rate and economic cost of transport are simulated as described in Chapter 4.

Example 7-1

A new vessel is to be placed in service in the next few months. During the past year, the sister ship has averaged the following speeds during voyages on the planned route:

18.6	19.3	20.7	21.0	17.9
18.3	20.5	19.2	18.9	19.7

What input would be appropriate for speed in a required freight rate simulation?

$$\text{Sample mean} = \sum_{i=1}^{n} x_i / n = 19.41$$

$$\text{Unbiased standard deviation} = \left[\sum_{i=1}^{n} \left(\frac{x_i - \bar{x}}{n - 1} \right) \right]^{1/2} = 1.0514$$

$$\text{Coefficient of variation} = .0542$$

Therefore, the input for the speed variable may be:

Minimum value 19.41 – 3(1.0514) = 16.2558

Median value = 19.41

Maximum value 19.41 + 3(1.0514) = 22.5642

Probability value = .135

or another suitable input for speed might be:

Minimum value 19.41 – 1.0514 = 18.3585

Median value = 19.41

Maximum value 19.41 + 1.0514 = 20.4514

Probability value = 15.8655

As Example 7-1 suggests, the simulation program user must use available statistical data or a rational method for estimating subjective data to develop each of the input distributions for RFRSIM. Example 7-2 presents another method for developing input information.

If a sample of an independent variable is changing over time, simple linear regression analysis is frequently useful in developing prediction estimates of the mean and variance needed for inputs to RFRSIM program. A least square fit of a straight line to a data sample is feasible when underlying theory or simple inspection supports the assumption that there is a linear relationship between x and y or $y = a + bx$. This fit requires the summation of the squared distances by which each data point deviates in the y direction from a line which minimizes the summation. The calculus is employed to find the best fit line, i.e., the line which provides the lowest value of the summation of the squares. If Q equals the sum of the squares, then:

$$Q = \sum_{i=1}^{n} (y_i - a - bx_i)^2$$

$$\frac{\partial Q}{\partial a} = -2 \sum_{i=1}^{n} (y_i - a - bx_i) = 0$$

$$\frac{\partial Q}{\partial b} = -2 \sum_{i=1}^{n} x_i (y_i - a - bx_i) = 0$$

Performing the indicated summations and substituting the estimated values of $\tilde{a}$ for a and $\tilde{b}$ for b and solving for $\tilde{a}$ and $\tilde{b}$ provides:

$$\tilde{a} = \bar{y} - \tilde{b}\bar{x}$$

$$\tilde{b} = \frac{\sum_{i=1}^{n} x_i y_i - n\bar{x}\bar{y}}{\sum_{i=1}^{n} x_i^2 - n\bar{x}^2}$$

Numerous texts on mathematical statistics demonstrate that any point y on the fitted line which corresponds to x has a normal distribution with the following mean and variance:

$$\bar{y}_o = \tilde{a} + \tilde{b}\bar{x}_o$$

$$\tilde{\sigma}_y^2 = \left[\frac{1}{n} + \frac{(x_o - \bar{x})^2}{\sum_{i=1}^{n} (x_i - \bar{x})^2} \right]$$

Example 7-2

A ship operator has a 16,000-dwt container vessel which, during the past eight years, has had the following annual port and terminal cost experience while engaged in a scheduled liner service.

Year	*Annual Cost*
1	917,950
2	974,917
3	1,000,069
4	1,036,154
5	1,068,200
6	1,082,944
7	1,135,200
8	1,160,000

Given this data, what mean and standard deviation should be used for the ninth year for input to RFRSIM for a similar vessel engaged in the same trade?

Applying regression we obtain the following estimates for a and b:

$$\tilde{a} = 898{,}191.2$$

$$\tilde{b} = 33{,}052.9$$

$$\tilde{\sigma} = 9{,}790.5$$

Then,

$$\tilde{y}_9 = 898{,}191.2 + 33{,}052.5\ (9)$$

$$\tilde{y}_9 = \$1{,}195{,}667.10$$

and,

$$\tilde{\sigma}^2_{y_9} = 95{,}853{,}498 \left[\frac{1}{n} + \frac{(x_o - 4.5)^2}{\sum_{i=1}^{n} (x_i - 4.5)^2} \right]$$

$$\tilde{\sigma}^2_{y_9} = 58{,}192{,}658$$

$$\tilde{\sigma}_{y_9} = 7{,}628$$

Therefore, the RFRSIM input for the estimated annual port and terminal costs for the simulation of the ninth year is:

low	1,188,039
median	1,195,667
high	1,203,295
probability	15.866

Example 7-3

The Mad River Shipping Company has accumulated the following information for a large container vessel. They intend to simulate the required freight rate and the economic cost of transport both of which will be expressed as probability distributions. The design speed is 20 knots.

Input Variable	*Low Value*	*Mode Value*	*High Value*	*Probability Input*
Round-trip, miles	5,800	6,000	6,200	.135
Sea speed, knots	17.5	20	20.5	.135
Port time, hours	60	65.28	75	.135
Operating days/year	350	354	358	.135
Vessel capacity	1,190	1,200	1,210	.135
Load factor, outboard	.85	.90	.95	.135
Load factor, inboard	.65	.80	.85	.135
Crew cost, annual	2,000,000	2,125,000	2,500,000	.135
Design speed fuel rate, ton/day	35.48	36.3	37.12	.135
Lube oil consumption, ton/day	.2	.25	.3	.135
In port fuel rate, ton/day	2	3	4	.135
Fuel cost, $/ton	84	100	108	.135
Lube oil cost, $/ton	600	750	800	.135
Annual M&R cost	1,700,000	1,790,000	1,850,000	.135
Insurance cost	450,000	465,000	475,000	.135
Port & terminal cost	1,188,039	1,195,667	1,203,295	15.8655
Stevedoring cost	7,200,000	7,800,000	7,900,000	.135
G & A cost	500,000	666,000	680,000	.135
Owner equity	13,800,000	14,000,000	14,200,000	.135
MARR	.135	.14	.145	.135
Mortgage payments	9,650,000	9,700,000	9,750,000	.135
Cargo value/unit	4,400	6,000	8,000	.135
Cargo interest rate	.08	.10	.14	.135

The ASCII input file for the above data prepared according to the instructions of Appendix A is shown in Figure 7-1 with some explanatory notes. After 600 stochastic simulations of the operating economics of the Mad River Shipping Company container vessel, the RFRSIM program produced an output file containing summary statistics and prediction intervals for the following quantities:

Voyages per year	Annual operating costs
Annual unit throughput	Annual total costs
Annual fuel cost	Required freight rate
Annual lube oil cost	Economic cost of transport

```
MAD RIVER SHIPPING COMPANY
BIG CONTAINER VESSEL
E. C. HUNT PE
20 ————————— Design speed is fixed
0 ——————————— No truncated variables
5,800
6,000
6,200
.135 ———————— High and low values are at three standard deviations
17.5
20
20.5
.135
60
65.28 ——————— Median value variable number three
75.
.135
350
354
358
.135
1,190
1,200
1,210
.135
.85
.9
.95
.135
.65
.8
.85
.135
2,000,000
2,125,000 ——— Median value variable number eight
2,500,000
.135
35.48
36.3
37.12
.135
.2
.25
.3
.135
2
3
4
.135
84
100
108
.135
600
```

Figure 7-1. ASCII input file to RFRSIM for Example 7-3.
(continued on next page)

Figure 7-1 (continued)

750
800
.135
1,700,000
1,790,000
1,850,000
.135
450.000
465,000
475,000
.135
1,188,039
1.195,667
1,203,295
15.866 ———— High and low values are at one standard deviation
7,200,000
7,800,000
7,900,000
.135
500,000
666,000
680,000
.135
13,800,000
14,000,000
14,200,000
.135
.135
.14 ———— Small variance approaching a constant
.145
.135
9,650,000
9,700,000
9,750,000
.135
4,400
6,000
8,000
.135
.08
.10
.14
.135

Figure 7-2 is the output file for the Mad River Shipping Company simulation. Reviewing this output under the assumption that rational methods were used to prepare the input data, we can make one or more of the following statements concerning the RFR for the Mad River vessel:

1. The mean and median value of RFR is $572.59 per container with a standard deviation of $15.91;

2. We are 90 percent confident that the RFR is between \$546.36 and \$598.83;
3. We are 95 percent confident that the RFR is between \$541.33 and \$603.86:
4. We are 99 percent confident that the RFR is between \$531.46 and \$613.74;
5. The distribution of the RFR is approximately normal.

```
**********************
*                    *
* REQUIRED FREIGHT RATE *
*                    *
* MONTE CARLO SIMULATION *
*                    *
*  Marine Engineering *
*   Economics And     *
*   Cost Analysis     *
*                    *
**********************
```

PREPARED FOR: MAD RIVER SHIPPING COMPANY
PROJECT: BIG CONTAINER VESSEL
PREPARED BY: E C HUNT PE
DATE: 08/09/1994
TIME: 16:12:57

	Voyages Per Year
EXPECTED VALUE	.231551E+02
VARIANCE	.188941E+00
STANDARD DEVIATION	.434673E+00
COEFFICIENT OF VARIATION	.187723E–01
GAMMA ONE, SKEWNESS	–.570134E+00
GAMMA TWO, KURTOSIS	.596277E+00
MAX. SAMPLE VALUE	.242878E+02
MIN. SAMPLE VALUE	.213221E+02
SAMPLE RANGE	.296565E+01
SAMPLE SIZE	600
PREDICTION LIMITS	
UPPER PREDICTION LIMIT 90% CONFIDENT	.238717E+02
LOWER PREDICTION LIMIT 90% CONFIDENT	.224384E+02
UPPER PREDICTION LIMIT 95% CONFIDENT	.240093E+02
LOWER PREDICTION LIMIT 95% CONFIDENT	.223008E+02
UPPER PREDICTION LIMIT 99% CONFIDENT	.242790E+02
LOWER PREDICTION LIMIT 99% CONFIDENT	.220311E+02

Figure 7-2. RFRSIM output file for the Mad River Shipping Company operating economics problem under risk. *(Continued on next page)*

Cargo Throughput, Units Per Year-Ship	
EXPECTED VALUE	.470514E+05
VARIANCE	.166306E+07
STANDARD DEVIATION	.128960E+04
COEFFICIENT OF VARIATION	.274082E-01
GAMMA ONE, SKEWNESS	–.463965E+00
GAMMA TWO, KURTOSIS	.503501E+00
MAX. SAMPLE VALUE	.505293E+05
MIN. SAMPLE VALUE	.426076E+05
SAMPLE RANGE	.792164E+04
SAMPLE SIZE	600
PREDICTION LIMITS	
UPPER PREDICTION LIMIT 90% CONFIDENT	.491776E+05
LOWER PREDICTION LIMIT 90% CONFIDENT	.449252E+05
UPPER PREDICTION LIMIT 95% CONFIDENT	.495858E+05
LOWER PREDICTION LIMIT 95% CONFIDENT	.445170E+05
UPPER PREDICTION LIMIT 99% CONFIDENT	.503860E+05
LOWER PREDICTION LIMIT 99% CONFIDENT	.437168E+05

Fuel Consumption, Tons Per Day	
EXPECTED VALUE	.358364E+02
VARIANCE	.407552E+01
STANDARD DEVIATION	.201879E+01
COEFFICIENT OF VARIATION	.563335E–01
GAMMA ONE, SKEWNESS	–.964791E+00
GAMMA TWO, KURTOSIS	.768191E+00
MAX. SAMPLE VALUE	.391011E+02
MIN. SAMPLE VALUE	.283558E+02
SAMPLE RANGE	.107453E+02
SAMPLE SIZE	600
PREDICTION LIMITS	
UPPER PREDICTION LIMIT 90% CONFIDENT	.391649E+02
LOWER PREDICTION LIMIT 90% CONFIDENT	.325079E+02
UPPER PREDICTION LIMIT 95% CONFIDENT	.398039E+02
LOWER PREDICTION LIMIT 95% CONFIDENT	.318690E+02
UPPER PREDICTION LIMIT 99% CONFIDENT	.410566E+02
LOWER PREDICTION LIMIT 99% CONFIDENT	.306163E+02

Fuel Cost, $ Per Year	
EXPECTED VALUE	.125472E+07
VARIANCE	.694160E+10

Figure 7-2. (continued)

STANDARD DEVIATION	.833163E+05
COEFFICIENT OF VARIATION	.664021E–01
GAMMA ONE, SKEWNESS	–.571383E+00
GAMMA TWO, KURTOSIS	.155152E+00
MAX. SAMPLE VALUE	.143123E+07
MIN. SAMPLE VALUE	.974362E+06
SAMPLE RANGE	.456870E+06
SAMPLE SIZE	600

PREDICTION LIMITS

UPPER PREDICTION LIMIT 90% CONFIDENT	.139209E+07
LOWER PREDICTION LIMIT 90% CONFIDENT	.111735E+07
UPPER PREDICTION LIMIT 95% CONFIDENT	.141846E+07
LOWER PREDICTION LIMIT 95% CONFIDENT	.109098E+07
UPPER PREDICTION LIMIT 99% CONFIDENT	.147016E+07
LOWER PREDICTION LIMIT 99% CONFIDENT	.103929E+07

Lube Oil Cost, $ Per Year

EXPECTED VALUE	.652528E+05
VARIANCE	.252859E+08
STANDARD DEVIATION	.502851E+04
COEFFICIENT OF VARIATION	.770620E–01
GAMMA ONE, SKEWNESS	.884580E–01
GAMMA TWO, KURTOSIS	–.236831E+00
MAX. SAMPLE VALUE	.798252E+05
MIN. SAMPLE VALUE	.517237E+05
SAMPLE RANGE	.281015E+05
SAMPLE SIZE	600

PREDICTION LIMITS

UPPER PREDICTION LIMIT 90% CONFIDENT	.735436E+05
LOWER PREDICTION LIMIT 90% CONFIDENT	.569620E+05
UPPER PREDICTION LIMIT 95% CONFIDENT	.751351E+05
LOWER PREDICTION LIMIT 95% CONFIDENT	.553705E+05
UPPER PREDICTION LIMIT 99% CONFIDENT	.782554E+05
LOWER PREDICTION LIMIT 99% CONFIDENT	.522502E+05

Operating Cost, $ Per Year

EXPECTED VALUE	.153440E+08
VARIANCE	.208705E+11
STANDARD DEVIATION	.144466E+06
COEFFICIENT OF VARIATION	.941513E–02
GAMMA ONE, SKEWNESS	–.261699E+00
GAMMA TWO, KURTOSIS	.278182E–01
MAX. SAMPLE VALUE	.157636E+08

Figure 7-2. (continued)

MIN. SAMPLE VALUE	.148866E+08
SAMPLE RANGE	.876923E+06
SAMPLE SIZE	600

PREDICTION LIMITS

UPPER PREDICTION LIMIT 90% CONFIDENT	.155822E+08
LOWER PREDICTION LIMIT 90% CONFIDENT	.151059E+08
UPPER PREDICTION LIMIT 95% CONFIDENT	.156280E+08
LOWER PREDICTION LIMIT 95% CONFIDENT	.150601E+08
UPPER PREDICTION LIMIT 99% CONFIDENT	.157176E+08
LOWER PREDICTION LIMIT 99% CONFIDENT	.149705E+08

Total Cost, $ Per Year

EXPECTED VALUE	.250449E+08
VARIANCE	.212570E+11
STANDARD DEVIATION	.145798E+06
COEFFICIENT OF VARIATION	.582147E–02
GAMMA ONE, SKEWNESS	–.229605E+00
GAMMA TWO, KURTOSIS	–.155422E–01
MAX. SAMPLE VALUE	.254538E+08
MIN. SAMPLE VALUE	.245888E+08
SAMPLE RANGE	.865056E+06
SAMPLE SIZE	600

PREDICTION LIMITS

UPPER PREDICTION LIMIT 90% CONFIDENT	.252852E+08
LOWER PREDICTION LIMIT 90% CONFIDENT	.248045E+08
UPPER PREDICTION LIMIT 95% CONFIDENT	.253314E+08
LOWER PREDICTION LIMIT 95% CONFIDENT	.247583E+08
UPPER PREDICTION LIMIT 99% CONFIDENT	.254219E+08
LOWER PREDICTION LIMIT 99% CONFIDENT	.246679E+08

Required Freight Rate, $ Per Unit-Year

EXPECTED VALUE	.574338E+03
VARIANCE	.239688E+03
STANDARD DEVIATION	.154819E+02
COEFFICIENT OF VARIATION	.269560E–01
GAMMA ONE, SKEWNESS	.577677E+00
GAMMA TWO, KURTOSIS	.803150E+00
MAX. SAMPLE VALUE	.631780E+03
MIN. SAMPLE VALUE	.536134E+03
SAMPLE RANGE	.956461E+02
SAMPLE SIZE	600

PREDICTION LIMITS

Figure 7-2. (continued)

UPPER PREDICTION LIMIT 90% CONFIDENT	.599863E+03
LOWER PREDICTION LIMIT 90% CONFIDENT	.548812E+03
UPPER PREDICTION LIMIT 95% CONFIDENT	.604763E+03
LOWER PREDICTION LIMIT 95% CONFIDENT	.543912E+03
UPPER PREDICTION LIMIT 99% CONFIDENT	.614370E+03
LOWER PREDICTION LIMIT 99% CONFIDENT	.534305E+03

Economic Cost of Transport, $ Per Unit-Year

EXPECTED VALUE	.600241E+03
VARIANCE	.260618E+03
STANDARD DEVIATION	.161437E+02
COEFFICIENT OF VARIATION	.268953E–01
GAMMA ONE, SKEWNESS	.505651E+00
GAMMA TWO, KURTOSIS	.688421E+00
MAX. SAMPLE VALUE	.659335E+03
MIN. SAMPLE VALUE	.560758E+03
SAMPLE RANGE	.985764E+02
SAMPLE SIZE	600

PREDICTION LIMITS

UPPER PREDICTION LIMIT 90% CONFIDENT	.626858E+03
LOWER PREDICTION LIMIT 90% CONFIDENT	.573624E+03
UPPER PREDICTION LIMIT 95% CONFIDENT	.631967E+03
LOWER PREDICTION LIMIT 95% CONFIDENT	.568514E+03
UPPER PREDICTION LIMIT 99% CONFIDENT	.641984E+03
LOWER PREDICTION LIMIT 99% CONFIDENT	.558497E+03

Figure 7-2. (continued)

Figure 7-3 is a left tail cumulative normal distribution plot of the RFR for the Mad River Shipping Company vessel. Plots of cumulative distributions are useful for the assessment of risk in comparing measure-of-merit for alternatives in ship design, ship purchases, and shipping ventures.

The required freight rate has a mean value of $572.60/container with a standard deviation of $15.91/container. If we assume that this is a normal distribution which is supported by the low values of gamma one and gamma two, then Figure 7-3 provides a very specific definition of the required freight rate expressed as a plot of a cumulative normal distribution.

Example 7-4

In an operation similar to Example 7-3, the Red River Shipping Company has a sistership container vessel, but their analysis has led to a different set of input data. They have determined that three of the input distributions must be truncated: round-trip distance, vessel speed, and number of operating days. The input for the data file is modified so that these independent variables are:

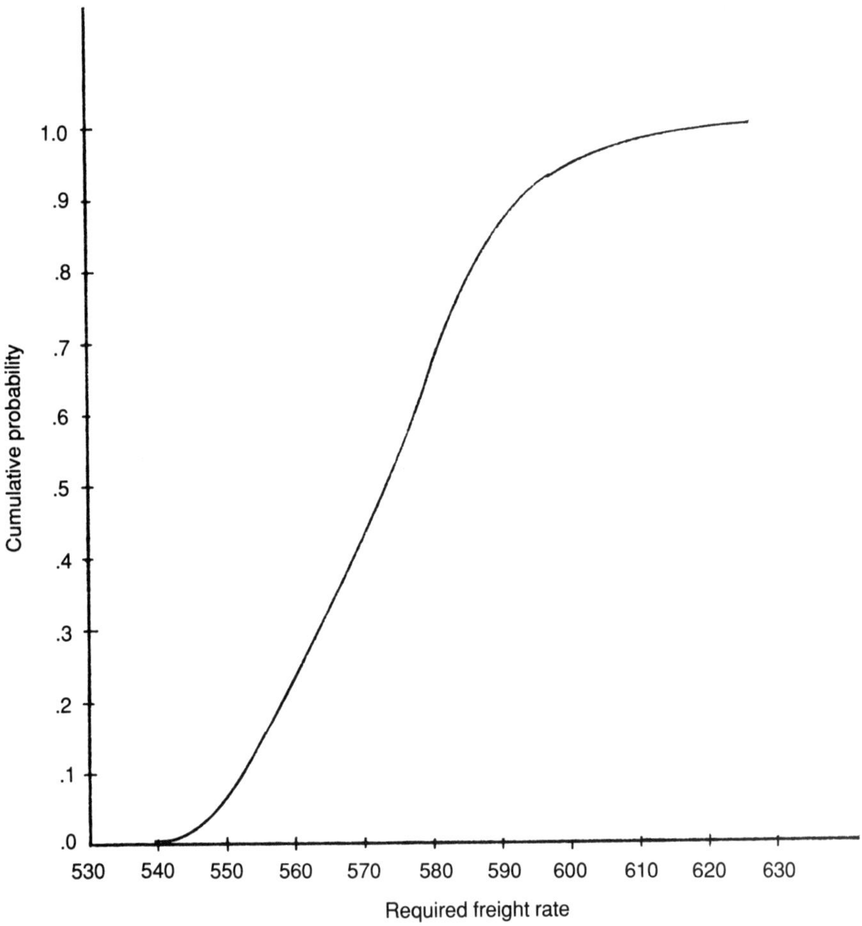

Figure 7-3. Left tail cumulative distribution of the simulated required freight rate for the Mad River Shipping Company problem

	Nautical Miles	*Speed, Knots*	*Operating Days*
Minimum value	5,800	17.5	352
Median value	6,000	20.0	354
Maximum value	6,200	21.5	356
Probability	.135	.135	15.866
Minimum truncation	5,850	10.0	340
Maximum truncation	6,150	20.5	365

For these three random variables, the shape is the normal distribution; however, no values above or below the truncation values are used in the simulation. Truncated variables are employed in simulation when there is a physical, contractual, business, or other rational reason to restrain the independent random variable from exceeding the truncated values. But, within the specified limits, the normal shape provides the best description of the random variable. Furthermore, some of the random variables are defined by specifying the specific probability in the distribution tails beyond the high and low values. Figure 7-4 shows the ASCII input file for the Red River Shipping Company simulation with some explanatory notes. Appendix A includes the detailed explanation for the preparation of the input file. Figure 7-5 is a portion of the output file for the Red River Shipping Company simulation showing the statistics of the required freight rate.

RED RIVER SHIPPING	
BIG CONTAINER SHIP	
E C HUNT CEng FIMarE	
20.	Design speed is a constant
3	Three truncated variables
1	
2	Truncated variables are 1, 2, and 4
4	
5800.	
6000.	
6200.	
.135	
5850.	First truncated variable
6150.	Low and high truncations
17.5	
20.	
21.5	
.135	
10.	Second truncated variable
20.5	Low and high truncations
62.	
65.28	
68.	
15.866	One standard deviation from median
352.	
354.	
356.	
15.866	One standard deviation from median
340.	Third truncated variable
365.	Low and high truncations
1190.	
1200.	

Figure 7-4. ASCII input file for the Red River Shipping Company RFR simulation. (Continued on next page)

1230.	
.135	Three standard deviations from median
.85	
.9	
.95	Two percent probability
2	in tails above and below
.65	high and low values
.7	
.85	Five percent probability
5	in tails above and below
2000000.	indicated values
2125000.	
2300000.	
.135	
30.	
34.	
38.	
.135	
.2	
.25	
.3	
.135	
2.	
3.	
4.	High and low values are
.135	three standard deviations
84.	from median values
100.	
110.	
.135	
600.	
750.	
800.	
.135	
1700000.	
1790000.	
1850000.	
.135	
450000.	
465000.	
475000.	
.135	
1188039.	
1195667.	
1203295.	High and low values
15.866	one standard deviation
7600000.	from median
7800000.	
7900000.	
.135	
620000.	
666000.	
680000.	
.135	

Figure 7-4 (continued)

13500000.
14000000.
14500000.
.135
1
.12 ——————— Median variable number twenty
14
.135
9600000.
9700000.
9800000
.135
4950.
6000.
7000.
.135
.08
.10 ——————— Median variable number twenty-three
.12
.135

Figure 7-4. (continued)

SIMULATION SAMPLE SIZE

The RFRSIM output provides confidence bands on all of the dependent random variables at 90 percent, 95 percent, and 99 percent confidence levels. These confidence upper and lower limits are based on the normal distribution. Prior to a simulation run, it is possible to estimate the sample size to yield a given confidence level that the simulation error will not exceed a specified amount. There are two useful approaches.

First, we can employ a procedure based on the normal distribution approximation of a binomial distribution. Let *p,* a binomial parameter, be the proportion of the simulation results which we require to have values within some specified error. Then *E* is the maximum allowable error in estimating p and $1 - \alpha$ is the confidence level, i.e., probability, that the estimated proportion $\tilde{p}$ does not differ from parameter p by more than $\pm E$. This is the classical problem of estimating a parameter of the binomial distribution presented in all texts on statistics, including Hahn and Bowker.

Assume that we wish to be 90 percent confident, i.e. $1 - \alpha = .90$, that the proportion of the RFR values between two specified values does not differ by more than E = .04 from the final estimated values, and the initial estimate of p is $\tilde{p} = .9$. Using a normal distribution approximation to the binomial distribution, we may estimate the number of trials of sample size as

$$n = \frac{\tilde{p}\,(1 - \tilde{p})}{E^2} Z^2_{1 - \alpha/2}$$

	Required Freight Rate, $ Per Unit-Year
EXPECTED VALUE	.585446E+03
VARIANCE	.258003E+04
STANDARD DEVIATION	.507940E+02
COEFFICIENT OF VARIATION	.867612E–01
GAMMA ONE, SKEWNESS	–.167681E+01
GAMMA TWO, KURTOSIS	.366809E+01
MAX. SAMPLE VALUE	.671361E+03
MIN. SAMPLE VALUE	.362173E+03
SAMPLE RANGE	.309188E+03
SAMPLE SIZE	600
PREDICTION LIMITS	
UPPER PREDICTION LIMIT 90% CONFIDENT	.669193E+03
LOWER PREDICTION LIMIT 90% CONFIDENT	.501699E+03
UPPER PREDICTION LIMIT 95% CONFIDENT	.685270E+03
LOWER PREDICTION LIMIT 95% CONFIDENT	.485623E+03
UPPER PREDICTION LIMIT 99% CONFIDENT	.716788E+03
LOWER PREDICTION LIMIT 99% CONFIDENT	.454104E+03

Figure 7-5. Portion of the output file showing the RFR statistics for the Red River Shipping Company simulation

where Z is the standardized normal variant tabulated in Appendix B. Then

$$n = \frac{(.1)(1-.1)}{(.04)^2}(1.65)^2$$

$$n = 153$$

A second approach avoids the estimation of the proportion p and instead employs the difference between the RFR parameter, μ, and the estimated mean, $\tilde{\mu}$, to establish an error for determining a sample size. In this case, E is the maximum allowable error in estimating μ and $1 - \alpha$ is the confidence level that $\tilde{\mu}$ does not differ from μ by more than $\pm E$. $\tilde{\sigma}$ is an estimate of the standard deviation. Then the approximate number of trials is

$$n = \left[\frac{z_{1-\alpha/2}\,\tilde{\sigma}}{E}\right]^2$$

If our estimated standard deviation is 15.6, the allowable error is 2.4 and $1 - \infty = .09$, then

$$n = \left[\frac{(1.65)(15.6)}{2.4}\right]^2 \approx 115$$

The key to either approach is the analyst's ability to adequately estimate the standard deviation $\tilde{\sigma}$ for the second case or the proportion $\tilde{p}$ for the first case.

REFERENCES AND BIBLIOGRAPHY

Hahn, G.J., and S.S. Shapiro. *Statistical Models in Engineering.* New York: John Wiley and Sons, 1967.

Bowker, A.H. and G.J. Lieberman. *Engineering Statistics.* Englewood Cliffs, N.J.: Prentice Hall, Inc., 1959.

QUESTIONS AND PROJECTS

1. What are the advantages and disadvantages of using simulation to calculate the required freight rate?
2. The program RFRSIM is based on the assumption that all the independent variables can be represented by a symmetrical (i.e., normal) distribution. Is this a reasonable approach?
3. The output of RFRSIM includes gamma one and gamma two. What do these values mean? How can they be used?
4. In Figure 7-3, the RFR is plotted as a left tail cumulative distribution. Plot this same distribution as a right tail cumulative distribution. Which plot is more useful? Why?
5. The prediction limits in RFRSIM output have a wider span as the confidence level increases. Why is this so? How would you use these prediction limits?
6. Given the distributions of two required freight rates for alternative shipping ventures, what values would you compare to make a selection between the alternatives? Why?
7. A sample input file for the Mad River Shipping Company is included on the floppy disk provided with this text. This input file does not include any truncated independent distributions. Select a variable which you believe should be truncated and revise the input files. Run the simulation program and compare the result.
8. If you are convinced that one or more of the input variables is best represented by a constant or deterministic value, how would you introduce this requirement into the simulation using RFRSIM?
9. If you have access to actual cost data for a fleet of vessels, determine the annual cost distributions for the following:

    1. Crew cost
    2. Maintenance and repair
    3. Insurance
    4. Port costs
    5. Stevedoring
    6. General and administrative costs

10. For a real shipping situation, use deterministic growth rates to predict the future values of the input mean and standard deviation for a five-year period. Plot the means and prediction limits of the RFR at each of the years. Note that expectation algebra provides for constant *a:*

$$E\,(ax) = \int ax\ f(x)\ dx = a\mu$$

$$E\,(ax - a\mu)^2 = \int (ax - a\mu)^2\ f(x)\ dx = a^2\ \sigma_x^2$$

11. Make several simulation runs at constant sample size of 500 and compare the results. Compare the means and variances of the RFRs for the several simulation results. Explain why they are not exactly the same result. Use the Mad River Shipping Company input file.
12. Make several simulation runs using the Mad River Shipping Company input file while increasing the sample size. Try n = 200, 400, and 600. Compare the means and variances of the RFR outcomes. Explain the differences.

Chapter 8

Ship Design Economics

Any solid lighter than a fluid will, if placed in the fluid, be so far immersed that the weight of the solid will be equal to the weight of the fluid displaced.

—*Archimedes*, On Floating Bodies I

INTRODUCTION

A ship is one of man's most complicated systems. Designed to provide a seaborne service under a wide range of dynamic environmental conditions and high-risk economic conditions, the commercial vessel requires careful consideration and reconsideration of all design parameters during the conceptual, preliminary, and detail design phases. The large capital cost of a modern merchant vessel suggests the need for exacting investigation and analysis of the many engineering and economic independent variables which will determine the vessel's life cycle ability to generate acceptable cash flow.

Ship design texts, such as Rawson and Taggart, usually include a design spiral. The spiral is an Archimedes spiral on which radial lines are superimposed. The necessary considerations in the design of a merchant vessel are listed sequentially on the radial lines so that the designer moves around on the spiral reconsidering past decisions as more information becomes available during the process of closing on a final design. A traditional ship design spiral might include the following items:

(1) Ship trading pattern
(2) Speed
(3) Dimensions
(4) Hull form
(5) Displacement
(6) General arrangement
(7) Freeboard
(8) Depth

(9) Volume
(10) Longitudinal strength
(11) Structure
(12) Powering
(13) Machinery
(14) Endurance
(15) Outfit
(16) Trim and stability
(17) Lightweight
(18) Deadweight

All of these design spiral factors contribute to the determination of the economic payoff of a ship venture. Furthermore, these factors are not independent of each other but are functionally related. For example, a sea speed decision is closely related to length of the hull. The hull form for a given speed and length is a major contributor to the economics of the ship venture. As suggested by the design spiral, these factors have traditionally been treated sequentially in the order listed. However, since the factors are all interrelated, it is efficient to develop computer algorithms which permit simultaneous consideration of most factors during the conceptual and preliminary design phases. An example of such an algorithm is included in this chapter. Figure 8-1 illustrates the functional relationships among ship size, speed, capacity, and the economic measure-of-merit, required freight rate.

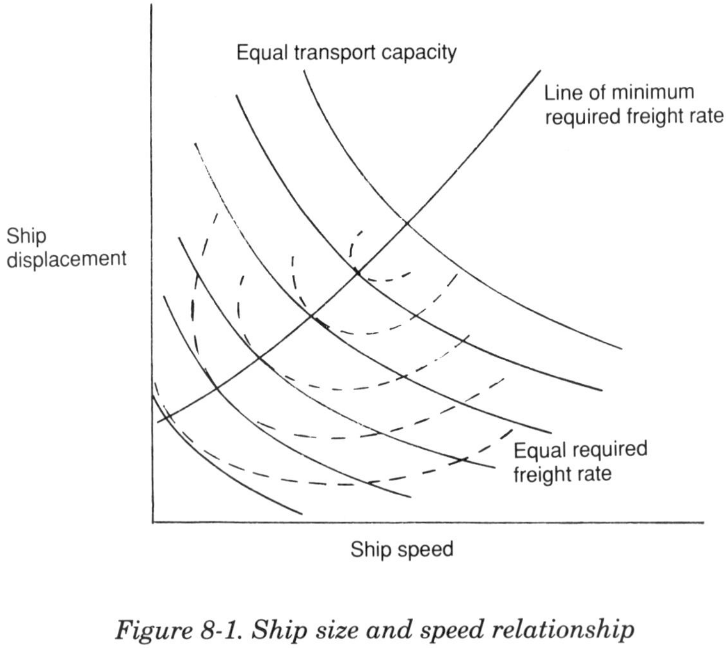

Figure 8-1. Ship size and speed relationship

PRELIMINARY SHIP DESIGN

During the conceptual design phase, the owner/operator requirements and economics are developed. A suitable economic measure-of-merit is selected and used to evaluate and compare a number of alternatives. If this work is done with a computer, the independent variables and alternatives to be considered may number in the hundreds. If a computer is not available, experience and judgment become the basis for the selection of a smaller number of alternatives for analysis.

The major engineering parameters considered by the ship designer during the conceptual design phase are:

Length, beam, depth, and draft
Displacement and deadweight
Block, prismatic, and midship section coefficients
Propulsion system performance and propeller RPM
Service allowance factor
Endurance
Sea speed
Crew size
Method(s) of cargo handling

The engineering parameters interact with the economic factors, resulting in a measure-of-merit such as required freight rate, net present value of lifetime cash flow, or the internal rate of return. The economic parameters include:

Ship price
Material costs
Labor costs
Productivity factors
Shipyard profit

Capital costs
Depreciation method and period
Amortization costs and period
Salvage value

Operating costs
Fuel
Lube oil
Labor
Maintenance and repair
Insurance
Taxes
Inflation rates

Operations
Port time
Revenue and load factor
Economic life

Ship Form Definition

The dimensions of a ship are determined early in the conceptual design stage and are then modified slightly as the design develops to meet suboptimization criteria for speed, stability, displacement, etc. Some elementary relationships among the hull dimensions are useful and are reviewed here.

Displacement Δ, the total weight of a ship, is a force defined as:

$$F = ma$$

$$\Delta = \frac{\rho V g}{g_c}$$

where ρ = density of water
V = volume of displaced water, ft^3
Δ = force due to displaced water, lb_f
g = acceleration of gravity, ft/sec^2
g_c = constant 32.2

Displacement includes weight of hull steel, machinery, fuel, stores, water, ballast, cargo, spare parts, crew, etc.

Several form coefficients are useful in representing the shape of the underwater hull. Block coefficient is:

$$C_B = \frac{V}{L \cdot B \cdot d}$$

where L = length between perpendiculars, ft
B = beam, molded, ft
d = depth equal to draft, ft

If the hull form is imagined to have been formed from a prism of length equal to the length of the ship and of constant cross section area equal to the immersed midship area, then we can define the prismatic coefficient as:

$$C_p = \frac{V}{\text{midship area} \cdot L}$$

The coefficient used to express the fullness of the midship section is the midship area coefficient, defined as:

$$C_m = \frac{\text{midship area}}{B \cdot d}$$

It should be noted that:

$$C_B = C_p \cdot C_m$$

The area enclosed by any particular waterline and the length times the beam at the waterline is called the waterplane area coefficient:

$$C_w = \frac{\text{waterplane area}}{L \cdot B}$$

C_w is usually defined at the full load draft for conceptual design studies. The following vertical prismatic coefficient is also defined at load draft:

$$C_{pv} = \frac{V}{\text{waterplane area} \cdot d}$$

and it can be shown that:

$$C_B = C_{pv} \cdot C_w$$

All of these coefficients will be used in this chapter to define a crude oil tanker hull form for conceptual design studies. Since these coefficients are nondimensional, they are useful in comparing geometrically similar ships while searching for optimum economic payoff.

In the conceptual design stages, all hull volumes and areas are based on the molded dimensions of the hull.

TANKER VENTURE ANALYSIS

A comprehensive example of a conceptual design and analysis is the computer program VENTURE (written in Fortran for the IBM Personal Computer or compatible and included with this text). The measure-of-merit for a tanker used in the program VENTURE is the internal rate of return.

The program VENTURE accepts data describing the characteristics of a tanker, i.e., handysize, capesize, VLCC, or ULCC; cost information; operating data; financial information; and shipbuilding information. The program defines the optimum characteristics and speed for a vessel of given length. The detailed program output includes:

Ship design particulars
Steel weight and cost
Machinery weight and cost
Outfit weight and cost
Hydrodynamic data
Propulsion power
Ship cost and price estimate
Economic summary
Annual cash flow for life cycle
Ship operating data
Construction payment schedule

The input data requirements are described in Appendix A and in the documentation for the program included on the accompanying floppy disk.

When the program has been calibrated for a given shipbuilding venue, it provides consistent and useful cost estimates for tanker construction. Propulsion is limited to the low-speed diesel in the current version. During the conceptual design phase, a shipowner or ship designer will find such a program invaluable when exploring deterministic engineering parameters, operating scenarios, and economic variables.

Parametric Studies Using VENTURE

During a conceptual design study, it is sometimes desirable to explore the relationships among one or more engineering and economic parameters and the economic measure-of-merit, i.e., the internal rate of return. Figure 8-2 is such a study indicating a series of similar tankers designed for

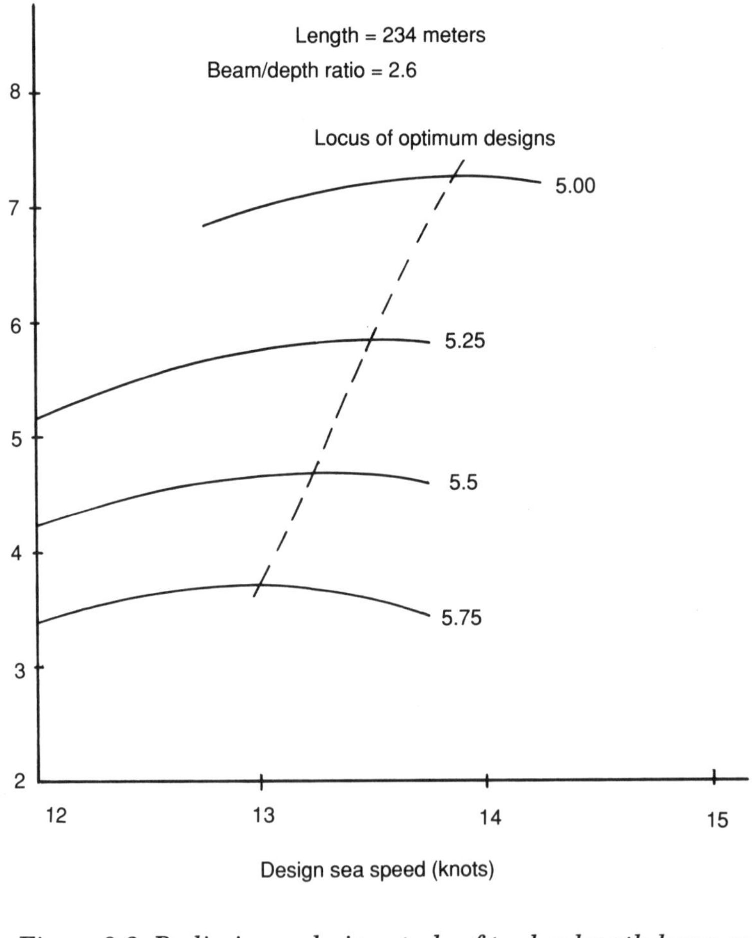

Figure 8-2. Preliminary design study of tanker length-beam ratio

length-beam ratio from 5.0 to 5.75 and sea speeds in the range of 12 to 14 knots. The optimum design speed at each length-beam ratio is indicated.

Bulk carrier midship section coefficients are generally greater than .99. Figure 8-3 illustrates the economic impact of increasing the midship section coefficient from .99 to .998 at three design sea speeds. Changes in internal rate of return are the results of small adjustments made by the

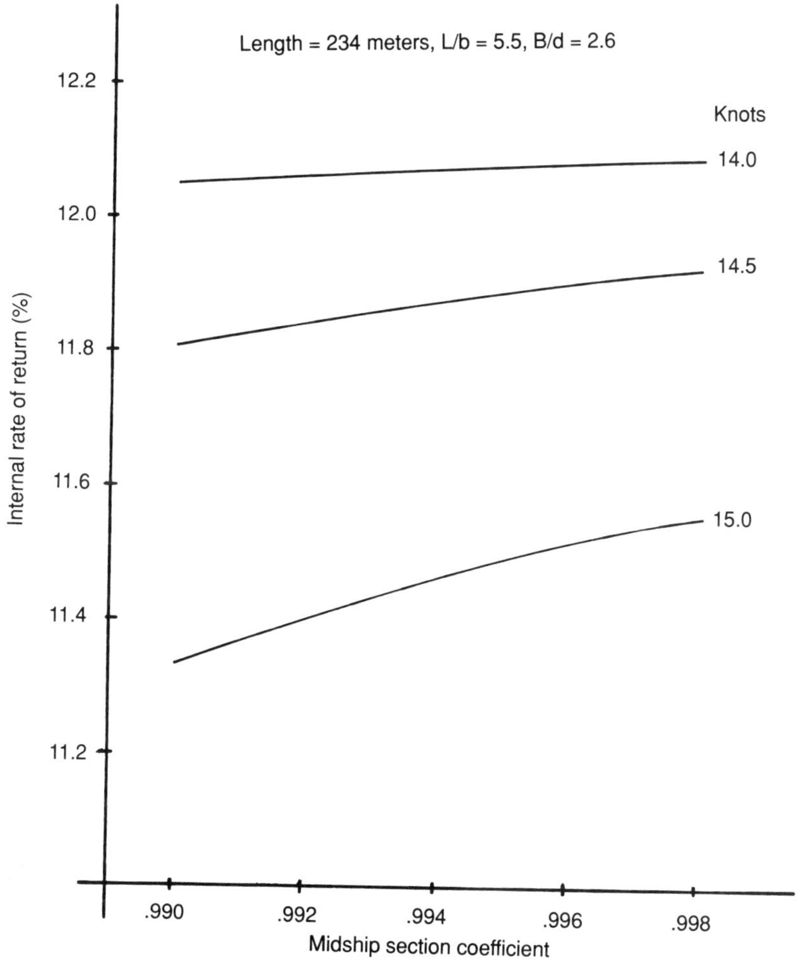

Figure 8-3. Midship section coefficient versus I.R.R.

VENTURE program to optimize the vessel for the given midship section coefficient and speed.

When a tanker is increased in size and there are no port or passage limitations, the measure of economic merit increases with size, reflecting the economy of scale principle. (Figure 8-4 shows a series of similar tankers

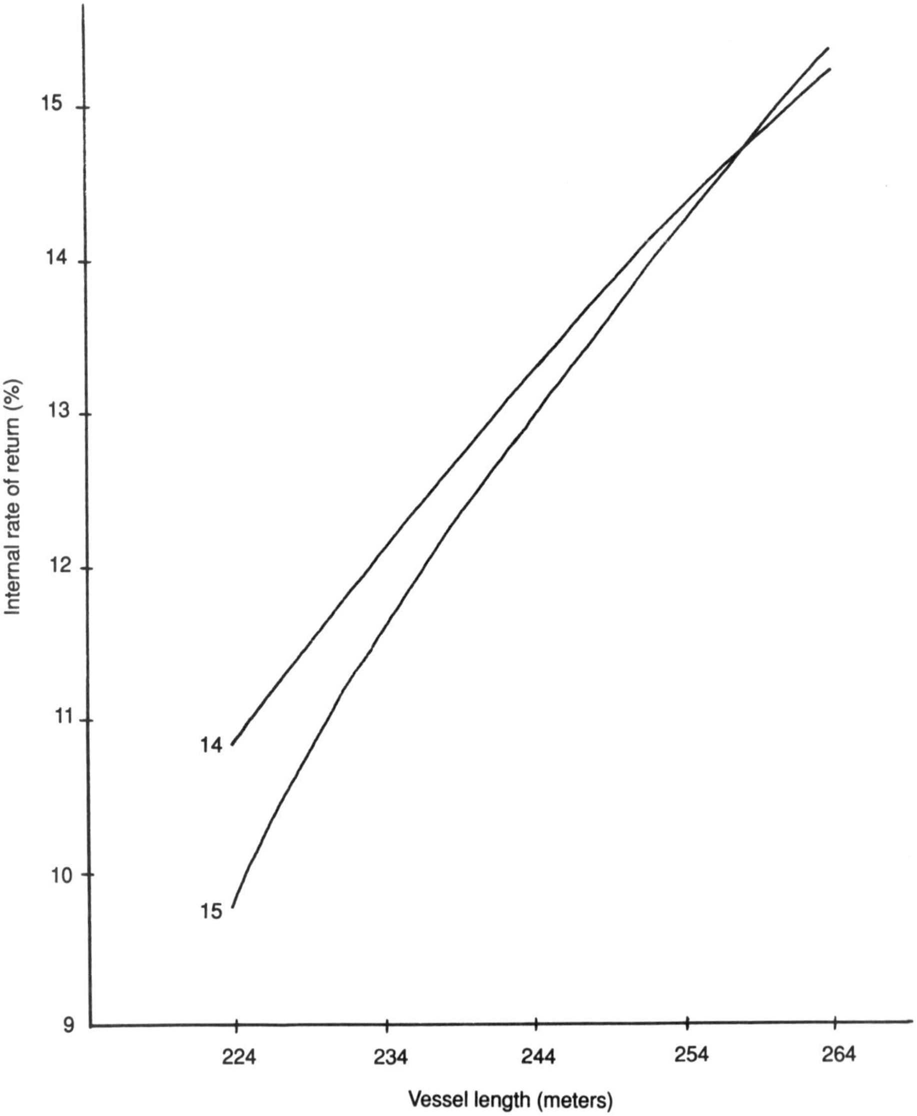

Figure 8-4. Ship length (size) versus internal rate of return for a series of similar tankers at 14 and 15 knots

ranging in length between perpendiculars from 224 meters to 264 meters at two speeds.) It is economy of scale that has resulted in the design and construction of the ultralarge crude oil carrier.

Recent events have renewed interest in protection of the environment relative to oil spills. This has resulted in laws mandating double-hull vessels in some ports. Figure 8-5 is a parametric study comparing a series

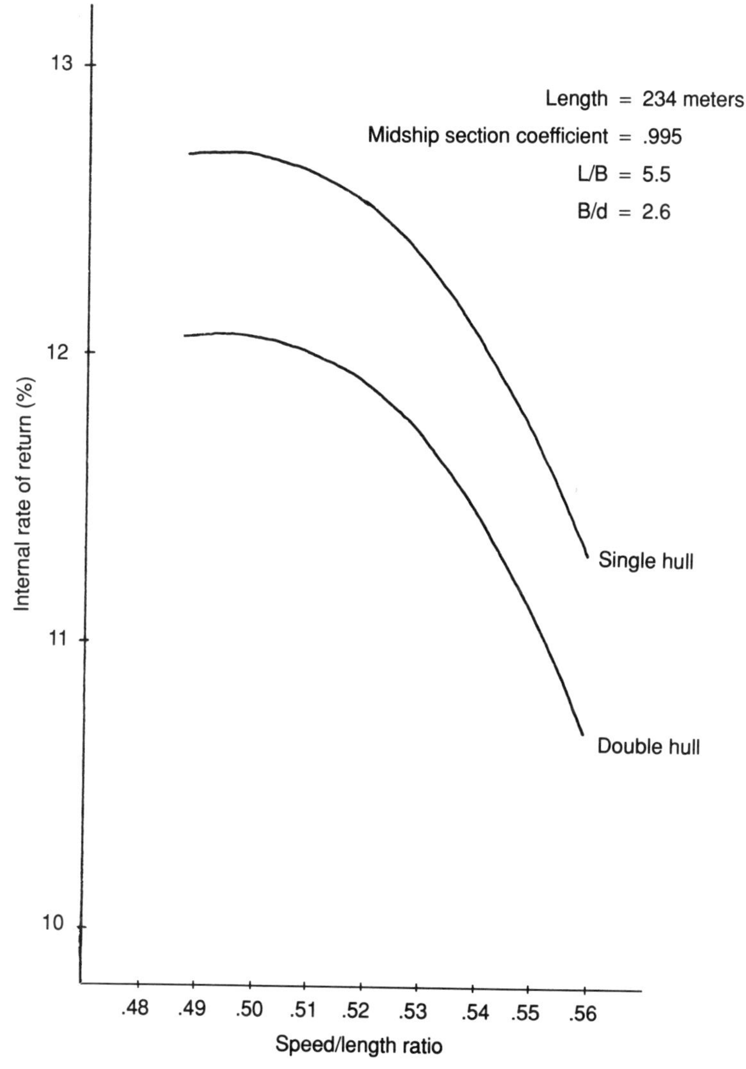

Figure 8-5. Internal rate of return versus speed / length ratio for a series of similar tankers with single or double hull

of similar tankers with single and double hulls over a speed-length ratio of .48 to .56. The difference between the economic merits is a result of the reduced cargo capacity and increased steel weight for a double hull versus a single hull. This study is suboptimal since the reduction in operating cost for the reduced oil spill risk has not been included in the life cycle costs of the vessels.

The reader should note that the program VENTURE provides a conceptual design of a crude oil tanker which is suboptimized for fixed independent variables, including length, length-beam ratio, and beam-depth ratio. Therefore, the illustrative parametric studies of Figures 8-2 through 8-6 represent crude oil carriers which are similar but have different displacements and cargo deadweight. Furthermore, the reader should not place undue significance in the absolute level for internal rate of return in the parametric studies since VENTURE accepts 51 input variables which are not explained in the illustrations. The reader is encouraged to become familiar with this program so that he or she can select input variables for meaningful parametric studies. A typical input data file suitable for use with the programs VENTURE and DESSEN is provided on the enclosed disks. The program DESSEN described in Appendix A is a variation of VENTURE. DESSEN is more efficient in exploring preliminary design and economic variables.

Where to Start

To prepare a ship design concept, the engineer must determine if the cargo for an individual ship will be unlimited, as in the case of crude oil tankers, or limited, as might be the case for a containership. In the unlimited cargo case, the larger vessel will have improved economy of scale. For the limited cargo, it will be necessary to find an optimum vessel for the average cargo deadweight anticipated. The limitation on vessel size for the crude oil carrier or bulk carrier will probably be established by a length, beam, or draft limit set by harbor, canal, or passage requirements. Within the dimensional limitations, the vessel should be as large as possible while maintaining a form which optimizes the construction and operating costs.

An interactive search for this form can be effectively carried out with a program such as VENTURE or more efficiently with DESSEN. Reasonable dimensional proportions should be used to start the search. A starting point for ocean vessel proportions suggested by Benford is:

Beam-draft ratio	3.00
Depth-draft ratio	1.43
Length-depth ratio	14.00
Length-draft ratio	20.00

If the vessel is a containership, passenger vessel, or some other type of vessel in which the cargo-carrying capacity is defined at the start, it will be necessary to use parametric studies and statistical data to close on a set of hull form dimensions which will provide the optimum life cycle cost for a given deadweight. The program VENTURE can be used for this purpose for an oil tanker. Other programs are available for various types of commercial vessels (see "Bulk Ocean Shipping" in the References). Before the advent of the computer, preliminary design was undertaken without the help of high-speed computation (see Taggart). As a result, a few alternatives were selected by judgment for consideration.

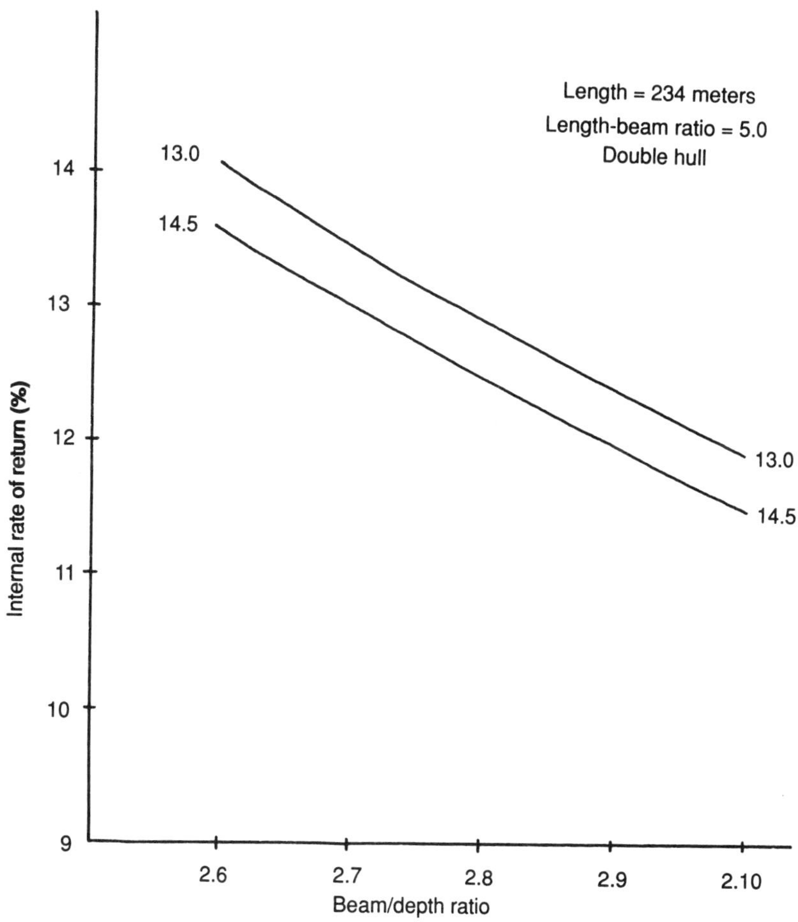

Figure 8-6. Beam/depth ratio versus internal rate of return for a series of similar tankers at 13 and 14.5 knots

A Conceptual Tanker Design for Decision under Certainty

In a decision under certainty, it is assumed that, initially, a variety of future economic and environmental conditions may exist. However, before the decision is made, we will know the specific future conditions and make the decision which will maximize the measure-of-merit. This eliminates the cost of uncertainty. In the following example, we will explore the length-beam ratio for an otherwise fully specified tanker in the face of the possibility of a range of economic variables—e.g., fuel cost, operating range, revenue—which will be determined before the decision is made.

Example 8-1

Given a 248-meter length between perpendiculars as a port limitation for a double-hull crude oil carrier, we have three alternative conceptual designs which are to be evaluated under an assumption of a discrete, exhaustive, and independent set of future events. The specific events which occur will be known before the decision.

The discrete design alternatives are:

Length-beam	4.5, 5.0, 5.5

The discrete future events are as follows:

Revenue, $/tonne-mile	.004, .006, .008
Fuel cost, $/tonne	80, 95, 110
Vessel range, nautical miles	2,000; 3,000

Since a ship design is generated and evaluated for each of the length-beam ratios and every combination of events, there are 54 payoff calculations. When it is known which event will occur, a selection can be made from among the three ratios.

The following ship requirements and economic events are some of the fixed independent variables:

Beam-depth ratio	2.6
Midship section coefficient	.995
Cargo density kg/m^3	915
Specific fuel consumption, kg/mw-hr	273
Steel cost, $/metric ton	481.26
Loan period, years	15
Interest rate on borrowed funds	8.5%
Owner equity	.25
Depreciation period, years	15
Port time per round trip, days	6
Inflation rate	.035
Operating days per year	351
Crew members	24

It is unnecessary to calculate the internal rate of return for all the alternatives since it becomes obvious that the L/B = 4.5 ship economic performance is better than either the L/B = 5.0 or the L/B = 5.5 under any of the future events; i.e., the L/B = 4.5 design dominates the decision. Until the problem is overwhelmed by the cost of increasing resistance, the approach of a higher block coefficient tanker is attractive. However, it is worthwhile to investigate the sensitivity of the selection of the L/B = 4.5 ship to the key future event parameters, i.e., one-way voyage distance, revenue rate, and heavy fuel cost.

For the L/B = 4.5 hull, there is a 16.7 percent improvement in internal rate of return for a vessel operating in a trade with a 3,000-nautical-mile one-way voyage distance compared to the vessel with a 2,000-nautical-mile voyage distance.

Figure 8-7 shows the change in internal rate of return as the revenue rate, increases from \$.004/tonne-mile to \$.008/tonne-mile.

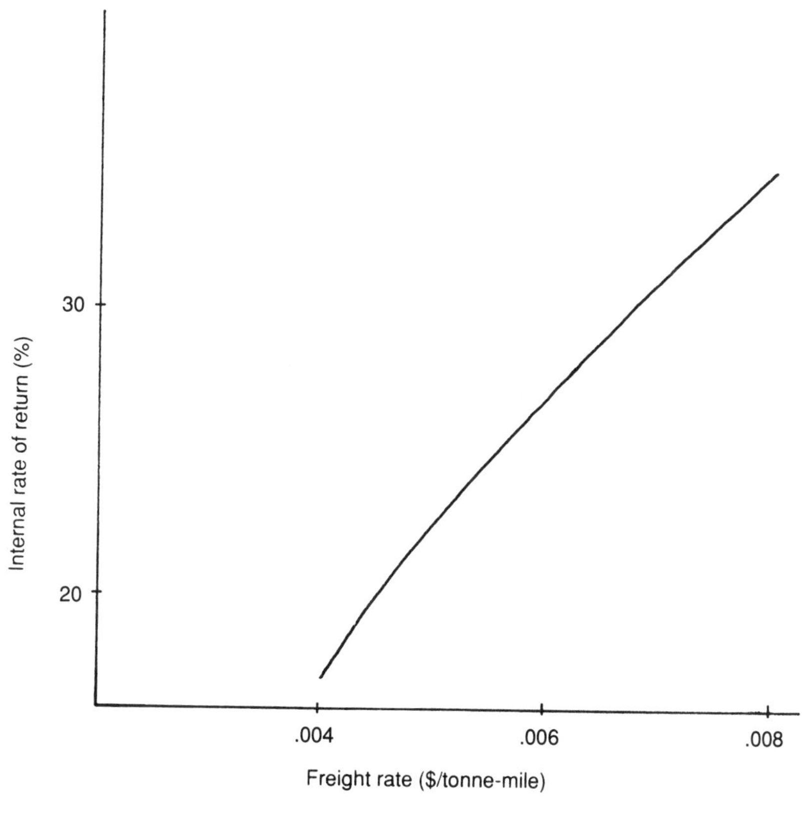

Figure 8-7. Sensitivity to revenue rate

Figure 8-8 shows, for the same double-hull tanker, the deterioration of internal rate of return with increasing heavy fuel cost from \$80 to \$110 per tonne.

The program DESSEN is useful for generating and evaluating a large number of tanker designs which vary in some combination of basic design

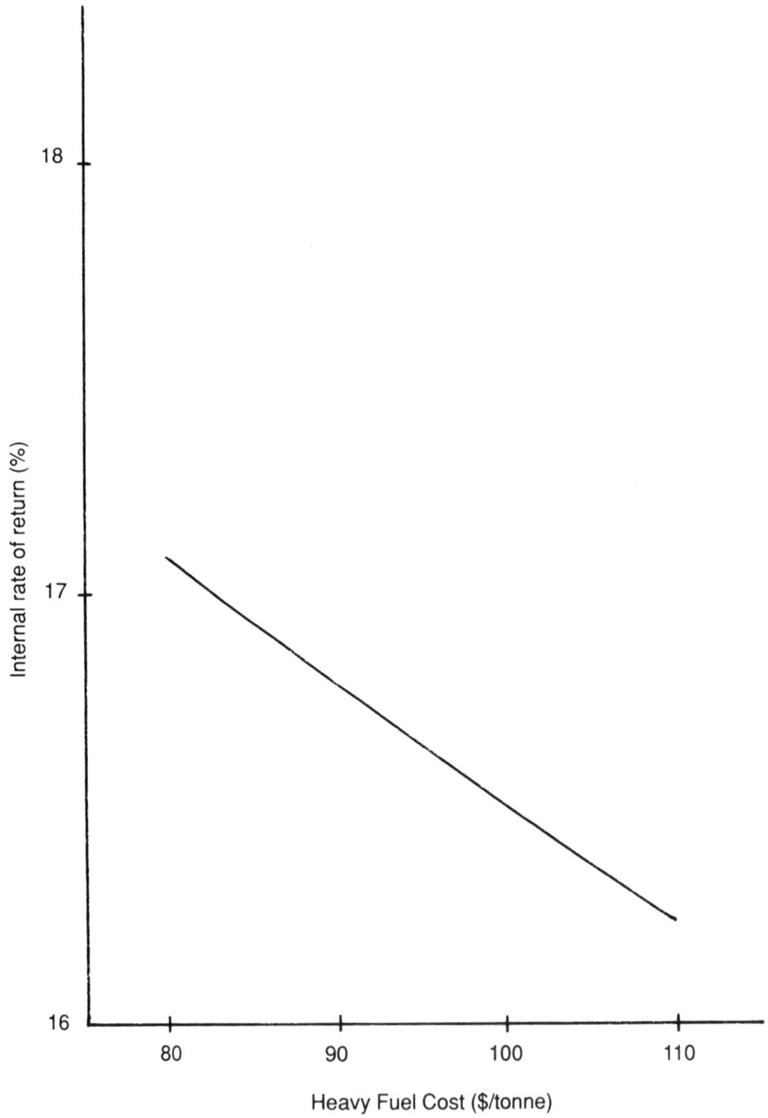

Figure 8-8. Sensitivity to fuel cost

parameters and/or economic and environmental factors. Figure 8-9 is a sample output of DESSEN used to explore a design parameter. Figure 8-10 is a sample output for study of economic independent variables. It is important to remember that DESSEN modifies the tanker design to provide the best vessel for a given set of economic and environmental factors. Therefore, to make a selection, we are assuming a deterministic future.

In a decision under uncertainty, the procedure requires that a measure-of-merit be calculated for a set of future conditions for each of the alternative designs under consideration. Since the alternative designs are not modified for optimization under the future states, the measure-of-merit will usually be poorer than the case for the deterministic design decision. A comprehensive example of a vessel selection decision under uncertainty is presented in Chapter 6 for a fast ferry. An approach for a cargo liner is described below.

DESIGN ECONOMIC EVALUATION

The evaluation and comparison of alternative ship designs or alternatives among other transportation facilities is a tedious process which generally includes the following steps:

1. Identify the design alternatives.
2. Prepare preliminary designs for the alternatives.
3. Determine costs for the alternatives.
4. Determine life cycle revenues and operating costs for the various operating scenarios.
5. Select, calculate, and compare a measure-of-merit.

A computer program like DESSEN, introduced earlier, which provides a suboptimized preliminary design for a set of future conditions or states, can be used to reduce the work of these steps for a tanker design selection in a decision or choice under certainty.

For the design selection under uncertainty or risk, the ship design alternatives are not suboptimized for a set of future conditions. In this case, it is necessary to define a set of alternatives and then evaluate the economic merit of these alternatives for each state of a collective and exhaustive set of future events. An appropriate decision rule in the uncertainty case or a probability distribution in the risk case is then applied to evaluate each alternative and select the one with the superior merit. This process recognizes that a fixed design alternative will have different economic performance given different future states which impact the performance. A computer program which is useful in calculating costs, revenues, and the measure-of-merit for a shipping venture is RFRDET, also included with this text. Unlike the programs VENTURE and DESSEN, RFRDET

* *

TANKER CONCEPTURAL DESIGN OPTIMIZATION AND
SENSITIVITY ANALYSIS FOR OPERATING VARIABLES

MARINE ENGINEERING ECONOMIC ANALYSIS

* *

PREPARED FOR Teekay Shipping
PROJECT ID Double Hull
PREPARED BY E C HUNT, PE
DATE: 09/12/1992
TIME: 15:39:35

LENGTH BP, M, FT	=	300.000	984.252
LENGTH-BEAM RATIO	=	5.000	
BEAM-DEPTH RATIO	=	2.600	
SERVICE ALLOWANCE	=	1.200	

DESIGN SEA SPEED VERSUS INTERNAL RATE OF RETURN

V/SQRL	CB	CM	CP	KNOTS
.446249	.873837	.996000	.877347	14.000
.458999	.865900	.996000	.869378	14.400
.471749	.857963	.996000	.861409	14.800
.484499	.850026	.996000	.853440	15.200
.497249	.842090	.996000	.845471	15.600
.509999	.834153	.996000	.837503	16.000
.522749	.826216	.996000	.829534	16.400
.535499	.818279	.996000	.821565	16.800

I.R.R.	DWT	DISPL	PWRMW	SHIP PRICE
.196249	201985.50	241860.4	21.2	101478600
.199852	201986.10	241762.3	21.8	101413700
.202984	202000.30	241683.2	22.6	101388000
.205705	202031.70	241628.3	23.5	101413700
.207852	202083.90	241603.3	24.7	101502700
.209323	202161.20	241614.1	26.2	101667100
.210010	202267.70	241667.2	28.1	101917600
.209797	202408.10	241769.1	30.5	102263900

Figure 8-9. Typical output of program DESSEN used to explore design variables

* *

TANKER CONCEPTUAL DESIGN OPTIMIZATION AND
SENSITIVITY ANALYSIS FOR OPERATING VARIABLES

MARINE ENGINEERING ECONOMIC ANALYSIS

* *

PREPARED FOR Teekay Shipping
PROJECT ID Double Hull
PREPARED BY E C HUNT, CEng., FIMarE
DATE: 09/12/1992
TIME: 15:42:27

RANGE, N.M.	=	3000.000
ANNUAL INFLATION RATE, DECIMAL	=	.03500
VOYAGE PORT TIME, DAYS	=	6.000
FREIGHT RATE, $/TONNE-MILE	=	.00350
COST OF HEAVY FUEL, $/TONNE	=	80.000
OPERATING DAYS PER YEAR	=	351.000

DESIGN SEA SPEED VERSUS INTERNAL RATE OF RETURN

V/SQRL	CB	CM	CP	KNOTS
.505278	.836251	.995000	.840453	14.000
.519714	.827273	.995000	.831431	14.400
.534151	.818296	.995000	.822408	14.800
.548587	.809318	.995000	.813385	15.200
.563024	.800340	.995000	.804362	15.600
.577460	.791362	.995000	.795339	16.000
.591897	.782384	.995000	.786316	16.400
.606333	.773407	.995000	.777293	16.800

I.R.R.	DWT	DISPL	PWRMW	SHIP PRICE
.041536	82494.79	101449.1	11.2	56140130
.041791	82555.87	101500.8	12.3	56506410
.041193	82633.69	101580.2	13.7	56956910
.039591	82730.75	101691.3	15.5	57496490
.036886	82849.74	101838.1	17.7	58128210
.032953	82993.52	102025.0	20.3	58853530
.027509	83165.19	102256.5	23.5	59672990
.020133	83367.86	102536.8	27.4	60585270

Figure 8-10. Typical output of program DESSEN used to explore design sensitivity to operating variables

does not develop a conceptual design, does not develop cost data for the design, and does not suboptimize the design for a given set of future states.

Because RFRDET has the limited function of calculating a measure-of-merit for a shipping venture, given the fixed design alternatives and a set of future events, it is a very flexible program with a wide range of applications. RFRDET can be used to:

1. Perform a financial analysis of an existing ship or ship design in a specified trade.
2. Prepare comparisons of alternative ship designs or ships for a single set of future economic and operating conditions, i.e., a decision under certainty.
3. Prepare comparisons of alternative ship designs for multiple sets of future economic and operating variables, some or all of which are defined by probability measures, i.e., a decision under risk.
4. Prepare comparisons of alternative ship designs for multiple sets of future economic and operating variables, for which the probability of occurrence is unknown, i.e., a decision under uncertainty.

In brief, RFRDET accepts a large amount of input data describing operating costs, capital costs, and trade scenarios, and then calculates either the required freight rate or, given a freight rate, calculates the present worth of the life cycle annual net cash flows. These measures-of-merit are calculated for three discount rates (MARR), seven speeds, and full displacement and three-fourths displacement conditions. The user must decide which measure-of-merit is the more pertinent for a given decision. Appendix A includes a description of the program RFRDET, including the input format and user instructions. An example of the use of RFRDET applied to a ship design decision under risk will be useful.

Example 8-2

A 16,000-dwt combination container and RO/RO cargo liner preliminary design has been developed for a 12,000-nautical-mile round-trip trade including a transit of the Panama Canal. An alternative to the basic design is a productivity package which will reduce the crew size for an incremental capital cost of $3.5 million. A second alternative is a propulsion enhancement package with a capital cost increment of $4 million. A third alternative is the combination of these packages for an incremental investment of $7.5 million. Therefore, the owner, Ajax Marine Transport Ltd., has identified the following four designs for consideration:

A. Basic design
B. Design for improved productivity
C. Design with propulsion enhancements
D. Design incorporating both 2 and 3 above

The following input data apply to these design alternatives.

	A	*B*	*C*	*D*
Ship capital cost, $ million	46	49.5	50	53
Crew size	18	12	18	12
Overhead, decimal	.25	.27	.26	.28
Cargo rate, LT/hr	300	350	300	350
Loaded displacement, LT	22,000	22,500	23,000	23,500
Ship availability, days	345	340	345	340
Full speed, kts	16	16	16	16
Freight rate, $/LT	100	100	100	100

The fuel rates for different speeds and displacements are shown in Figure 8-11. All additional input information required for RFRDET are common for the four alternatives and are tabulated in Appendix A.

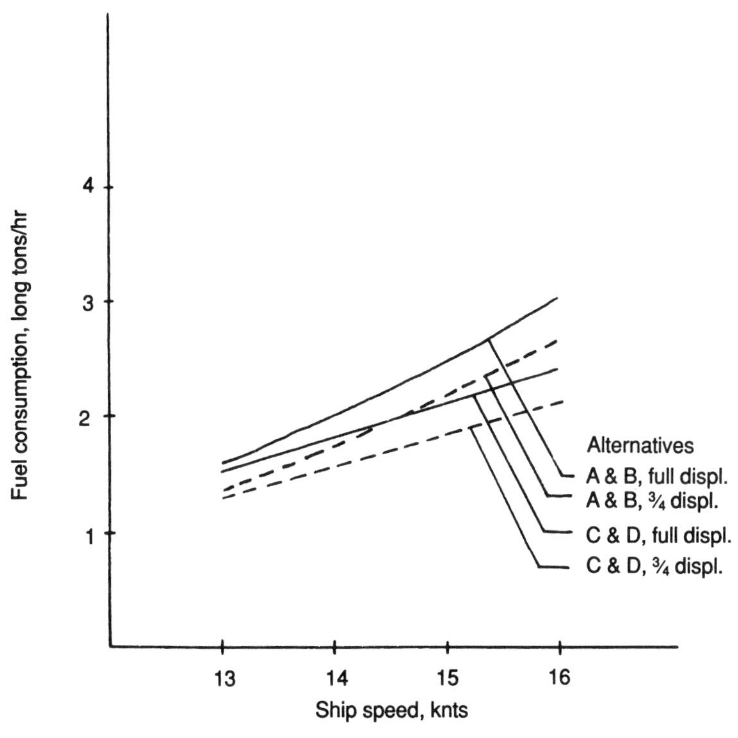

Figure 8-11. Ship speed (knts) versus fuel consumption (lt/hr). Courtesy Ajax Marine Transport Ltd.

The shipowner has identified two variables as the key risks for a cargo liner in the planned trade. These are the cargo load factor and the need for reduced speed due to weather conditions on the trade route. Based on the owner's long-term experience in the trade, the following discrete probability distributions have been developed to represent the risk from these random variables.

Speed	*Probability (speed)*
12	.05
13	.10
14	.16
15	.19
16	.50

Displacement	*Probability (displacement)*
full	.65
3/4	.35

The convolved distribution for these two independent random variables is:

State	*Probability (state)*
12 kts - full	.0325
13 kts - full	.0350
14 kts - full	.1040
15 kts - full	.1235
16 kts - full	.3250
12 kts - 3/4	.0175
13 kts - 3/4	.0350
14 kts - 3/4	.0560
15 kts - 3/4	.0665
16 kts - 3/4	.1750

Using the program RFRDET with the above input information, the present worth of the sum of annual net cash flows for each of the design alternative and future state combinations (i.e., 40 present-worth summations) was calculated. An example of a typical output from RFRDET is shown in Figure 8-12 for alternative *A* at full displacement. Figure 8-13 includes a summary tabulation of the 40 net present-worth summations for the Ajax Marine Transport Ltd. example. Rearranging this information into a payoff matrix where the rows represent the design alternatives or decision maker acts and the columns represent the 10 discrete future states, it is possible to apply the decision analysis program DECIANA to evaluate the four design alternatives under the probability distribution provided by the shipowner. Figure 8-14 shows the result of this analysis. Alternative *B* has the highest present worth of annual net cash flows. The

RESULTS OF ANALYSIS

(LOADING CONDITION = FULL-LOAD)

SPEED	ANNUAL TONS HAULED	SERVICE FREQUENCY	DISCOUNT RFR	RATE = .10 NPV	DISCOUNT RFR	RATE = .12 NPV	DISCOUNT RFR	RATE = .14 NPV
(KTS)	(LONG TONS)	(TRIPS/YR)	($/LT)	($)	($/LT)	($)	($/LT)	($)
12.0	96407	6.89	100.00	9618856	100.00	5773806	100.00	2874288
13.0	102996	7.36	100.00	11101250	100.00	7081779	100.00	4039745
14.0	109406	7.81	100.00	12093710	100.00	7969736	100.00	4840697
15.0	115642	8.26	100.00	12517520	100.00	8370428	100.00	5218887
16.0	121714	8.69	100.00	12541130	100.00	8427894	100.00	5299127
17.0	127626	9.12	100.00	12084350	100.00	8073660	100.00	5022132
18.0	133384	9.53	100.00	10692300	100.00	6918754	100.00	4051171

Figure 8-12. Typical output of RFRDET analysis program

Future States		Alternative Designs			
Knots	Displ.	A, $	B, $	C, $	D, $
12	full	9618856.	11168610.	641789.	6585176.
13	full	11101250.	12699600.	8621948.	8790706.
14	full	12093710.	13721010.	10642260.	10848490.
15	full	12517520.	14148770.	12464180.	12708190.
16	full	12541130.	14158010.	14374900.	14662750.
12	¾	763472.	1626055.	–2196228.	–2508481.
13	¾	2050514.	2984131.	–363378.	–632292.
14	¾	2972760.	3968814.	1381180.	1158163.
15	¾	3415646.	4461479.	2952178.	2777220.
16	¾	3600836.	4689065.	4657472.	4532921.

Figure 8-13. Table of payoffs for Ajax Marine Transport net present worth, 20-year life

* *

ELEMENTARY DECISION ANALYSIS
MARINE ENGINEERING ECONOMIC ANALYSIS

PREPARED FOR AJAX MARINE TRANSPORT LTD.

PROJECT ID 16000 DWT CARGO LINER

PREPARED BY E C HUNT

DATE: 08/19/1992

TIME: 15:04:57

MEASURE OF MERIT IS PW OF NET CASH FLOW DOLLARS

* *

PROBABILISTIC DECISION ANALYSIS

ACT	EXPECTED PAYOFF	EXPECTED OPPORTUNITY LOSS	COST OF IRRATIONALITY
1	.90226E+07	.15780E+07	.14140E+07
2	.10437E+08	.16404E+06	.00000E+00
3	.91256E+07	.14750E+07	.13109E+07
4	.92253E+07	.13753E+07	12113E+07

VALUE OF DECISION UNDER CERTAINTY IS .10601E+08

COST OF UNCERTAINTY IS .16404E+06

Figure 8-14. Analysis of Ajax Marine Transport alternatives

four design alternatives can be arranged in the order of decreasing merit as follows:

Alternative or Act	*Description*	*Payoff, $*
2 (B)	Productivity package	10,437,000
4 (D)	Productivity & propulsion	9,225,300
3 (C)	Propulsion enhancement	9,125,600
1 (A)	Basic design	9,022,600

From Figure 8-13, it can be observed that if the selection decision were made on the assumption that the vessel would operate only at the design speed of 16 knots and full displacement, then alternative *D* would be the correct decision. Also it should be noted that alternative *B* is always preferred to alternative *A* regardless of the future state; i.e., alternative *B* dominates alternative *A*.

The cost of uncertainty in Figure 8-14 suggests that if it is possible to improve the information, i.e., reduce the uncertainty in the future states through further study, then it is justifiable to spend less than $164,000 on this activity to remove the risk inherent in the probability measure. Furthermore, a decision to select alternative *A* given the information in this analysis, would result in an additional expected loss of net present worth equal to $1,414,000, i.e., the cost of irrationality for selecting alternative *A*.

THE MARINE ENGINEERING PERSPECTIVE

Working in support of the naval architect's overall ship design objectives, the marine engineer is confronted with a wide range of economic decisions involving main and auxiliary machinery and systems located within the hull. A basic design requirement classification for these systems has been suggested by Kane:

Marine Propulsion System
Shaft horsepower
Propeller rpm
Specific fuel consumption
Bunker capacity
Space and weight
Adaptability to ship configuration
Central control and operation
Data collection and analysis

Auxiliary Ship Systems
Electric distribution
Electric generation
Auxiliary steam
Heating, ventilation, and air-conditioning
Compressed air

Firefighting, bilge, and ballasting
Potable water system and evaporators
Fuel transfer
Refrigeration
Lubricating oil storage and transfer
Hydraulics
Plumbing
Sewage treatment
Waste disposal

Hull Engineering Systems
Anchor handling
Steering engine and telemetering control
Cargo-handling gear
Crane systems
Elevators
Bulk cargo systems, self-unloaders, etc.
Container systems
Tanker cargo systems
WT door operation
Stabilization

Electronic and Navigation Systems
Interior communications
Exterior communications
Navigational aids, radar, loran, etc.
Shipboard computers: stability, cargo, stores
Computer-based maintenance

In every design decision there are numerous economic factors which must be quantified for the alternatives under consideration. For example, consider the selection of main propulsion machinery for a crude oil carrier.

Since all of the crude oil carriers currently under construction in the world have been designed with the long-stroke low-speed diesel propulsion system, it would seem obvious that this is the most economic propulsion system. However, arriving at that conclusion requires a lengthy, time-consuming, and expensive analysis. The key costs which must be evaluated and compared for the propulsion systems under consideration over the planned economic life include at least the following:

Total crew cost
Cost of crew training
Maintenance costs
Special maintenance tool requirements
Spare parts requirements and on-board storage
Reliability of the system
Impact of propulsor performance
Fuel cost for planned operating profile
Lubricating oil cost

Fuel treatment
Performance deterioration during economic life
Salvage cost
Installed capital cost
Insurance cost

After selection of an appropriate interest rate, escalation rate, and economic life, these costs can be combined into a measure-of-merit such as equivalent annual cost or present worth for comparison among the alternatives.

While the long-stroke low-speed diesel engine is the choice for most modern tankers and cargo liners, it can be shown that two medium-speed diesel engines geared to a contrarotating propeller have an economic advantage when applied to a vessel which must operate a portion of the time at reduced speed due to weather or other navigational conditions. The simple-cycle gas turbine has economic promise for application to cruise liners which travel short distances, and a modern geared steam turbine is the economic winner for a large crude carrier which operates in Arctic ice.

The selection of a propulsion system has a major impact on the ship's overall design and lifetime performance. For this reason, the propulsion plant alternatives are frequently defined as different ships for comparison of the economic merit. Therefore, if the hulls are identical for the alternative ships, the lift capacity will be different due to differences in machinery weight, volume, and bunker requirements. Conversely, if lift capacity is maintained constant among the vessels, then the hull sizes will be different to accommodate the differences in plant weight and volume.

Machinery Economic Evaluation

All the machinery and systems installed in a ship contribute to the overall lifetime economics of the vessel. Size, capacity, efficiency, effectiveness, weight, reliability, availability, usage, performance, and installed cost of each machinery item—each has an effect on the economic performance of the vessel. Very often, several competitive products with different performance levels and physical characteristics are available to provide the same shipboard function. It is necessary to undertake evaluation studies to make a design selection. For example, there are numerous electric motors on a modern cargo vessel ranging in size from fractional horsepower to ratings of 300 horsepower or more. Montgomery has discussed the practice of major manufacturers to provide induction motors for these auxiliary services at a *conventional* efficiency level and at an *energy efficient* level. At the three-fourths rated load, these efficiencies are typically 91 percent and 95 percent respectively. The energy efficient motors are priced about 20 percent higher than the conventional induction motors to reflect the higher costs of the extended ferrous cores of the low-loss laminations and the larger conductor sizes used. To make a decision between a conventional and an energy efficient motor requires the calculation of an economic measure-of-merit for each of the two alternatives. The following factors must all be considered in this calculation:

Installed cost of motors
Normal operating load (possible random variable)
Annual operating hours (possible random variable)
Motor efficiencies
Cost of fuel (possible random variable)
Electric plant fuel rate

Example 8-3

A 15 HP potable water pump is available for installation in a cargo liner with a conventional rated load efficiency of .875 or as an energy efficient design with a .93 efficiency. The installed cost differential is $374. The following data are pertinent to the selection of the induction motor:

Operating hours	8,736 per year
Fuel rate	.4613 lbs per kwhr
Fuel cost	$.0802 per lb
MARR	.10 per annum
Economic life	20 years

Assuming a deterministic problem, calculate the annual net saving for the selection of an energy efficient induction motor for this application.

$$\text{Change in efficiency} = 1/.875 - 1/.93 = .0676$$

$$\text{Change in kw load} = (15 \text{ hp})(.0676)(.7457 \text{ kwhr/hp-hr}) = .756$$

$$\text{Annual capital cost} = (\$374)(A/P, .10, 20) = \$43.9$$

$$\text{Annual fuel saving} = (.756 \text{ kw})(8{,}736 \text{ hrs})(.4613 \text{ lbs/kwhr})(\$.08/\text{lb}) = \$243.7$$

$$\text{Net annual savings} = \$244.1 - 43.9 = \$200.2$$

The break-even annual operating hours for which the net annual saving would be zero are:

$$\text{Break-even operating hours} = 43.9/(.756)(.4613)(.08) = 1573.5 \text{ hours}$$

Modern diesel-propelled vessels have several large, high-utilization induction motors which are used to drive fans or pumps at reduced capacity for significant portions of time. The reduced capacity operation is usually determined by environmental conditions, such as sea temperature, air temperature, etc., or the operating power of the main propulsion engine. The reduced capacity can be obtained by throttling the flow (with the consequent energy losses, which are a period cost) or by controlling the speed of the motor (which requires an increase in the installed cost for the control system). Kusko has undertaken an extensive study of the application of voltage source inverter (VSI) and current source inverter (CSI) variable-speed drives to induction motors. The key to the economic merit of VSI/CSI speed control applications aboard a ship is the induction motor utilization factor. For example, despite the large power ratings of windlasses and winches, the utilization factor for these equipments is too low to justify VSI/CSI variable-speed control compared to the less costly step controller. Table 8-1 is a summary of typical adjustable-speed machinery application opportunities for general marine use. VSI controllers are usually selected for large high-utilization motors. On a typical motor ship, these may include:

	Horsepower
Main saltwater circulator	60
Main freshwater circulator	250
Jacket cooling pump	40
Machinery space exhaust fan	25

Figure 8-15 shows the typical characteristics for the voltage source inverter. The controller maintains constant motor volts per hertz. In Figure 8-15 the motor torque speed characteristics are shown for two frequencies, *f1* and *f2*. The figure also shows a typical centrifugal fan load characteristic. A two-to-one reduction in the inverter frequency produces a two-to-one reduction in speed, while the torque reduces by a ratio of four to one due to the fan torque-speed characteristic. The general approach to the economic evaluation of a VSI speed controller to a shipboard fan or pump is as follows:

1. Prepare a head-flow curve for the fan or pump and system at various speeds and increasing system head due to throttling.
2. Determine the required fan or pump flow capacity for the various ship operating and environmental conditions.
3. Analyze environmental conditions on the ship trade route and prepare a probability distribution to describe the pertinent environmental random variables such as air temperature, sea temperature, etc.
4. Calculate and compare the kw load for operation at constant speed with flow throttling and variable speed control.
5. Calculate the expected value of the equivalent annual saving or the net present worth of annual cash flows resulting from the selection of the VSI.

Example 8-4

It is necessary to select a flow control for an engine room exhaust fan. The engine room heat load has a mean value of 12,860 Btu/min. For the planned trade route, the atmospheric temperature can be represented by the following distribution expressed as percent of time in a year:

TABLE 8-1
Typical Shipboard Adjustable-Speed Machinery Application Opportunities

Application	*Range of HP Rating*	*Range of Motor Speed, RPM*	*Percent Speed Range*	*No. of Quadrants*	*Torque Characteristics*	*Power Characteristics*
C.T. winch	50-150	600-1,800	±300	4	Max. at low N	Limited
Cargo winch	50-150	600-1,800	±200	4	Max. at low N	Limited
Capstan	25-150	600-1,800	±400	2	Max. at low N	Limited
Windlass	50-150	600-1,800	±300	4	Max. at low N	Limited
Feed pump	100-500	7,500	10-100	1	$T \alpha N^2$	$P \alpha N^3$
Circ. pump	50-200	600-1,200	10-100	1	$T \alpha N^2$	$P \alpha N^3$
Cond. pump	50-100	600-1,200	10-100	1	$T \alpha N^2$	$P \alpha N^3$
L.O. pump	25-50	1,200-1,800	10-100	1	$T \alpha N^2$	$P \alpha N^3$
F.O pump	25-50	1,200-1,800	10-100	1	$T \alpha N^2$	$P \alpha N^3$
Ballast pump	50-150	1,800-3,600	10-100	1	$T \alpha N^2$	$P \alpha N^3$
Cargo pump	50-3,000	600-1,800	10-100	1	$T \alpha N^2$	$P \alpha N^3$
F.D. fan	50-200	600-1,800	10-100	1	$T \alpha N^2$	$P \alpha N^3$
Vent. fan	10-50	600-1,800	10-100	1	$T \alpha N^2$	$P \alpha N^3$
Thruster	200-2,200	600-1,200	±100	4	$T \alpha N^2$	$P \alpha N^3$
Compressor	50-500	600-1,800	10-100	1	T constant	$P \alpha N$
Steering	50-200	600-1,200	±200	4	Max. at low N	Limited

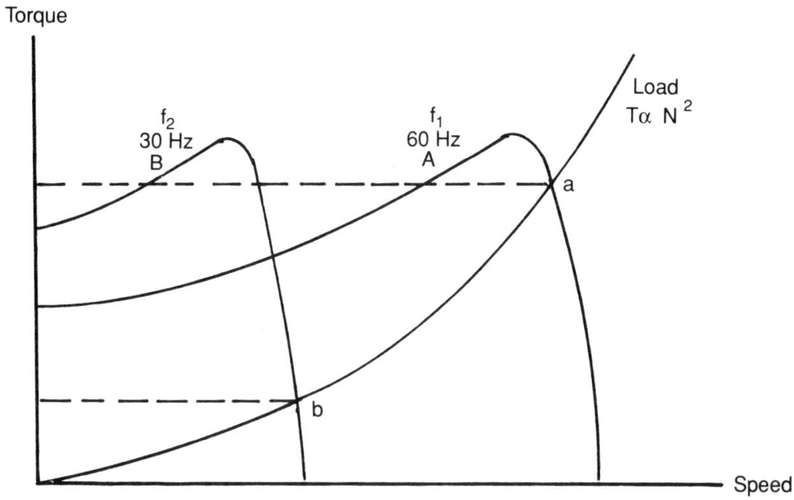

Figure 8-15. Torque-speed characteristics for centrifugal ventilating fan with VSI speed controller

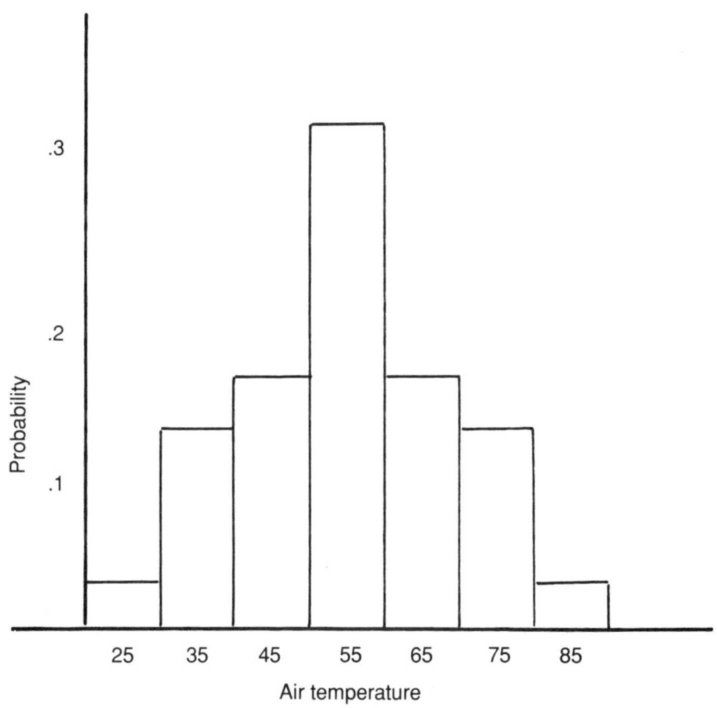

Figure 8-16. Probability distribution of air temperature in the ocean trade for Example 8-4

Percentage time	*Temperature °F*
3.0	25
13.7	35
16.8	45
33.0	55
16.8	65
16.7	75

This temperature data is converted to a discrete probability distribution in Figure 8-16.

A centrifugal fan is required with the flow capacity to handle the mean heat load for the full range of environmental temperatures. The fan flow requirement as a function of outside temperature is plotted in Figure 8-17.

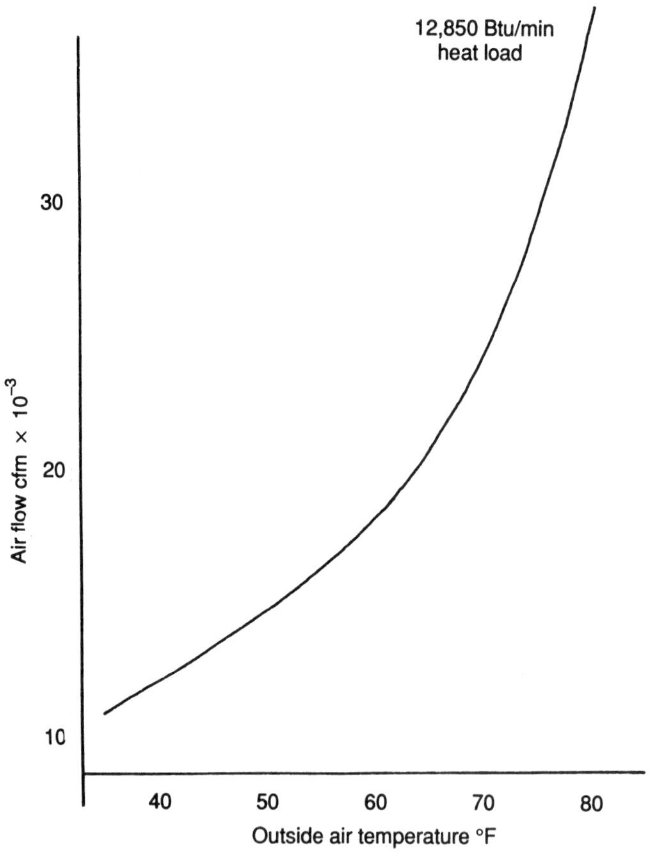

Figure 8-17. Engine room ventilation flow requirement versus outside air temperature

The head-flow characteristic for the selected fan is shown in Figure 8-18 for five speeds and four system heads. The incremental cost for the installation of a VSI speed control system instead of a simple flow throttle is \$6,400. Additional needed information is tabulated on the next page:

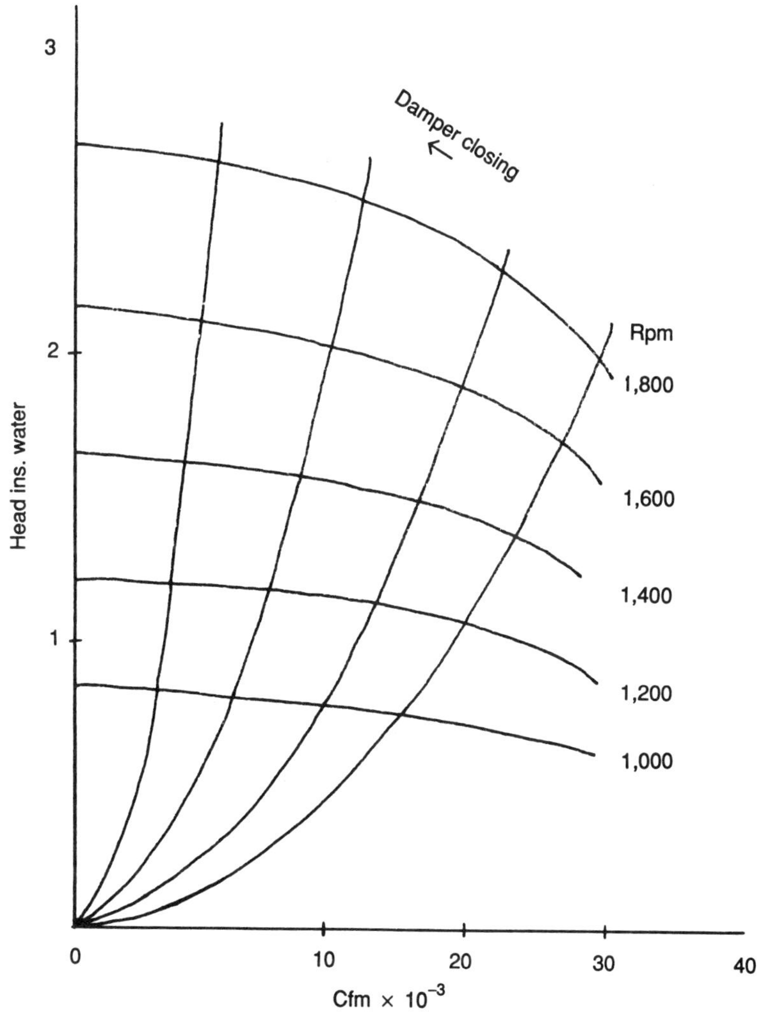

Figure 8-18. Engine room ventilation fan and system head-flow characteristics

MARR	12%
Economic life	20 years
Generator fuel rate	.4613 lbs/kwhr
Fuel cost	$.0802/lb

Should the VSI control system be selected?

Figure 8-19 is a plot of the kw requirement of the centrifugal fan with damper flow control and variable speed. Using this information and the flow-time distribution, it is possible to calculate the annual kwhr savings as follows:

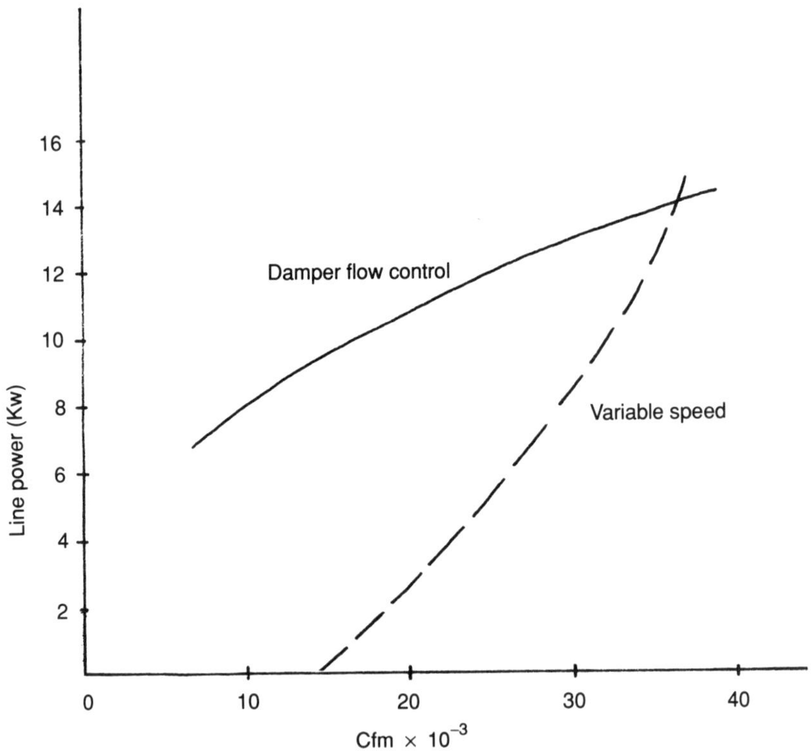

Figure 8-19. Engine room ventilation fan line power requirement (kw) versus flow

Flow, CFM	*% Time*	*Kw Saving*	*Hours*	*Kwhr Saving*
37,000	3.0	-0.7	262.8	(184)
29,600	13.7	4.8	1,200.1	5,761
21,100	16.8	7.6	1,471.7	11,185
16,500	33.0	8.8	2,890.8	25,439
13,400	16.8	9.0	1,471.7	13,245
11,100	16.7	9.3	1,462.9	13,605
				69,051

Annual cost saving = (.4613 lb/kwhr) ($.08/lb) (69,051 kwhr/year) = $2,548/year

Equivalent annual capital cost = (A/P, .12, 20) ($6,400) = (.1339) ($6,400) = $856.96

Net annual saving = $2,554 – $857 = $1,697

So, the incremental present worth of the decision to select VSI speed control is:

Incremental present worth = (P/A, i, n) ($2,554) – $6,400

= (7.469) ($2,552) – $6,400

= $12,660

REFERENCES AND BIBLIOGRAPHY

Benford, H. *General Cargo Ship Economics and Design.* Ann Arbor: University of Michigan, 1968.

———. "Optimal Life and Replacement Analysis for Ships and Shipyards." RINA *Transactions,* Vol. 114 (1972).

———. "The Rational Selection of Ship Size." SNAME *Transactions,* Vol. 75, (1967).

Benford, Harry. "Of Dollar Signs and Ship Designs." SNAME *Star Symposium,* August 1975.

———. "The Blacksmith Ship Economist." Report 270. Ann Arbor: University of Michigan, January 1983.

———. "Bulk Cargo Inventory Costs and Their Effect on Design of Ships and Terminals." SNAME *Marine Technology,* Vol. 18, No. 4 (October 1981).

———. "A Second Look at Measures of Merit for Ship Design." Report No. 223. Ann Arbor: University of Michigan, Department of Naval Architecture and Marine Engineering, August 1980.

Bulk Ocean Shipping—Computer Optimizations and Economics, "Fast Ferry Analysis, "Ship Analysis," New York: Transtech Marine Company, 1985.

Buxton, Ian. "Engineering Economics Applied to Ship Design." RINA *Transactions,* Vol. 114 (1972).

———. "Application of Engineering Economics to Ships and Fleets." WEGEMT; September 1978.

———. "Economic Comparisons of Ships and Their Propulsion." Stone Manganese Marine Conference, 1974.

———. "Ships in the Marine Economic Environment." WEGEMT; 1978.

Carreyette, J. "Preliminary Ship Cost Estimation." RINA *Transactions* Vol. 120 (1978).

Chapman, Kenneth R. "Economics and Ship Design." *North East Coast Institute of Engineers and Shipbuilders Transactions,* Vol. 83 (1967).

Erichsen, Stian. "Optimizing Containerships and Their Terminals." SNAME *Spring Meeting Papers* (1972).

Fisher, K.W.; "Economic Optimization Procedures in Preliminary Ship Design"; RINA *Transactions* ; Volume 114, 1972.

———. "The Relative Costs of Ship Design Parameters." RINA *Transactions,* Vol. 114 (1972).

Gallin, C., and O. Hiederich. "Economical and Technical Studies of Modern Ships." *Shipbuilding and Marine Engineering International,* (April 1983).

Goss, Richard. "Economic Criteria for Optimal Ship Design." RINA *Transactions,* Vol. 107 (1965).

Hutchinson, B. "Application of Probabilistic Methods to Engineering Estimates of Speed, Power, Weight and Cost." *Marine Technology* (October 1985).

Johns, R. A., and P. Marshall. "Energy-Saving Induction Motors". *Marine Engineers Review;* December 1983

Kane, J.R."General Considerations in Marine Engineering." Chapter 1, *Marine Engineering.* Harrington, R., ed. New York: SNAME, 1976.

Kerlen, H. "How Does Hull Form Influence the Building Cost of Cargo Vessels?" Proceedings of Second International Marine Systems Design Conference; Danish Technical University. Lyngby: May 1985.

Kusko, A. "The Application of Adjustable-Speed AC (CSI/VSI) Drives to Shipboard Machinery and Systems"; Vols. I and II, Maritime Administration, NTIS, December 1981.

Lamb, T. "A Ship Design Procedure." *Marine Technology,* (October 1969).

Mentz, Paul B. "What Price Speed?" SNAME Chesapeake Section; September 1975.

Montgomery, D. C. "Evaluation of Energy Efficient Motors." Proceedings of Textile Industry Electric Conference, (May 1979).

Nowacki, H., F. Brusis, and P.M. Swift. "Tanker Preliminary Design—An Optimization Problem with Constraints." SNAME *Transactions*, Vol. 78 (1970).

Parsons, M.G. "Optimization Methods for Use in Computer-Aided Ship Design." SNAME Star Alpha, August 1975.

Rawson, K.J., and E.C. Tupper. *Basic Ship Theory.* London: Longman Group, 1978.
Reinertsen, W. "Economic Trends in Ship Design"; *Motor Ship* (May 1982).
Taggart. *Ship Design and Construction.* New York: SNAME, 1980.

QUESTIONS AND PROJECTS

1. As vessels increase in size and where there are no dimensional port or canal limitations and cargo is available, what changes can we expect in the venture measure-of-merit? Discuss the reasons.
2. Discuss the differences in design decision making under the following two conditions: (a) A suboptimized design is generated for each future state in a collectively exhaustive set of states. (b) A finite set of designs is evaluated for economic performance under a collectively exhaustive set of future events.
3. Discuss the application and importance of a computer in conceptual design decisions for shipping ventures.
4. Study the documentation and computer input file structure for the tanker conceptual design program VENTURE. The example input is for a 281.7-meter L.b.p. (length between perpendiculars) vessel. Select another length for the vessel without changing other design and economic parameters. Run program VENTURE for the new vessel and compare the results to the 281.7-meter vessel.
5. Develop a complete set of design and economic independent variables required for the VENTURE program input.
6. Use the input data from question 5 with the program DESSEN to investigate the independent variable, midship section coefficient. Present the results in a useful plot and comment on the changes in economic performance which result from the variations in the midship section coefficient.
7. For the best combination of vessel speed and midship section coefficient determined in question 6, run program VENTURE for this single condition and print out design data and annual cash flows. Comment on the results.
8. Using the same input file developed for question 5, modified to include best midship section coefficient, run the program DESSEN to quantify the impact on the suboptimized design for several of the key economic variables:

(a) Freight rate	\$/tonne-mile
(b) Fuel cost	\$/tonne
(c) Operating days	days/year

9. As a design engineer, how would you select a measure-of-merit for a specific owner? What are the considerations?

10. List the sources which should be used to collect input data for economic analysis of ship design alternatives.
11. Benford suggests the following proportions as a start for deep water ship design:

Beam/draft	3.10
Depth/draft	1.43
Length/depth	14.00

    Does the program DESSEN confirm these as desirable tanker proportions?
12. Redo the Ajax Marine Transport Ltd. problem using RFR as the measure-of-merit at 10 percent MARR. Does this require a reconsideration of the decision of the text example?
13. Redo the Ajax Marine Transport Ltd. problem using 12 percent and 14 percent MARR. Does this change the result of the text example?
14. For the Ajax Marine Transport Ltd. problem, calculate the equivalent annual cash flows for the expected payoff for each alternative. Discuss the meaning and potential use of a measure-of-merit calculated in this manner.
15. For the Ajax Marine Transport Ltd. problem, assume the probability distribution for the 10 future states is unknown. How would you analyze the payoff matrix? Analyze the payoff matrix without the future event probability distribution and compare your results to the text solution.
16. Develop a complete set of input variables for the program RFRDET. Current technical papers or trade journals are a source of typical information. Run the program RFRDET with the set of input variables and vary the discount rate, i.e., MARR, and the interest rate on borrowed funds, and the inflation rates.

PROBLEMS

1. A 60 HP air compressor induction motor drive must be selected for a diesel-propelled vessel. The designer must decide between a "conventional" efficiency motor and an "energy efficient" motor. Given the following data, what selection would you make?

Conventional motor efficiency	.91
Energy efficient motor efficiency	.954
Annual operating hours	874
Cost difference	$814
Generator fuel rate, lbs/kwhr	.46
Fuel cost, $/tonne	185

Economic life, years	20
MARR	.12

2. Solve problem 1 assuming the annual operating hours are properly represented by a random variable with the following discrete distribution:

Hours	*P (hours)*
1,000	.25
850	.35
750	.30
600	.10

3. Solve problem 1 assuming the annual operating hours and the fuel cost are best described by the following random variables:

\$/pound	*P (\$/pound)*
.081	.30
.072	.50
.063	.20

Hours	*P (Hours)*
1,000	.25
850	.35
750	.30
600	.10

4. An engineer is required to make a selection between a constant speed drive flow throttle control and a VSI variable speed drive for two 250 HP main freshwater circulating centrifugal pumps. The system heat load is defined as follows:

% of Time	*Heat Load, millions Btu/hr*
20	16.7
50	25.0
30	33.4

The annual seawater temperature on the specified trade route is adequately defined by a normal distribution with a mean of 68.72 and a standard deviation of 1.17°F. Figure 8-20 shows the pump head-flow characteristics at five speeds and the system head-flow characteristics for four flow throttling conditions. Figure 8-21 shows the pump variable speed operating characteristics, including flow, controller efficiency, motor and line kw, and discharge head over the nominal operating speed range. Figure 8-22 shows the flow requirements for each heat load over the range of seawater temperatures. The installed incremental cost for controls on the two freshwater cooling pumps is \$28,000. The MARR is 10 percent and the economic life is 20 years. What is the economic gain from the application of VSI speed control versus throttle flow control of the cooling water?

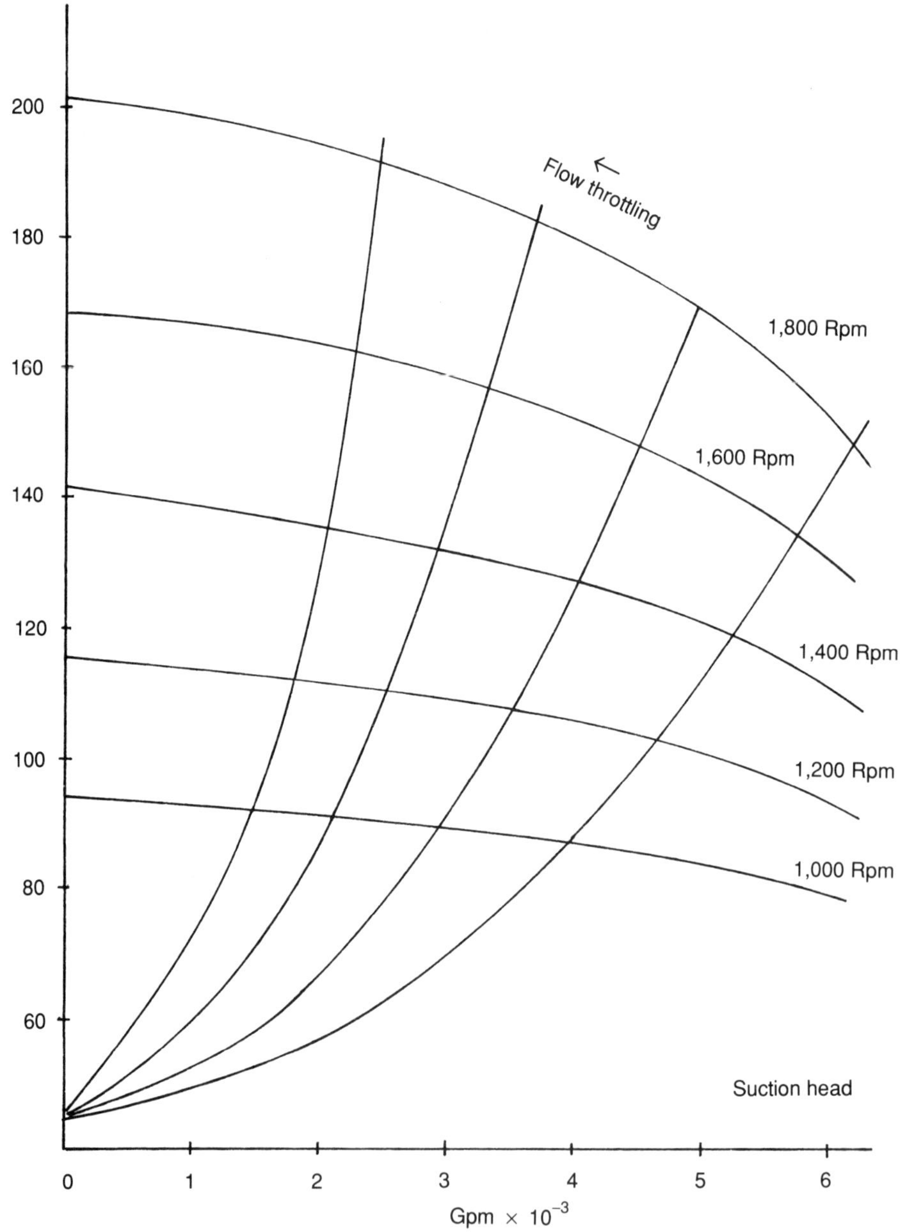

Figure 8-20. Fresh-water cooling pump and system head-flow characteristics

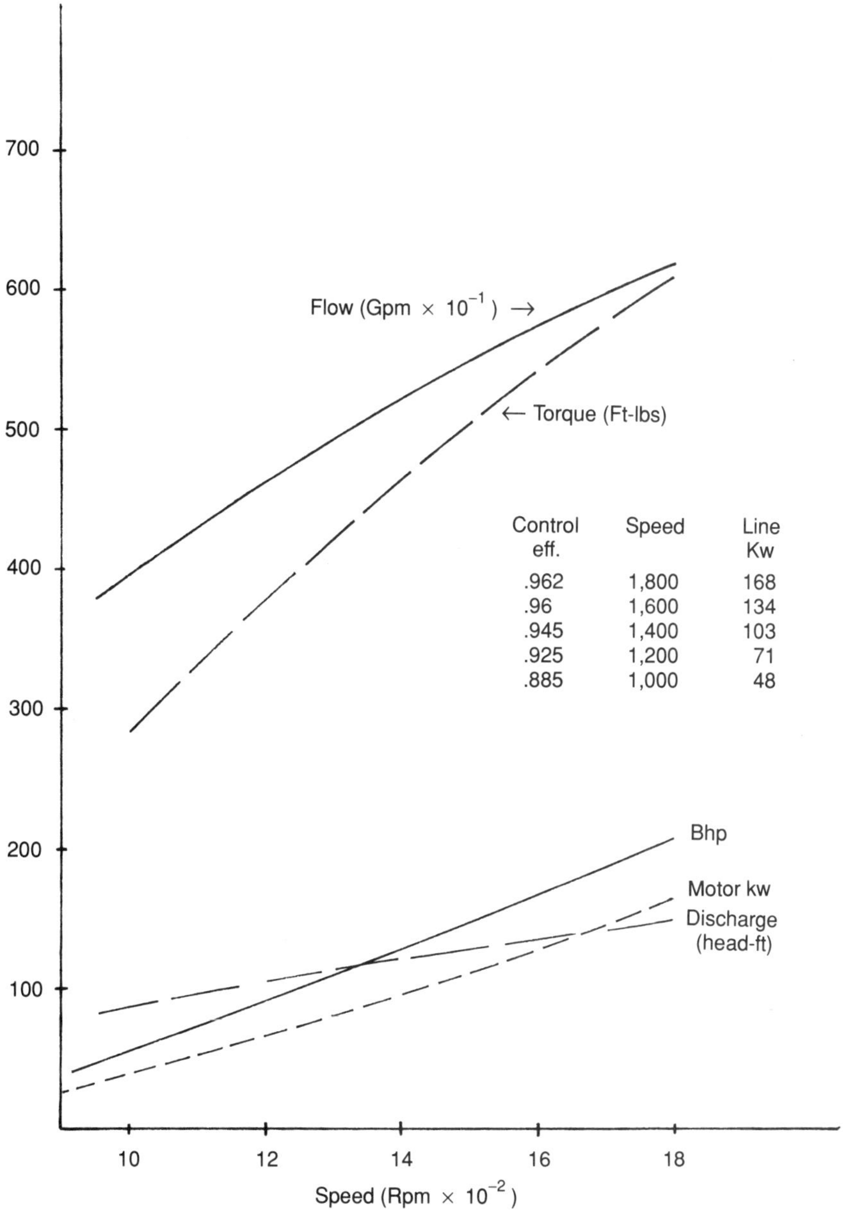

Figure 8-21. Central cooling pump variable speed operating characteristics

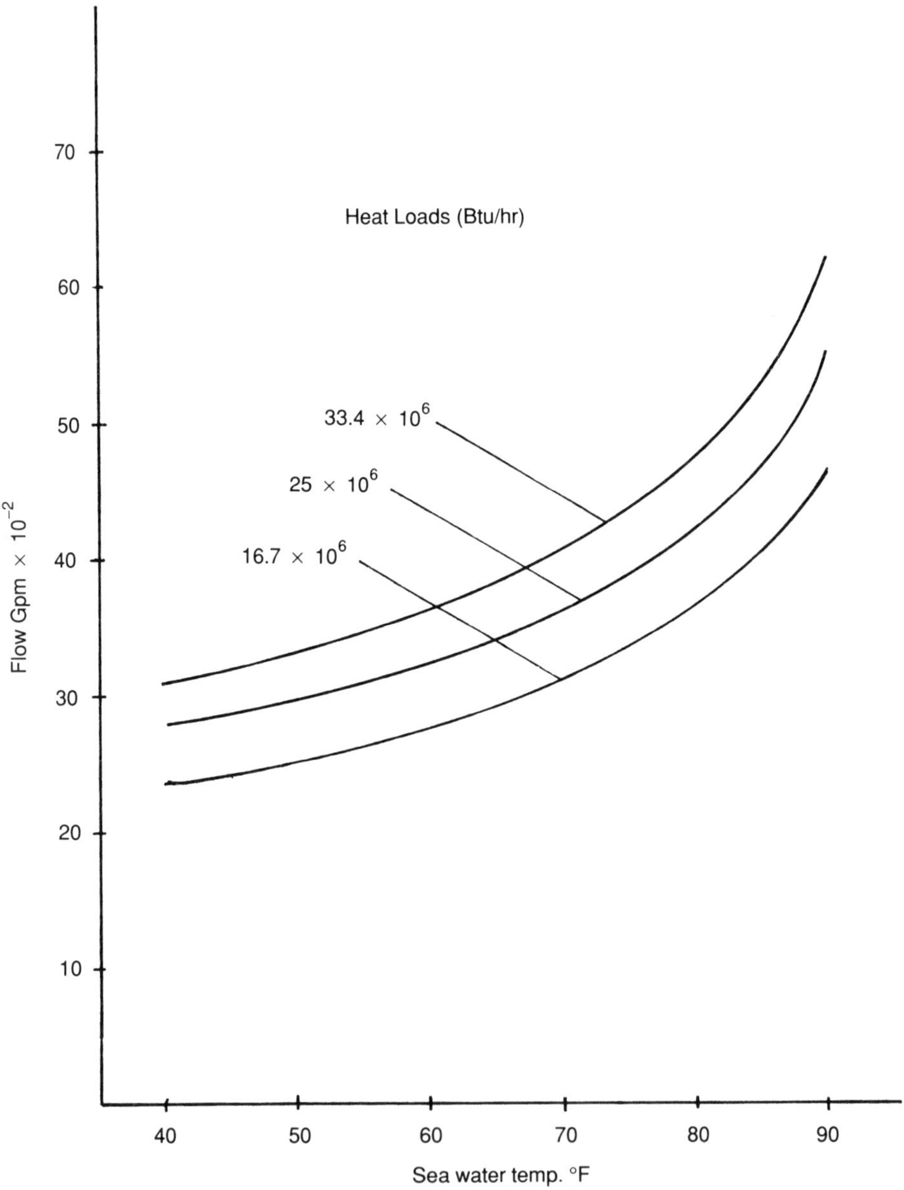

Figure 8-22. Central fresh-water cooling pump flow versus seawater temperature

Chapter 9

Cost Analysis and Estimating

The Shipyard estimator, like the Flying Dutchman, is forever at sea. His course passes constantly between those dangers, Scylla and Charybdis.
—L.S. Summers, Marine Technology, *Vol. 10, No. 1*

Building a modern ship is a large and risky construction project which may require hundreds of millions of dollars to conclude successfully. Even ship repair projects may involve many millions of dollars. The shipbuilder is required to quote a price for such work and, if awarded the contract, to deliver the ship for the contract price in accord with all the contract provisions, which may be quite simple or very complicated. The basis for the price offer is the cost estimate, usually prepared by a cost estimating or cost engineering department within the shipyard.

Shipbuilding estimators are usually engineers or designers who have extensive experience in collecting and evaluating shipbuilding returned costs. These are used to estimate the labor and material costs of proposed vessels which may be very similar or only slightly similar to the vessels for which the costs were accumulated.

Another category of shipbuilding cost estimates is derived from the preliminary estimate based on functional relationships between the cost and the design and weight parameters. Such relationships are developed by ship designers to facilitate design optimization studies and preliminary design costs. These estimates are internally consistent but, in the absolute sense, they lack the accuracy expected of an estimate used to prepare a shipbuilding bid. Appendix A describes the computer programs DESSEN and VENTURE, which rely on functional relationships to prepare preliminary design cost estimates for ships. Various researchers, including Benford, Buxton, Carreyette, and others, have applied, with considerable success, regression analysis to the problem of estimating shipbuilding labor and material costs given the preliminary parameters of a ship, including length, breadth, depth, draft, block coefficient, speed, and power.

The general relationship, $C = aX^n$, is very useful when fitted to actual shipbuilding cost data.

The following is a summary of the popular relationships derived for preliminary cost estimating.

Steel work direct labor cost:

$$C_{SL} = K_1 \frac{W_S^{2/3} L^{1/3}}{C_b}$$

Outfit direct labor cost:

$$C_{OL} = K_2 W_O^{2/3}$$

Machinery installation direct labor cost:

$$C_{MI} = K_3 P^{.82}$$

Steel material cost:

$$C_{SM} = K_4 W_S^{1.0}$$

Outfit material cost:

$$C_{OM} = K_5 W_0^{.95}$$

Machinery material cost:

For diesel propulsion, $C_{MM} = K_6 P^{.82}$

For steam propulsion, $C_{MM} = K_7 P^{.6}$

where K_j = constant which depends on shipbuilder, system of units, and cost per unit
W_s = steel weight
L = length between perpendiculars
C_b = block coefficient
W_O = outfit weight
P = service propulsion power

Shipbuilding and ship repair costs are divided into categories for accumulation purposes and for use in estimating or predicting the costs of a future project or contract. These categories include:

Direct Labor includes the cost of all labor expended directly and wholly on the ship or project. Appendix C contains a comprehensive list of direct labor cost accounts for a typical commercial shipyard.

Indirect Labor includes labor expended in support of all or some of the ships or projects underway during a given year. Such labor may include safety workers, production planners, security workers, facility repair workers, and many other categories. Cost accountants are normally responsible for determining a rational basis for allocating the indirect labor costs to ships and projects during an accounting period such as a fiscal year. This cost called burden or overhead is applied as a percent of direct labor.

Direct Material includes all materials which can be assigned directly to a ship or project. Machinery, steel plate, pipe, fittings, furniture, etc. are in this category. Appendix C contains a comprehensive list of direct material cost accounts for a typical commercial shipyard.

Indirect Material includes materials, services, and supplies which are difficult to account for on a ship or project basis. Such items may include compressed air, welding rod, electric power, cleaning materials, sand blasting material, etc. The cost of these items is included in the overhead rate charged to all projects.

Overhead Costs includes the indirect labor and materials, depreciation expenses, interest costs, and all other costs of the business which cannot be allocated directly to a ship or project. In shipbuilding and ship repair, engineering expenses are usually charged directly to the ship or project for which the work is undertaken. Marketing and general management costs are sometimes broken out in a special overhead or burden category called General and Administrative Expenses.

COST ESTIMATES FOR DESIGN STUDY

Scher and Benford (see the References) have developed typical functional relations between design and weight parameters and ship costs for bulk carriers. Earlier in this text, Figure 3-2, taken from the Scher and Benford study, showed a relationship in which the dependent variable is the required freight rate for a bulk carrier and the independent variables are the block coefficient and the speed-length ratio. Such a plot requires knowledge of building, operating, and financing costs. To obtain the shipbuilding labor and material costs, the Scher-Benford study used the following typical relationships which have been modified only as needed to make them relevant for costs in the 1994 period.

Steel weight (long tons):

$$W_S = 1.185\,(10^{-3})\,E^{1.36}\,[.65 + .5C_b + (1 - C_b)(.8D - T)/6T]$$

where $E = L(B + T) + .85L(D - T) + .75 \sum L_e H_e$

C_b = block coefficient
D = depth
T = draft
L = length between perpendiculars
B = beam
L_e = length of houses
H_e = height of houses

Ship steel material cost ($):

$$C_{SM} = W_S\,(1.167 - .117\ C_B)\ (493.7\ \$/\text{ton})$$

Ship steel labor cost (labor-hours):

$$MH_S = 157\ W_{si}^{.9}$$

At $16.76 per labor-hour, the steel labor cost is:

$$\text{Labor cost} = 16.76\ MH_S$$

Outfit weight (long tons):

$$W_O = .0290\ LB - 1.59\ 10^{-5}\ L^2\ B$$

Outfit labor and material costs ($):

At 270 labor-hours/ton of outfit, the outfit labor cost is;

$$\text{Labor cost} = [\ 270\ \text{mh/ton}\ ]\ [\ 16.76\ \$/\text{mh}\ ]\ W_O$$

At $7,290/ton of outfit, the outfit material cost is;

$$\text{Material cost} = \$7{,}290\ W_O$$

Low speed diesel propulsion machinery weight (long tons):

$$W = .124\ (MCR)^f\ (RPM)^{-.167} + .555\ (MCR)^{.7}$$

$$\begin{aligned} \text{where } MCR &= \text{maximum continuous rating HP} \\ f &= 1.299\ (\log_{10} RPM)^{-.44}\ ;\ f \leq 1 \\ RPM &= \text{engine speed, rpm} \end{aligned}$$

Diesel propulsion costs ($) :

$$\text{Machinery cost} = 13{,}672\ (MCR)^{.7} + 760{,}500$$

Shipboard electronics are $504,000 installed.

Total ship estimated costs:

$$\text{Ship cost} = \text{material} + \text{labor} + \text{overhead}$$

where overhead is 90% of the direct labor cost.

Example 9-1

Given a single hull bulk carrier with the following characteristics, estimate the total cost using the above functional relationships:

Length	820.21 ft.
Beam	107.2 ft.
Depth	65.2 ft.
Draft	44.5 ft.
House volume	18,297.5 cu. ft.
Steel cost	$493.7/ton
Labor rate	$16.76/labor-hour
Propulsion	low-speed diesel
MCR	20,600 HP
Engine speed	90 rpm

$$E = 820.21\ (107.2 + 65.2) + .85\ (820.21)\ (65.2 - 44.5) + .75\ (18{,}297)$$

$$E = 169{,}558.9$$

$$W_S = .001185\ (169{,}558.9)^{1.36}\ [\ .65 + .5\ (.853) + (1 - .853)\ (52.16 - 44.5)\ /\ 6\ (44.5)]$$

$$W_S = 15{,}331.8\ (1.08072) = 16{,}569.3 \text{ tons}$$

$$\text{Steel material cost} = 16{,}569.3\ [1.167 - .117\ (.853)\ ]\ (493.7)$$

$$\text{Steel material cost} = \$8{,}729{,}969$$

$$MH_S = 157\ (17{,}682.7)^{.9} = 1{,}043{,}985 \text{ labor hours}$$

$$\text{Steel labor cost} = 1{,}043{,}985\ (16.76) = \$17{,}497{,}153$$

$$W_O = .029\ (820.21)\ (107.2) - .0000159\ (820.21)^2\ (107.2)$$

$$W_O = 1{,}403.2 \text{ tons}$$

$$\text{Outfit material cost} = 6{,}264\ (1{,}403.2) = \$8{,}789{,}645$$

$$\text{Outfit labor cost} = (270)\ (1403.2)\ (16.76) = \$6{,}334{,}606$$

$$f = 1.299\ (\log_{10} 90)^{-.44} = 1.26$$

$$f = 1.0$$

$$W_m = .124\ (20{,}600)^{f}\ (90)^{-.167} + .555\ (20{,}600)^{.7}$$

$$W_m = 1{,}785.6 \text{ tons}$$

$$\text{Machinery total cost} = 13{,}672\ (20{,}600)^{.7} + 760{,}500$$

$$\text{Machinery total cost} = \$15{,}067{,}188$$

$$\text{Electronics total cost} = \$504{,}000$$

Summary

Hull	$26,226,849
Outfit	$15,124,245
Machinery	$15,067,188
Electronics	$504,000
Overhead	$21,484,577
Total	$78,406,859

This total represents a crude estimate which would be useful for comparing alternatives during a preliminary design study.

Engineers involved in cost studies always use the most reliable information available. If the only information available is not reliable and suitable to the need, the study should not be undertaken.

COST ESTIMATE FOR CONTRACT BIDDING

Shipbuilding Estimate

The shipbuilder receives a set of contract plans and specifications to prepare a formal bid for a shipowner. The contract plans and specifications may be prepared by the owner's naval architect and marine engineer or they may be prepared by the yard's advance design group if the shipbuilder is the ship designer. Mack-Forlist (see References) has offered the following steps for a shipyard cost-estimating organization to prepare a formal shipbuilding estimate.

1. Determine material quantities from contract plans and specifications. This activity results in detailed lists of steel plate and shapes, machinery, electronic, and outfit components.
2. Estimate the cost of material. Current prices are used for steel, pipe, paint, etc. Price quotations are requested from suppliers of major machinery components.
3. Obtain the returned labor-hours from shipbuilding cost accounts for the most similar ship(s) previously built. The most recent information is the most valuable.
4. Develop an overall plan for the construction of the vessel. Define major subassemblies, move methods, locations for work, etc.
5. Evaluate the differences between the previously built ship(s) and the proposed ship.
6. Adjust the returned cost of labor-hour units for the differences in items 4 and 5.
7. Review the conditions of performance, e.g., productivity, experienced with the previously built ship(s).
8. Adjust labor-hour units for the difference between current and previous conditions which affect productivity.

9. Anticipate and consider changes in conditions which may occur during the construction period that could affect productivity and adjust the labor-hour units to compensate. There is considerable uncertainty in this activity, which can be formally accommodated by a method to be described later.
10. Establish labor-hours and material quantities for the first ship of a multiship contract using the foregoing information.
11. Develop a learning curve for labor-hours of the subsequent ships of a multiship contract using the shipyard historical experience.
12. Estimate the dollar-per-hour labor rate on the basis of existing labor agreements and the changes expected during the construction cycle. There are important elements of uncertainty to be considered here.
13. Estimate the overhead rate per labor-hour for the proposed ship(s). This is an activity wherein a cost accountant devises a rational basis for allocating all nondirect costs to the ships under construction and to other shipyard work.
14. Review the make-or-buy alternatives and make decisions using corporate criteria to achieve income objectives.
15. Collect material, labor, and overhead costs into a total cost estimate for each ship in the contract. Present to management for a price decision.

Ship Repair Estimate

Ship repair estimators have a difficult task, since the exact nature of the required work, the work methods to be employed, and the quantity of work required are sometimes poorly defined despite a shipowner's detailed repair specifications for the work. Repair estimates are usually based on the estimator's experience and shipyard standards for units of repair work. Standard labor units may include the labor-hours to repair weld one foot of ¾-inch plate, labor per cubic foot to water wash a boiler fireside, and to chip and paint (.003-inch coat) one square foot of deck plate. If a large data base is available, it is possible to fit curves to such data and develop a computer-based method of generating labor estimates for ship repair.

The ship repair estimate procedure is similar to that for ship building but contains more uncertainty. For ship repair, the shipowner provides work specifications along with the needed drawings, instruction manuals, etc. For complicated machinery or electronic components, the shipowner usually requires the shipyard to retain a service engineer from the manufacturer of the equipment. The estimating procedure is as follows.

1. Review the owner's specifications and drawings.
2. Develop a production plan for each repair task. The plan will define location of the work, required tools, number and type of craft workers, support services, labor-hours for each craft, materials and sup-

plies required, quality control and classification society requirements, and environmental and safety requirements. The estimator will use preplanned methods and standard labor rates to the maximum extent possible.

3. Prepare a material list for the project. Use material cost standards or supplier quotations to develop the total material cost for the project.
4. Using the estimator's production plan or labor standards for repetitive tasks, estimate the labor-hours for each repair task. Use current or expected labor rates to estimate this cost for the proposed contract.
5. Use an appropriate overhead rate for ship repair work to calculate an estimated overhead cost.
6. Sum material, labor, service, and overhead costs for a total cost which will be presented to management for a price decision.

In shipbuilding, a production planning and control group prepares and issues detailed work packages for each shipping task. These packages authorize labor, specify methods, set quality requirements, provide working drawings, and control the local and overall schedule. In ship repair, the bid cost estimator frequently completes much of this work in preparing the bid estimate.

COST ACCOUNT SYSTEM

Account Structure

Shipbuilding cost accounts are used to record the returned labor-hours and material costs for ships under construction. The accounts are identified by the particular ship systems for which they are used to accumulate and record labor and material. For example, typical labor and material cost accounts may be labeled as follows:

Accommodation ladders and gangways
Metal sheathing
Light metal bulkheads
Winches and capstans

To assure consistency in accumulating labor and material to such accounts, the accounts have detailed descriptions of the types of costs which are chargeable to each account. The descriptions are readily available to all users of the accounts. Appendix C contains the detailed descriptions of a typical cost account structure utilized by a commercial shipbuilder. There are usually more than 100 such cost accounts to cover all the systems of a commercial ship. The cost of engineering and yard services is also accumulated in these cost accounts for each ship.

Typical Cost Account Analysis

Appendix C also includes a complete fractional allocation analysis to the account structure of the labor-hours and the material costs for a 119,054-dwt double-hull crude oil carrier and a 14,090-dwt RO/RO cargo vessel. Summing these fractional account values into major categories provides the following:

Account Grouping	*RO/RO Material*	*RO/RO Labor*	*Tanker Material*	*Tanker Labor*
Hull	.405672	.443480	.242559	.441251
Mechanical	.342279	.150022	.404712	.210720
Electrical	.053107	.020845	.113591	.047586
Outfitting	.072169	.043871	.122498	.068855
Engineering	.036542	.130139	.004093	.043182
Yard Services	.064339	.200050	.054507	.181313
Other	.025892	.011593	.058150	.007093
Total	1.000000	1.000000	1.000000	1.000000

Shipbuilding Estimates by Accounts

The shipbuilding labor and material estimate for a bid is prepared using the cost account structure. The shipbuilder's discipline of carefully accumulating the same cost elements to the same cost accounts on all contracts makes it possible to use these accounts for the database employed by the shipbuilding cost estimators to estimate the labor and material for a new contract bid.

Total Cost and Price

Once the labor-hours and material costs which can be directly charged to a specific hull or other contract have been determined, the remaining costs (called overhead) must be distributed among the hulls and projects which the shipbuilder anticipates during an accounting period. Typical expenses which are handled as overhead or burden include executive compensation and services such as employee relations, financial services, security, facility maintenance, and depreciation of capital equipments, etc. It is common practice to sum these fixed and semivariable expenses and to allocate the total to shipbuilding products in proportion to the direct labor-hours required by a product.

Finally, the profit is applied to the total cost to arrive at the product price. The profit is determined by the market conditions which prevail.

SHIPBUILDING LABOR AND MATERIAL INDICES

A useful index of the direct labor and material cost trends experienced by American shipbuilders has been compiled as of January each year since

1939 by the Maritime Administration. Table 9-1 is a listing of the published indices since 1939. The data in this table may be used to update old cost estimates or returned costs for delivered vessels in the absence of more specific and accurate information. Nonlinear regression analysis of these data will provide a useful prediction of cost growth for short periods into the future.

TABLE 9-1

Index of Estimated Shipbuilding Costs in the United States

Year	*Material*	*Labor*	*Total*	*Year*	*Material*	*Labor*	*Total*
1939	55	45	100	1966	125	192	318
1940	56	45	101	1967	130	201	331
1941	57	48	105	1968	135	208	343
1942	60	59	119	1969	139	220	359
1943	61	66	127	1970	148	231	379
1944	61	71	132	1971	158	241	399
1945	61	74	135	1972	168	250	418
1946	62	69	131	1973	175	268	443
1947	81	77	158	1974	187	283	470
1948	90	85	175	1975	249	309	558
1949	102	87	189	1976	263	330	593
1950	96	90	186	1977	282	354	636
1951	100	98	198	1978	299	379	677
1952	101	111	212	1979	329	414	743
1953	102	120	222	1980	363	448	811
1954	107	125	232	1981	399	493	892
1955	108	130	238	1982	431	548	979
1956	117	141	258	1983	434	590	1024
1957	127	143	270	1984	447	621	1068
1958	133	152	285	1985	457	625	1082
1959	133	159	292	1986	458	645	1103
1960	133	162	295	1987	456	665	1121
1961	127	170	297	1988	478	660	1138
1962	124	175	299	1989	507	654	1151
1963	123	180	303	1990	515	673	1188
1964	124	187	311	1991	523	687	1210
1965	124	189	313				

The index is developed primarily from data published by the Bureau of Labor Statistics, Department of Labor. The material portion of the index is determined from three BLS wholesale price subgroup indices, Groups 10-1, Iron and Steel; 11-4, General Purpose Machinery and Equipment;

and 11-7, Electrical Machinery. These BLS indices are weighted 45 percent, 40 percent, and 15 percent respectively, to represent the material index for merchant ship construction.

The material index represents 50 percent of the total MarAd index. The other 50 percent is the labor portion, which is based on a BLS publication titled “Indexes of Change in Straight-time Average Hourly Earnings for Selected Shipyards for Steel Vessel Construction.”

Example 9-2

For a cargo vessel completed in 1980, the total returned labor-hours were 1,289,319 and the total material costs were $13,745,574. The average rate in 1980 for shipbuilding crafts was $10.50 per hour. Estimate the labor and material cost for delivery in 1990.

1990 material index	515	1990 labor index	673
1980 material index	363	1980 labor index	448
Increase	152	Increase	225
Percent increase	41.87%	Percent increase	50.22%

$$\text{1990 labor estimate} = 10.50\ (1.5022)\ (1{,}289{,}319) = \$20{,}336{,}557$$

$$\text{1990 material estimate} = \$13{,}745{,}574\ (1.4187) = \$19{,}500{,}845$$

Example 9-3

A particular shipyard has an average craft rate of $16.60 per hour in 1991. There is no contract agreement which establishes craft rates for the near future. An estimate is needed for the average rate in 1995. In the absence of better information, it is decided to use the labor index to predict the craft average rate in 1995.

Since the slope of the material index since 1983 has been almost constant, the prediction will be based on a linear regression of the labor index from 1983 to 1991.

x	y	x	y
83	590	88	660
84	621	89	654
85	625	90	673
86	645	91	687
87	665		

Using available curve-fitting algorithms, it is found that the best fit of this data is the following curve:

$$y = a + b/x + c/x$$

where a = -5257
b = 1,104,000
c = 51,330,000
Correlation factor = .9216

Based on this fit to the data, the prediction for the near future for the labor index is:

1992	680.79
1993	681.46
1994	680.76
1995	678.77

If a fit to a straight line is used, the correlation factor is .8776 and the prediction is as follows:

1992	698.08
1993	708.37
1994	718.65
1995	739.22

A plot of the data may reinforce the judgment that the straight line is the prudent choice in these circumstances.

INTERNATIONAL SHIPBUILDING LABOR COMPENSATION

In ship design studies and shipbuilding studies, it is frequently desirable to make cost comparisons among costs of major shipbuilding countries. Table 9-2 provides the history of the hourly compensation costs, including fringe benefits, in U.S. dollars since 1982.

TABLE 9-2

Shipbuilding Hourly Compensation Costs

Country	*1982*	*1983*	*1984*	*1985*	*1986*	*1987*	*1988*	*1989*	*1990*
U.S.	13.48	13.81	14.21	14.57	14.47	14.26	14.33	14.77	15.50
Germany	11.60	11.64	10.68	11.00	15.26	19.35	20.89	20.16	26.50
Norway	11.59	11.19	10.90	11.16	14.15	18.50	19.88	19.63	24.36
Denmark	9.56	9.26	8.46	8.62	11.70	15.55	16.99	16.23	21.86
Netherlands	10.02	9.57	8.63	8.61	12.61	15.23	15.87	—	21.70
France	8.95	8.97	8.86	8.92	11.66	13.97	14.51	14.09	18.60
Italy	8.32	8.63	8.02	8.58	11.53	14.18	14.62	15.10	19.22
U.K.	6.97	6.47	6.11	6.37	7.73	8.91	9.89	10.06	12.55
Japan	6.96	7.68	7.76	8.13	11.60	13.00	14.83	14.67	15.80
Korea	2.08	2.21	2.13	2.19	2.53	2.92	4.40	6.35	10.00

Using regression analysis or other curve-fitting programs, it is possible to determine trends in the growth of labor compensation. In the absence of better information or contradictory information, such trends may be used to forecast the shipbuilding labor compensation for a given year and country. The forecast can then be applied to cost estimating or used as input for the programs DESSEN and VENTURE.

Example 9-4

In a tanker design study, the analyst plans to include among the independent variables the choice of building the ship in Japan or the United States. If the major portion of the labor will be scheduled for the year 1995, what shipbuilding compensation rates should be used in the study?

Inspection of Table 9-2 shows that the Japanese compensation rate is increasing much faster than the U.S. rate. Predicting these rates five years into the future is always risky. However, let's assume that there is no reason to believe that the trend will not continue. Therefore, a straight line may be fit to the data. (Many other curves were fit to the data of Table 9-2. Some resulted in higher correlation coefficient; however, there was no rationale in the economic sense to assume that a complicated function was a better fit than the straight lines shown below.)

Year	*U.S. Labor Rate Prediction*	*Japan Labor Rate Prediction*
1990	15.11	16.18
1991	15.32	17.43
1992	15.53	18.69
1993	15.74	19.94
1994	15.96	21.20
1995	16.19	22.45
Percent increase, 1990-95	7.14	38.75
Correlation coefficient	.7156	.9235

An analyst may be reluctant to predict the 38.75 percent increase in Japanese shipbuilding labor rate between 1990 and 1995. However, the Japanese rate between 1982 and 1990 increased 127 percent and there is no indication of a leveling-off at this time.

ACCOMMODATING UNCERTAINTY IN LABOR ESTIMATES

Estimation is the process of making a decision or judgment concerning the approximate value of certain parameters when the decision is weighted or influenced by all available information. It is natural for knowledgeable estimators to give more weight to factors which tend to increase the task labor-hour estimate. This practice assures that they do not produce a low estimate with severe consequent problems for the shipbuilding corporation. Estimates which tend to be high and estimates which tend to be low are not as useful to management decision makers as an estimate which provides a mean labor-hour estimate and a measure of the variance about the mean. Experienced estimators know with some precision what future events are likely in any shipbuilding or ship repair task and the effect of these events on the labor required to complete these tasks. The estimator needs a methodology to permit the quantification of his or her understanding of the range, mode, low value, and high value of the task labor-hour estimate in a tractable form.

The inherent uncertainty in such estimates can be expressed as a subjective probability distribution or, if a significant number of ships have been constructed, the labor-hours for each task can be corrected for learning and other differences to prepare a consistent sample from which the parent distribution parameters can be estimated.

One of the authors experienced success in applying subjective probability distributions to quantify the uncertainty in ship repair task labor estimates. The task labor-hour probability distributions were summed to provide a probability distribution of the total project labor estimate. In this case, the estimators were asked to provide three estimates of labor for each task:

1. Estimate a low labor value *LL* and the estimate of the probability *P(LL)* that the actual labor is equal to or less than the low value.
2. Estimate the mode of the labor, i.e., the most likely value.
3. Estimate a high labor value *HL* and the probability *P(HL)* that the high value is equal to or greater than the actual task labor estimate.

Figure 9-1 is a sketch of a typical task probability distribution showing the above values. A short training program is needed to teach estimators to encode their knowledge in such distributions so that consistent results are produced.

The next step is to fit each of the three estimate sets to a Weibull or Normal probability distribution. The selection of the distribution is based on a goodness-of-fit criterion. The Central Limit Theorem, found in most statistics texts, allows us to sum a large number of the labor task distributions to a single Normal distribution which describes the uncertainty in the total estimate of project labor.

LABEST PROGRAM

In the case of shipbuilding and ship repair, it would be tedious to fit 100 to 200 sets of data to task distributions and then sum them to a total distribution. To replace this prohibitive effort, a computer program, LABEST, has been prepared to check the input file for errors, to fit each of the labor

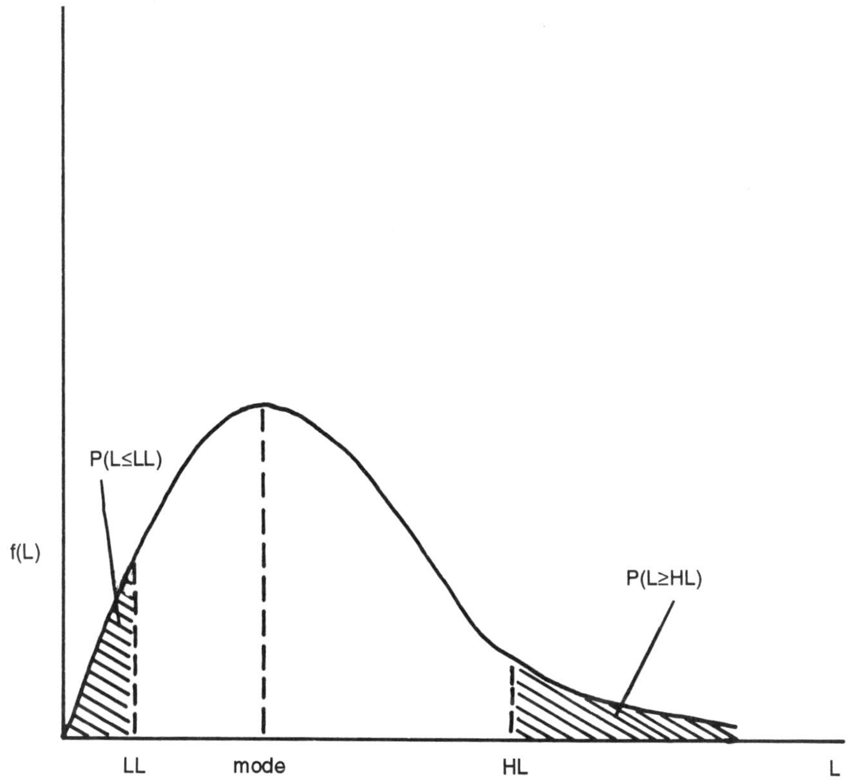

Figure 9-1. Probability distribution of task labor-hours

task estimates to an appropriate distribution, and to calculate the Normal distribution of the total estimate. Table 9-3 shows the output of the LABEST program for shipbuilding labor-hours for a 14,090-dwt RO/RO cargo liner. Figure 9-2 is a plot of a left tail cumulative probability distribution of the total labor-hour estimate. A review of the plot and the program output shows that management has much more information available compared to a single value for the total estimate. For example, the expected value of the total labor estimate is 1,301,370 labor-hours. Furthermore, there is a 10 percent chance that the total estimate will equal or exceed 1,325,007 labor-hours and a 10 percent chance that the total estimate will equal or be less than 1,277,733. User information for the program LABEST is included in Appendix A.

TABLE 9-3

LABEST Program Output for Shipbuilding Labor Estimate: Total Labor Estimating Program for Marine Engineering Economics and Cost Analysis

PREPARED FOR: Red River Shipping Proposal
PROJECT ID: 14090-dwt RO/RO Carrier
PREPARED BY: Cost Engineering Department
DATE: 02/23/1993
TIME: 12:40:12

Labor Task Estimates

*Task**	*LL*	*P(LL)*	*Mode*	*HL*	*P(HL)*	*Mean*	*Sigma*	*Dist.*
714	562,667	.100	564,896	600,878	.050	574,017.90	13,909.91	N
713	10,112	.100	10,876	11,530	.100	10,842.19	554.57	N
721	3,458	.300	3,715	4,945	.300	4,466.63	1,631.03	N
722	1,942	.300	1,995	2,059	.300	2,000.26	114.86	N
723	916	.250	966	988	.300	954.42	58.76	W
724	2,217	.300	2,444	2,855	.300	2,593.21	613.66	N
725	178	.200	188	202	.250	191.58	17.80	N
734	200	.020	6,960	8,604	.300	6,936.17	3,186.42	W
744	4,422	.150	5,069	6,006	.200	5,304.03	836.02	W
745	351	.250	386	503	.200	423.64	96.23	W
748	1,095	.100	1,207	1,262	.200	1,200.70	78.26	N
750	2,818	.150	2,858	2,992	.150	2,888.55	85.74	N
752	406	.050	420	430	.050	419.35	6.76	N
755	5,159	.250	5,403	5,634	.200	5,392.90	317.39	N
756	1,014	.100	1,054	1,110	.100	1,061.20	35.65	N
758	273	.150	303	353	.150	312.50	39.07	N
765	26,005	.150	28,049	29,510	.100	27,569.38	1,600.59	N
780	367	.200	440	524	.200	446.61	90.12	W
783	5,050	.100	5,081	6,659	.150	5,608.65	862.12	N
784	105	.200	136	231	.200	161.87	72.01	N
785	1,513	.100	1,581	1,810	.150	1,646.46	154.00	N
789	246	.100	290	391	.100	303.32	56.13	N
790	982	.200	1,009	1,037	.200	1,008.47	33.73	N
791	1,303	.100	1,439	1,540	.200	1,451.61	114.78	N
792	2,348	.200	2,406	2,723	.200	2,536.52	247.14	N
793	1,658	.250	1,710	1,754	.300	1,706.70	84.51	N
794	5,997	.150	6,339	6,521	.250	6,333.15	308.78	N
795	630	.200	731	814	.200	721.88	106.50	W
796	154	.250	183	225	.250	191.09	50.64	W

TABLE 9-3 *(continued)*

797	930	.100	1,212	1,320	.200	1,167.31	177.57	W
331	6,710	.150	7,265	7,923	.200	7,414.52	680.45	N
811	445	.150	502	1,568	.020	744.28	311.04	W
817	488	.200	518	612	.200	552.21	80.61	N
840	1,311	.150	1,409	1,565	.150	1,422.79	122.14	N
841	1,812	.050	1,900	2,000	.050	1,904.72	53.61	N
842	11,173	.100	14,954	16,317	.100	14,100.65	2,153.74	N
844	6,452	.200	6,457	6,699	.150	6,546.90	137.99	N
855	80	.050	179	204	.050	159.65	40.88	N
856	12,190	.100	12,664	13,032	.100	12,633.40	334.55	N
858	5,045	.200	5,349	5,690	.200	5,335.86	379.07	N
869	16,763	.150	19,736	25,712	.150	21,205.83	4,240.99	W
877	21,583	.150	24,155	25,679	.150	23,652.84	1,978.23	W
879	10,190	.330	10,283	10,378	.330	10,294.09	227.09	N
880	8,205	.300	8,460	11,041	.100	9,116.16	1,387.04	N
882	4,227	.250	4,232	4,331	.300	4,300.58	106.57	N
890	6,481	.150	6,839	8,393	.050	7,082.19	649.21	N
893	364	.100	450	491	.050	433.56	44.92	N
894	22,044	.120	22,963	26,608	.060	23,562.64	1,646.00	N
900	20,312	.150	21,533	22,080	.150	21,243.67	846.29	N
901	15,593	.200	15,881	16,498	.200	16,015.03	559.19	N
902	2,009	.100	2,200	2,300	.200	2,190.93	132.44	N
903	409	.150	435	596	.070	468.95	69.56	N
904	5,445	.200	5,728	6,104	.150	5,758.38	354.87	N
906	10,253	.150	11,367	11,861	.150	11,065.44	807.88	W
907	5,943	.200	6,025	6,439	.200	6,162.91	306.89	N
914	1,936	.250	2,291	2,400	.300	2,159.34	402.50	W
915	9,804	.330	9,908	10,009	.250	9,871.08	183.01	N
916	5,486	.150	7,383	8,791	.250	7,501.77	1,884.99	W
917	1,395	.050	1,480	1,500	.200	1,469.85	38.71	N
918	10,793	.100	12,040	12,300	.250	11,773.33	738.93	W
919	12,235	.150	12,803	13,324	.100	12,740.31	502.34	N
920	10,998	.050	12,071	12,288	.200	11,883.82	483.42	W
921	10,857	.120	12,073	13,644	.230	12,582.71	1,593.59	N
927	60	.150	65	67	.150	63.59	3.53	N
930	2,230	.100	2,268	2,293	.200	2,266.92	27.89	N
938	1,246	.200	1,290	1,607	.100	1,366.30	169.91	N
939	28,935	.150	31,180	37,015	.050	31,511.46	2,818.34	N
940	10,320	.150	10,645	10,987	.150	10,641.72	332.20	N

TABLE 9-3 (*continued*)

941	7,845	.150	8,523	9,647	.200	8,784.80	1,006.97	N
942	2,014	.120	2,227	2,285	.160	2,180.95	125.16	N
943	838	.200	1,061	1,095	.150	952.03	182.92	N
946	430	.150	460	595	.070	489.83	65.77	N
950	17,299	.050	18,585	21,313	.060	18,972.96	1,313.42	N
952	66,628	.100	69,703	76,833	.100	71,020.73	4,285.36	N
957	36,156	.150	37,717	38,356	.100	37,264.07	1,069.49	N
958	19,746	.150	22,384	24,188	.100	21,800.35	1,982.45	W
959	14,438	.100	14,553	14,636	.100	14,542.82	77.90	N
960	36,829	.120	37,133	37,943	.180	37,394.07	626.92	N
962	14,669	.100	16,310	16,906	.250	16,133.72	1,028.56	N
965	777	.120	852	1,082	.160	924.04	162.37	N
956	17,744	.100	21,940	28,789	.080	22,518.96	3,957.74	N
440	285	.100	500	727	.100	505.16	169.16	W
700	1,379	.150	2,383	2,670	.200	2,083.06	681.37	W
701	2,713	.020	10,000	10,600	.150	8,901.61	2,255.69	N
702	2,460	.020	7,661	12,083	.100	8,048.70	3,032.66	W
703	3,133	.100	8,466	9,000	.200	7,071.74	2,482.84	N
800	8,235	.100	11,403	14,056	.250	12,053.13	3,031.43	W
935	3,076	.100	5,248	6,742	.150	4,997.40	1,606.90	W
953	1,047	.200	1,060	1,185	.100	1,097.72	64.46	N
975	3,252	.150	3,594	6,125	.200	4,573.60	1,684.29	N

* The task numbers refer to the shipbuilding account structure described in Appendix C.

Distribution of Total Labor Estimate

Labor Expectation, X′	*Left Tail Cum Probability, P(X <= X′)*
1277733	.100
1296697	.400
1301370	.500
1306043	.600
1313810	.750
1325007	.900
1331707	.950
1344277	.990
1358366	.999

Standard Deviation of the Total Labor Estimate = .1844E + 05
Ratio Maximum Labor Task Variance to Total Variance = .569

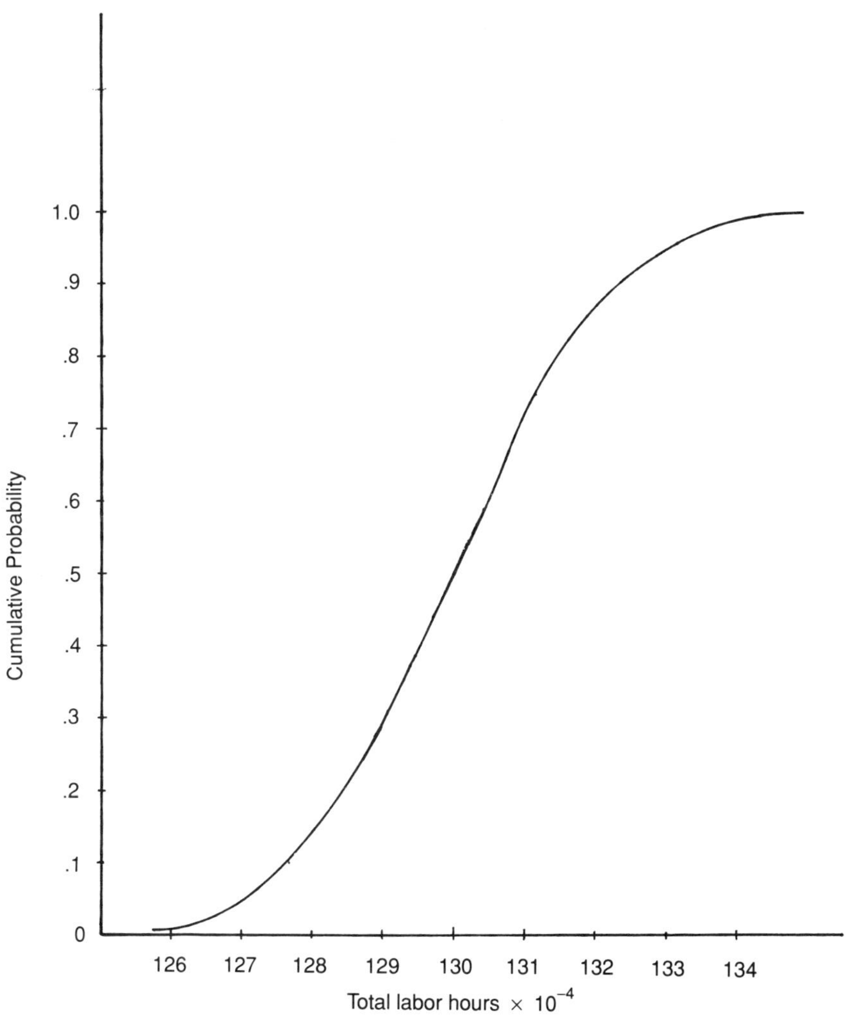

Figure 9-2. Left tail cumulative probability of labor estimate

THE LEARNING CURVE

Learning Principle

In shipbuilding and other manufacturing activities, it has been observed by Hancock and Bayha in Salvendy (see References) that repetitive manufacture of identical products will improve the performance of individual workers and the overall manufacturing organization. This characteristic may be employed to successfully predict the labor-hours required for building a series of same class vessels given the labor-hour estimate for the first vessel.

The factors that affect the rate at which people and organizations learn to do repetitive jobs include complexity of the project, capabilities of people and the organization, and time between the repetitive tasks.

Figure 9-3 shows the direct labor-hours curve for a series of 10 identical vessels. The curve indicates a continuous improvement in the labor-hours required per hull. The diminishing rate for productivity improvement suggests an exponential relationship between the labor-hours per hull and the cumulative hulls constructed. When plotted on log-log paper in Figure 9-4, the same data show as a straight line. Therefore, the functional form is

$$L = K V^{C}$$

where L = man power per vessel
K = man power for first vessel
V = number of vessels
C = constant slope of learning curve

Taking the logarithms of both sides provides:

$$\log L = \log K + C \log V$$

which is the equation of a straight line in the form:

$$y = cx + k$$

If actual shipbuilding data are plotted on log-log paper and a straight line is fitted, c will be the slope and b will be the intercept of the fitted line.

A useful property of the learning function is that every time N, the number of vessels, is doubled, then L, the direct labor-hours per vessel, decreases by a fixed percentage. If the learning curve is a 92 percent curve (which is not uncommon in shipbuilding), each time the number of hulls doubles, the direct labor will be 92 percent of the prior value. Mack-Forlist suggests that the shipbuilding learning rates are between 85 and 95 percent.

To calculate the average labor for vessels N_a to N_b, where N_a is the first vessel of the series in average, N_b is the last vessel of the series in average,

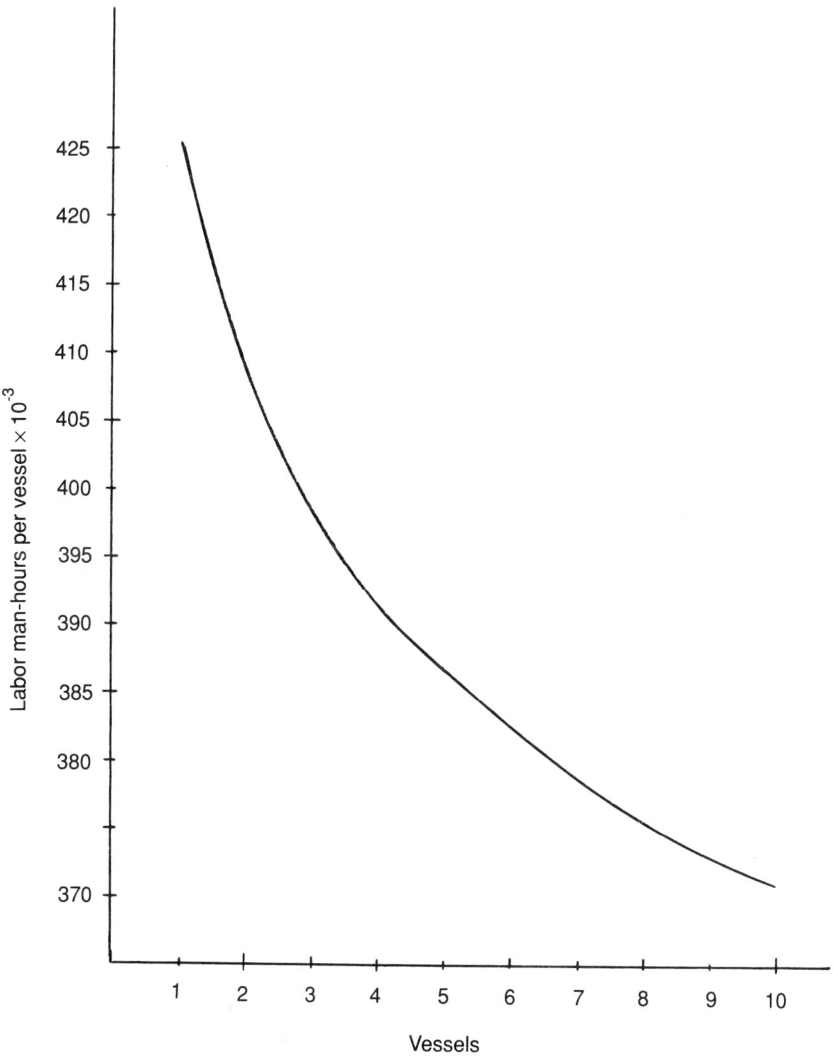

Figure 9-3. Ninety-six percentage learning curve for ten ships program

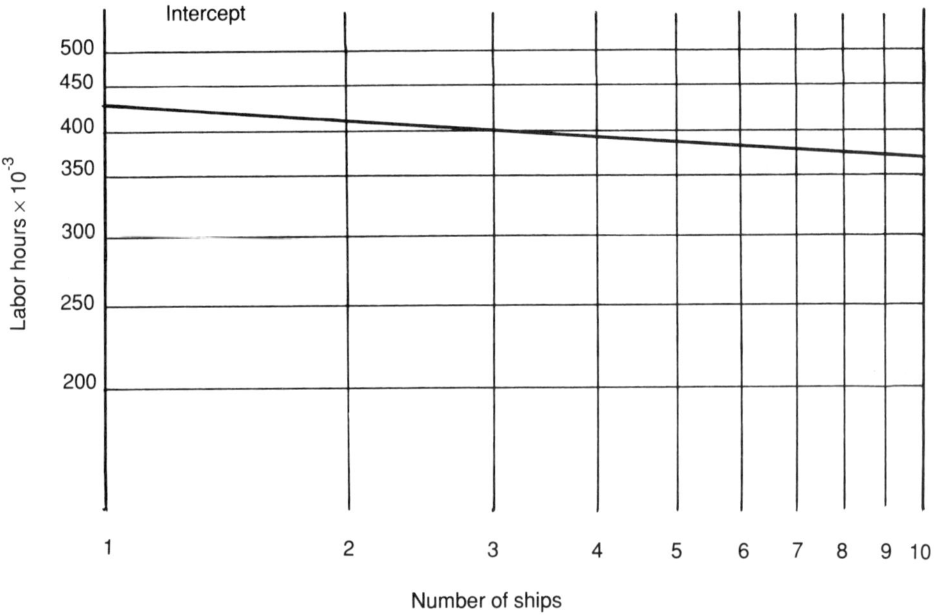

Figure 9-4. Log-log plot of learning curve for ten ships, 96% learning

and L is the average direct labor for included vessels, it is necessary to sum the learning function over the vessels of interest and divide by the number of vessels as follows:

$$\bar{L} = \frac{\int_{N_a-.5}^{N_b+.5} KV^{-C}\,dV}{(N_b+.5)-(N_a-.5)} = \frac{\left[\dfrac{K(N_b+.5)^{1-C}}{1-C}\right] - \left[\dfrac{K(N_a-.5)^{1-C}}{1-C}\right]}{N_b - N_a + 1}$$

$$\bar{L} = \frac{K\left[(N_b+.5)^{1-C} - (N_a-.5)^{1-C}\right]}{(1-C)(N_b - N_a + 1)}$$

Example 9-5

A shipbuilder with experience which suggests a learning rate slope constant of .08 plans to build a series of 10 same class vessels for several owners. What is the total estimated direct labor-hours for the second and last vessels if the first vessel is 423,509 labor-hours?

$$L = K V^{-C}$$

$$L_2 = 423509 \ (2)^{-.08}$$

$$L_2 = 400664 \text{ labor–hours}$$

$$L_{10} = 423509 \ (10)^{-.08}$$

$$L_{10} = 352259 \text{ labor–hours}$$

Example 9-6

Use the data of Example 9-5 to calculate the average labor-hours for vessels 5 through 10.

$$\overline{L}_{5-10} = 423509 \frac{\left[(10.5)^{1-.08} - (4.5)^{1-.08} \right]}{(1 - .08)(10 - 5 + 1)} = 361{,}337 \text{ labor–hours}$$

Example 9-7

The shipbuilder estimates that 361,337 labor-hours will be the standard labor for continuous manufacture of this vessel. How many vessels must be constructed with learning to reach the standard?

$$\frac{423{,}509}{361{,}337} = 1.1721$$

$$L = 1.1721 \, V^{-.08}$$

$$\log V = \frac{\log 1.1721}{.08}$$

$$V = 7.28 \text{ vessels}$$

Example 9-8

If the shipbuilder starts the program and has the following experience for the first three vessels, what adjustment should be made to the assumed learning curve?

Vessel	*Labor-hours*
1	423,600
2	381,240
3	358,454

The learning curve has a slope constant of –.1520, which could be determined by linear regression of the logarithms of the labor-hours and cumulative vessels or by plotting the above data on log-log paper.

$$\frac{423{,}600}{381{,}240} = 1.111$$

$$1.111 = (\tfrac{1}{2})^{C} = .5^{C}$$

$$\log\ 1.111 = C\ \log\ .5$$

$$C = -.152$$

The general relationship for calculating the slope constant of the learning curve, given a decimal ratio of learning for doubled production, is:

$$-C = \frac{2 - \log\ R_p}{\log\ 2}$$

where R_p is the percentage learning ratio. Here are some slope constants and percentage learning ratios in the typical shipbuilding range:

Percentage Learning Ratio	*Slope Constant*
100	0.0
98	-.0291
96	-.0589
94	-.0893
92	-.1203
90	-.1520

A relationship for the slope, given a decimal learning ratio R_D is

$$C = \frac{\log\ R_D}{\log 2}$$

or

$$C = \frac{\ln\ R_D}{\ln\ 2}$$

INFLATION AND LABOR COST

Given the learning rate, it is feasible to predict future labor-hours for identical units of manufacture. If the labor rate average for all trades is known, the product of labor-hours and dollars per hour will provide the direct labor cost for a vessel. However, there is usually an annual inflation in labor costs. When it is required to make comparisons of labor costs subject to both productivity improvement and inflation, the present worth of the costs in terms of zero time dollars is the appropriate measure. It may be desirable to make such comparisons to permit decisions among production methods, management methods, facility utilization, etc.

To deal with inflation and an interest rate simultaneously, it is necessary to calculate an equivalent interest rate which incorporates both factors and provides a present worth or annual equivalent cost in terms of zero time dollars. An inflated dollar is worth less in zero time dollars so the discount rate must increase with inflation.

$$\text{Equivalent } i = (1 + i_n)(1 + i_f) - 1$$

$$i = i_n + i_f + i_n i_f$$

where i = equivalent interest
i_n = interest without inflation
i_f = inflation

PRODUCTION PLANNING AND CONTROL

The traditional approach to shipbuilding production planning and control employs ship systems as the bases for labor and material allocation and reporting. A more effective method allocates and controls labor and material by ship zones irrespective of the type or number of systems included in a zone. Recent shipbuilding experience suggests that zone-oriented shipbuilding has a potential for a labor reduction of 40 percent compared to the system-planning method. This reduction in direct labor is obtained with a high initial cost for the expanded and detailed engineering, drafting, and planning required before construction starts. This increases the cycle time and the engineering labor cost for the series of vessels. The learning curve is a useful method for the comparison analysis of economics of ships constructed under these two methods.

Example 9-9

A shipbuilder expects to be awarded contracts for 10 vessels of the same class to be manufactured at the rate of two per year for a five-year period. He is planning the use of the zone construction method which, it has been estimated, will require 200,000 additional labor-hours of engineering effort before construction starts. The current average rate for engineering labor is \$23 per labor-hour. The current average rate for shipbuilding labor is \$14 per labor-hour. The anticipated inflation rate during the next five years for shipbuilding labor is 2 percent per year. The shipbuilder's cost of money is 8 percent. The labor content of the first vessel is 425,000 labor-hours. The shipbuilder has experienced a learning rate of 96 percent on previous multiship contracts. In current dollars, i.e., corrected for inflation, what is the present worth of the economic gain for using the zone method compared to the system planning method?

Calculate the equivalent interest rate with inflation:

$$i = i_n + i_f + i_n\, i_f$$

$$i = .08 + .02 + (.08)(.02) = .1016$$

Calculate annual labor based on 96 percent learning:

$$\text{For zone method } L = KV^{-C} = 255{,}000\ V^{-.0589}$$

$$\text{For system method } L = KV^{-C} = 425{,}000\ V^{-.0589}$$

Vessel	*Year*	V^{-C}	*Zone Method Direct Labor*	*System Method Direct Labor*
1	1	—	255,000	425,000
2	1	.960	244,800	408,000
3	2	.937	238,935	398,225
4	2	.922	235,110	391,850
5	3	.910	232,050	386,750
6	3	.900	229,500	382,500
7	4	.892	227,460	379,100
8	4	.885	225,675	376,125
9	5	.877	223,635	372,725
10	5	.873	222,615	371,025

Present worth of labor = \$14 (labor–hours) (P/F, i, n)

		Zone Method Labor		*System Method Labor*	
Year	*(P/F,i,n)*	*Labor-hours*	*PW, \$*	*Labor-hours*	*PW, \$*
1	.9077	499,800	6,351,358	833,000	10,585,597
2	.8241	474,045	5,469,247	790,075	9,115,411
3	.7481	461,550	4,833,998	769,250	8,056,663
4	.6791	453,135	4,308,136	755,225	7,180,226
5	.6164	446,250	3,850,959	743,750	6,418,265
Present worth of labor			24,813,698		41,356,162

Difference in labor	\$16,542,464
Incremental engineering	– 4,600,000
	\$11,942,464

Therefore, the present worth of economic gain in zero year dollars is \$11,942,464, or the annual equivalent gain is \$11,942,464. (A/P,.1016,5) = \$3,163,559 for the use of the zone planning and control method for shipbuilding when inflation is 2 percent and the cost of money is 8 percent during the five-year building cycle for the 10 vessels.

REFERENCES AND BIBLIOGRAPHY

Buxton, I. L. "Engineering Economics Applied To Ship Design" RINA *Transactions*, (1972).

Carreyette, J. "Preliminary Ship Cost Estimation". RINA *Transactions,* (1978).

Hagan, H. H. "Hull Cost Estimating and Cost Control." *Shipbuilding and Shipping Record* (February 1936).

Houlden, G. H. "Ship Tendering and Factors Influencing Building Costs." NECIES *Transactions*, Vol. 71 (1955).

Katsoulis, P.S. "Optimising Block Coefficient by Exponential Formula." *Shipping World & Shipbuilder* (February 1975).

Mack-Forlist, M., and R. Goldbach. "Bid Preparation In Shipbuilding." SNAME *Transactions*, Vol. 84 (1976).

McNeal, James K. "A Method of Comparing Costs of Ships Due to Alternative Delivery Intervals and Multiple Quantities." SNAME *Transactions,* Vol. 77 (1969).

Newhaus, Hebert M. "Ship Cost and Labor Estimating." *Marine Engineering and Shipping Review,* Vol. 50, Nos. 7 and 8 (July 1945).

Salvendy, Gavriel. Chapter 4.3, *Handbook of Industrial Engineering*. New York: John Wiley and Sons, 1982.

Scher, Robert M., and Harry Benford. *Some Aspects of Fuel Economy in Bulk Carrier Design and Operations.* No. 228; Ann Arbor: University of Michigan, September 1980.

Summers, Lewis . "The Prediction of Shipyard Costs." *Marine Technology,* Vol. 10 (January 1973).

QUESTIONS AND PROJECTS

1. Use the VENTURE program to prepare a cost estimate for a typical single hull tanker. Use the functional relations in this chapter to calculate the labor and material of a ship with the same characteristics. Are the estimates the same? Can you suggest reasons why they may be different?
2. In estimating, why is it important to have consistent cost structures for reporting returned costs?

3. Why is a stable workload needed to control costs and improve productivity in a shipyard?
4. What do we mean by the statement that shipbuilding is a labor-intensive activity?
5. Suggest a list of productivity improvement sources in shipbuilding which could contribute to the learning process.
6. If a shipyard experience in productivity improvements for a series of vessels does not fit the traditional exponential learning curve, what would you do to quantify and predict learning?
7. What interest rate would you use to find a future worth if time value of money and inflation are both included?
8. In a shipbuilding series of 12 vessels, a redesign is implemented after the fifth vessel to change the propulsion plant from medium speed diesel to gas turbine. Comment on the validity of the learning curve experience gained on vessels 1 through 5.

PROBLEMS

1. Given the following labor estimates for five tasks of a ship repair proposal, use the program LABEST to calculate the expected labor. Comment on the standard deviation. What is the significance of the ratio of the maximum task variance to the total variance? Does this appear to be a risky bid? What could be done to reduce the risk?

LL	*P(LL)*	*Mode*	*HL*	*P(HL)*
1,000	.100	3,000	5,000	.150
1,200	.050	2,000	7,000	.010
2,000	.080	4,000	6,000	.080
1,300	.010	7,000	9,000	.200
500	.200	1,000	1,500	.200

2. Given a 97 percent learning experience, calculate the learning curve slope.
3. Given a .965 learning ratio, calculate the learning curve slope.
4. The average labor rate is $13.80/hour. The first vessel in a series requires 360,000 labor-hours of direct labor. If the shipbuilder learning experience is 96 percent, what will be the estimated labor cost for the fourth vessel in the series?
5. Use the information of problem 3 to calculate the average labor cost for vessels 5, 6, and 7 in the construction series.
6. You are provided the direct labor-hours for the first four vessels in a series of 10. The hours are:

Vessel	*Direct Labor-Hours*
1	425,000
2	418,000
3	405,000
4	400,500

Use linear regression analysis to calculate the slope of the learning curve.

7. If the inflation rate is 2 percent per annum and the time value of money is 8 percent, calculate the interest rate you would use to find (P/F, i, n).
8. *Motor Ship* periodically publishes an international ship repair price index. A recent issue contained the following indices for the years 1985 to 1990.

Year	*Ship Repair Price Index*
1985	100
1986	95
1987	95
1988	95
1989	110
1990	129
1991	115

Use these data to predict the rate of ship repair cost increases for years beyond 1991 for a ship design study.

Chapter 10

Economics of Shipyard Operations

It must be remembered that the principal object of detail costing is to supply the cost estimating department with the information that will allow it to anticipate the cost of a proposed vessel.
—*H.H. Hagan,* Shipbuilding and Shipping Record, *Vol. 47, No. 8*

CLASSIFICATION OF SHIPYARDS

Shipyards may be classified based on their markets, capacity, types of ships, etc. Functional classification recognizes:

1. The yards which are primarily involved in ship construction and also in major overhaul and refitting
2. Those involved in both ship construction and repairs
3. Ship repair yards with a complete service, e.g., all American naval shipyards and many commercial yards
4. Topside repair enterprises
5. Specialized repair plants and shops, carrying out regular and emergency repairs of the ship's main and auxiliary machinery

This classification is subject to modification due to factors such as changing market conditions which lead to permanent or temporary shifts from one classification to another. For instance, many ship construction yards have found themselves involved in repairs due to a lack of new shipbuilding orders. In 1990 only 116 American yards remain operational, and only 20 of them were ship construction yards.

Manufacturers of marine components should be considered an integral part of the shipyard system. These enterprises include manufacturers of boilers, diesels, steam and gas turbines, deck machinery, pumps, valves and other piping components, marine electronics, marine instrumentation, etc. Many of these enterprises act as subcontractors or as vendors; however, their role in the ship construction and repair process is somewhat similar to that of the craft shops of shipyards.

Also, there are providers of certain marine materials such as ship construction steels and other metals, joiner materials, piping, paint, welding wire, and electrodes. Although they are not shipyards, most of the methods of economic analysis discussed in this book are applicable to them as well.

Shipyards differ widely in their capacity as measured by the number of ships they are capable of building or repairing during a period of time and also by the maximum size of ships they can handle. Based on the annual report entitled "U.S. Shipbuilding Industry" (see "Report on Survey" in the References), only 19 American yards are capable of building ships of 400 feet long.

SHIPYARD RESOURCES

The principal resources of a shipyard system include the following groups: production facilities, labor force, and production materials.

Production Facilities

Based on their role in the shipyard process, the production facilities might be subdivided into two groups: the first group includes buildings and installations, and the second one contains production equipment.

Buildings and installations. Shipyards are among the most expensive manufacturing enterprises, primarily because the unique production buildings and installations make them different from other manufacturing companies and add substantially to their cost. Examples include:

1. Waterfront structures such as seawalls and deep draft anchorages
2. Shiplifting facilities like graving and floating dry docks, synchrolifts, marine ways, and ship-lifting cranes
3. Outfitting or repair piers equipped with cranes, power and energy lines, and various auxiliary installations needed to complete ship construction after launching or to perform topside ship repairs
4. Inclined or horizontal building ways: concrete paved areas with rails, keel blocks, and heavy-lifting carriages
5. A variety of cranes, including those capable of lifting and installing pre-outfitted ship modules with a weight up to 500-700 tons
6. Warehouses and storage areas for materials and equipment
7. Special transportation means to carry ship modules from assembly sites or painting areas to the erection area on building ways
8. Buildings intended to assemble, weld, paint, and outfit large ship modules and blocks (and in many cases even the entire ship) indoors

In addition to a high initial cost, most of the above facilities are intended to be used for a very long period of time, in many cases over 50 years. Therefore, the following should be considered when dealing with shipyard facilities: high initial cost, and future modernization or expansion costs; likelihood of a costly decision to discontinue use of these installations before the end of their expected economic life; and high cost of dismantling and demolition if future conversion of shipyards into other types of manufacturing facilities is planned.

Another cost factor related to a shipyard is the premium cost of land occupied by the yard due to its waterfront location—the shipyard has to compete with other waterfront activity. Economic decisions regarding future shipyard development or construction of a new shipyard should consider the possible alternative use of land.

Production equipment. The production equipment of shipyards includes the following groups:

1. Production machinery, including various welding and metal-cutting machines, metal forming, woodworking, pipe-bending equipment, etc.,
2. Energy sources and equipment such as electrical generators and motors, air compressors, steam generators
3. Means of transportation and material handling such as cranes, transporters, trucks and lorries, tugboats and barges
4. Measuring and quality control equipment and devices
5. Universal and specialized tooling.

Labor Force

Depending on their functions and method of allocation of their cost, the shipyard workforce may be divided into the following groups: direct production workers, indirect production workers, non-production workers, and management and administrative personnel.

Direct production worker utilization is normally recorded in the time sheets for a specific contract or project, and the cost is included in operational cost as a separate direct item. While the American naval yards use a standard list of uniform crafts, the distribution and titles of worker trades differ among commercial shipyards. The principal shipyard crafts are shipfitters or ironworkers, pipefitters, welders, machinists, painters, carpenters, and electricians.

The *indirect production worker* category includes the labor force of departments other than direct workers; i.e. their cost cannot be directly allocated to specific contracts at the yard. Repair and maintenance personnel, certain transport workers, and personnel of the shipyard power plant are in this category.

The *nonproduction worker* category includes safety and fire prevention personnel and warehousing labor.

As discussed in Chapter 9, the cost of direct production labor is generally understood as a dollar rate per labor-hour. The high cost of labor is one explanation for the decline of U.S. commercial shipbuilding during the 1980s. However, due to a sharp increase in labor costs in Japan and Europe during the late 1980s, America is no longer the highest-paying country in the world. While the average labor rate in 1989 for American shipyards was $15.64 per labor-hour and in Japan $15.82, six European countries had labor rates substantially higher, especially West Germany and Norway with rates over $21 per labor-hour (see Frankel in the References).

The issue of labor productivity is closely related to labor cost. Productivity is normally measured by amount of work done per unit of time or amount of time needed for a unit of work. For example, how much actual work is done during one labor-hour, or, conversely, how many labor-hours are needed to perform a certain job? Another productivity unit is the total amount of labor-hours required per ship. Frankel shows that an average U.S. yard in 1990 required 2.31 labor-hours of direct labor to build an 80,000- dwt tanker for every single labor-hour at a typical Japanese yard. This means that American shipbuilding productivity was 43 percent of Japanese productivity in 1990.

Considering this difference in productivity, the actual direct labor cost in American shipyards, adjusted by the ratio of productivities, appears to be 2.28 times the rate in Japan and perhaps the highest in the world. This example demonstrates that both productivity and labor rate must be considered in the economic evaluation of any endeavor involving change in labor consumption. The productivity difference is normally reflected in the change of labor consumption among the alternatives considered in an analysis.

The costs of labor categories other than direct production labor are included in overhead or labor burden.

Production Materials

Based on their roles in the production process, there are three principal groups of production materials used at shipyards:

1. Direct materials and ship equipment, including all materials and equipment listed in the contract specifications and drawing: construction steel, piping, cables, paint, ship propulsion plant, power plant, electronic, radio and auxiliary equipment, outfitting and accommodation materials and equipment, etc. The cost of basic materials and equipment is directly included in the cost calculation of specific contracts as a separate item. For example, out of $91.2 million cost to build an 89,000-dwt tanker at a typical U.S. Atlantic

Coast shipyard, the direct material cost is over $45 million or approximately 50 percent.

2. Ship-related auxiliary materials, like electrodes, brazing and soldering components. This cost is normally estimated based on the average cost per one labor-hour of the corresponding trade or one running foot of a welding seam, or some similar way. As a result this cost is allocated to a specific contract in proportion to another direct expense, direct labor, or materials. However, when the consumption of an auxiliary material changes as a result of a proposal to be evaluated, the economic analysis should include direct estimation of changes of the cost of those auxiliary materials. For example, if the plasma metal-cutting process is considered to replace the acetylene-based gas cutting, costs of acetylene and electricity per one running foot of cut should be included in the estimates.
3. Non–ship related materials are those used for repair and maintenance of shipyard equipment and facilities, and also fuel and water for the shipyard steam- and power-generating plant, outside electrical energy, etc. The cost of these materials and services is normally included in overhead. However, if affected by a proposal under evaluation, direct cost estimates would be required.

BIDS AND COST ESTIMATING

As discussed earlier, bidding is the usual method for a shipyard to obtain a contract for ship construction or repair. A bid is a price and delivery offer prepared by a shipyard in response to a request for proposal (RFP) or an invitation for bid (IFB) issued by ship operators or shipowners. The contract is awarded based on the prior announced criteria: either to the lowest-price bidder, or to a combination of the price and duration, or any other set of conditions where the price is normally the primary component.

Bids result from a detailed analysis and evaluation of a set of specifications and contract drawings. Normally, the contract documentation—preliminary specifications and drawings—is produced by an engineering firm hired by the shipowner. However, in some cases the shipowner retains a shipyard to provide the preliminary ship design. Obviously, any design activities should be included in the shipyard's cost estimates.

Cost estimations for bids are carried out in accordance with the procedures established by each shipyard. Some American shipyards have adopted the cost classification system developed by MarAd. This classification consists of direct cost of 42 ship systems plus indirect expenses, engineering, overhead and profit charges. The systems are grouped into (1) steel (systems 1 through 3), (2) outfit (systems 4 through 24), and (3) machinery (systems 25 through 42). The cost estimates are developed by creating unit rates for material and labor for each system or group of systems from all information available. Appendix C includes a detailed description of a typical commercial yard cost account system which is used for both cost accounting and cost estimating. Chapter 9 includes a discussion on the estimating method employing shipbuilding cost accounts.

According to the MarAd classification, the definitions for the above direct cost of categories are as follows.

Steel: All structural steel including, in addition to hull steel, all superstructure steel ⅛-inch thick and over, masts, kingposts, rudder, integral hull casting, and foundations.

Outfit: All hull attachment equipment, furnishings, piping outside the machinery space, and all of the electrical generation and distribution systems.

Machinery: The main propulsion machinery, shafting, propeller, and all other equipment, piping, motors in the engine room, except electrical generation and distribution systems, and air-conditioning and refrigeration machinery.

Indirect expenses: Costs not to be confused with overhead, as they are directly chargeable to a specific ship but are not chargeable in the usual sense to the production department (i.e., pipe fitters, shipfitters, electricians, etc.). These charges, which most of shipyards call "direct expenses" or "direct operating expenses" include such items as insurance, classification fees, bonds and royalties, launching expenses, dry dock charges, ship cleaning, watchmen, general material handling, and trial expenses.

Overhead: Three major categories of expense: the first is fixed plant charges, such as maintenance and operation of facilities; second is direct labor charges not associated directly with production work, such as management salaries and shipyard maintenance; the third category is cost associated with labor but not paid to the worker directly, such as Social Security, F.O.A. benefits, unemployment insurance premiums, etc.

Engineering: All engineering services associated with production work, such as engineers' and draftsmen's salaries, plan reproductions, models, and materials purchasing. Laying down and fairing lines is included in the mold loft.

Profit: An arbitrary allowance whose value depends upon the use to which the estimate is going to be put.

In the MarAd system which is intended strictly for bid estimates, overhead is applied as a percent of the direct, indirect, and engineering labor costs. In order to allow for more thorough cost analysis, most of the shipyards use a modified cost-estimating system which might be represented by a formula:

$$C = [\, M + DE + ES + DL \times LR\,(1 + {}^{OH}\!/_{100})\,] \times (1 + {}^{P}\!/_{100}),$$

where C = shipyard price estimate, \$
M = cost of materials for the 42 systems, \$
DE = direct expenses or "indirect" as per MarAd classification, \$
ES = cost of direct engineering services, \$
DL = direct production labor for 42 systems, labor-hours
LR = weighted labor rate, \$/labor-hour
OH = percent of overhead expenses to direct production labor cost, %
P = profit as a percent of total cost estimate, %

Material cost in this case covers only direct charges for material procurement, transportation, and possible storage, but does not include any installation costs. Examples of these material items are steel plates, pipes, and paint, as well as manufactured or partially fabricated pieces of equipment such as main engines, boilers, pumps, and shafting. All of them are embraced by the three groups of systems in the MarAd methodology—steel, outfitting, and machinery. Material charges can also represent subcontract cost involving both materials and subcontract labor. An example would be joiner work, refrigerated cargo space, or air-conditioning control.

The labor cost is estimated separately based on the specific labor-hour unit rates multiplied by the amount of work. These unit rates might be expressed in labor-hours per ton of steel or per pump. Total direct production labor cost is found by applying labor rates to the total labor-hours. In turn, labor rate is an average straight-time hourly rate. Two different labor rates are used, one for the shipyard production workforce and another for engineering. In the MarAd publication (see *Relative Cost of Shipbuilding* in the References), the average hourly labor rates at American naval yards in 1990 were in the range between \$10.72 and \$14.01. If total labor required to build a PD-337 commercial cargo ship is estimated to be 1,971,000 labor-hours, total direct labor cost would be in the range of \$21 million and \$27.5 million out of total production cost of \$123 million, or between 17 percent and 22 percent.

The bulk of expenses is distributed indirectly. The procedure of this distribution is as follows: Based on the historical data and the projected workload for a planned period, normally a year, the shipyard estimates the amount of indirect expenses per planned period and also the predetermined direct labor cost for the same period. A simple ratio gives the percent of indirect expenses or overhead, which in practice means a uniform distribution of overhead expenses proportional to direct labor. For example, if the overhead at a shipyard is 250 percent and the direct labor per a contract is \$500,000, the overhead expenses allocated to this contract would be found from the expression:

$$\text{Overhead amount} = \text{direct labor } (\text{overhead } \%/100)$$
$$= 500{,}000\ (250/100) = \$1{,}250{,}000$$

Example 10-1

A pleasure boat manufacturer establishes the cost of its product based on the following actual costs:

Basic materials (as per specification)	$40,000
Direct operating expenses	1,800
Direct labor	35,000
Subtotal direct cost	$76,800
Overhead (130%)	45,500
Total production cost	$122,300
Profit (15%)	18,345
Total price	$140,645

SHIPYARD CONTRACTS AND PRICES

There are three principal types of shipyard contracts: (a) fixed price, (b) fixed price plus changes, and (c) "time and material."

A fixed-price contract sets the price firmly, allowing for just a few changes which have been agreed upon at the time of the contract award. This type of a contract does not consider any substantial change of work scope, therefore it requires a thorough development of contract documents and bid estimates. For instance, the scope of a standard repair of a certain unit is fixed, regardless of the actual condition of components. The only criterion is the number of hours the unit was in operation.

The most widely used ship construction and overhaul contract is one that provides compensation to the shipyard for contract changes in addition to the original fixed price. The price for changes is established by the shipyard with the owner's approval based on the agreed labor rates and certain unit prices. Examples of the unit prices are cost per one ton of installed metal plate, cost per one running foot of electric cable or pipes, and cost to install certain fittings, valves, joiners, and equipment.

The "material and time" contract is mainly used for emergency repairs and is based on calculations of the actual expenses incurred by the yard. Unit prices might be used if agreed by the yard and the shipowner. The profit rate should also be agreed before the contract commences.

Most shipbuilding contracts include provision to make payments for escalation of labor and material costs during the life of the contract.

JUSTIFICATION OF SHIPYARD EXPANSION AND DEVELOPMENT

The principal directions of capital investments at shipyards are:

1. Development and improvement of buildings and facilities, including construction of new covered ways and docks for ship assembly; improvement of building ways to allow ship movement and access for heavy duty transporters; various specialized facilities like painting areas; renovation and modernization of graving and floating docks.
2. Modernization or replacement of production equipment and systems. Some of the examples might include installation of computer-controlled machine tools; high capacity cranes; production lines for steel preparation, pre-fabrication and painting; automated quality control and testing facilities; and computerized lofting stations. On a larger scale, relative to ship construction yards, most of the proposed changes would be intended to support modern ship construction methods.
3. Improvement of nonproduction facilities such as storage, transportation, and safety equipment.
4. Improvement of technology, organization, and management of specific contracts in ship construction and repairs. These topics are discussed in Chapters 11 and 12.

DEVELOPMENT OF FACILITIES AND BUILDINGS

When a proposal for new facilities or buildings is evaluated, the following considerations should be emphasized. First of all, the initial cost represents the total construction expenses plus profit. This cost might also include, if applicable, demolition costs, repair, and overhaul. Secondly, most shipyard installations are long-life facilities. Therefore, it is vital to properly establish the overall and remaining service life in order to assure a correct decision. Next, in most cases, the impact on operating expenses will be reflected in the overhead, and it is important to identify the specific overhead components which will be affected.

Example 10-2

A new pipe-bending machine is considered to replace a manual, semi-mechanized process. Based on the annual load of the piping shop, total reduction of 5,000 hours would be achieved. Assuming the labor rate is $14.00 per labor-hour, total labor cost saving is expected to be $70,000. However, certain indirect expenses would increase: energy by $2,300 and maintenance and repairs by $3,700. Total production cost savings is $70,000 − ($2,300 + $3,700) = $64,000. It should be noted that a serious error would be committed if the overhead rate were applied to the total labor savings to estimate the savings. At the overhead rate of 130 percent, that would give a result of $70,000 (1 + 130/100) = $161,000, or an error of almost $100,000. In most cases involving new production equipment, certain indirect expenses increase while the labor costs decrease.

Finally, shipyard production time for construction and/or repair might also change as a result of the proposal and should be reflected in calculations.

Example 10-3

In order to enhance utilization of a large old graving dock, it is required to subdivide the dock into two separate work areas and install an intermediate lightweight caisson with integral flooding ducts and valves.

The total construction cost, including building the new caisson and cutting new groove, is estimated at \$7.2 million. Additional operating expenses include the cost of increased docking time (additional six hours per docking), for a total of \$12,000 a year, and additional maintenance and energy, estimated at \$360,000. Assuming 25 percent required rate of return before tax and remaining useful life of 20 years, what minimum additional benefits (profit) would justify the construction?

Using the rate of return method the following equation is applied:

$$\frac{\text{PW benefits}}{\text{PW costs}} = 1$$

or

$$\text{PW benefits} = \text{B (P/A, .25, 20)}$$

$$\text{PW costs} = \text{IC} + (\Delta\text{MC} + \Delta\text{OC})\ \text{(P/A, .25, 20)}$$

where PW benefits = present worth of benefits (profit) during the service life, \$

PW costs = present worth of initial and operation costs during the service life, \$

B = annual benefits (increased profit)

P/A, .25, 20 = present worth factor = 3.9539

IC = first cost = \$7,200,000

ΔMC = additional annual maintenance and energy expenses = \$360,000

ΔOC = additional annual operating cost = \$12,000

$$\text{B} = \frac{\text{PW costs}}{\text{(P/A, .25, 20)}} = \frac{7{,}200 + 372\ (3.9539)}{3.9539} = \$2.193\ \text{million}$$

Based on the results of the above analysis, the suggested modification of the dry dock would be justified only if the additional revenues exceed the additional operational costs by \$2.193 million.

In many cases of economic evaluation the first iteration provides a foundation for further investigation of the other alternatives, which might either improve the financial results or simply make attractive the previously rejected proposal.

Example 10-4

Further analysis of the data in Example 10-3 reveals another alternative which might reduce the required profit and revenues. The alternative considers reduction of annual maintenance by two-thirds during the first 10 years, one midlife overhaul, and an increase of maintenance cost by 50 percent in the remaining nine years. Total cost of maintenance and repairs in this case would exceed that in the example above.

Item	*Cost Value*
Capital cost, \$000	7,200
Additional annual operating expenses, \$000	12
Additional annual maintenance & energy cost first 10 years, \$000	120
Additional annual maintenance & energy cost after 10 years, \$000	540
Midlife overhaul, \$000	1,500
Rate of return, %	25
Remaining service life, years	20

Using the same formula as in the above example, the present worth of costs is:

$$\text{PW costs} = 7{,}200 + 12\ (P/A, .25, 20) + 120\ (P/A, .25, 10) + 1{,}500\ (P/F, .25, 10) + 540\ (P/A, .25, 9)\ (P/F, .25, 11)$$

$$= 7{,}200 + 12(3.9539) + 120\ (3.5705) + 1{,}500\ (0.1074) + 540\ (3.4631)\ (0.0859)$$

$$= \$7.997 \text{ million}$$

$$B = \frac{7{,}997}{(P/A, .25, 20)} = \frac{7.997}{3.9539} = \$2.02 \text{ million}$$

The alternative appears more attractive than the one in the above example because of the reduction of the expenses during first 10 years when the value of money is much higher.

Economic life of shipyard facilities normally exceeds 20 years, therefore the impact of its increase or reduction is not as substantial as it might be expected, and in very many cases its increase might have a negative result.

Example 10-5

Assume that the remaining service life in the previous example is increased from 20 to 30 years by carrying out another overhaul of the caisson in year 20 at a total cost of \$2 million. Obviously, further increase of maintenance cost should be expected. Assume a 20 percent growth of the additional annual maintenance expenses after the second overhaul to the level of \$650,000 (no increase in the year of the overhaul). It is also assumed that the dry dock might also be operated for another 30-year period. Note that the additional 10 years of extra operational expenses ($12,000 a year) should be included in the calculation of the present worth of the cost.

$$\text{PW costs} = 7{,}997 + [650\ (P/A, .25, 9)\ (P/F, .25, 1) + 12\ (P/A, .25, 10) + 2{,}000\ ]\ (P/F, .25, 20)$$

$$= 7{,}997 + [650\ (3.4631)\ (0.8000) + 12\ (3.5705) + 2{,}000]\ (0.1074)$$

$$= \$7.997 + (1{,}801 + 43 + 2{,}000)\ (0.1074)$$

$$= \$7.997 + 413$$

$$= \$8.410 \text{ million}$$

$$\text{PW of benefits} = B\ (P/A, .25, 30) = B\ (3.995)$$

$$B = \frac{8.410}{3.995} = \$2.105 \text{ million}$$

Although the new alternative provides a 50 percent increase of the service life, the financial result is poorer.

Under tight market conditions the shipyard might consider a revision of the minimum acceptable rate of return.

Example 10-6

If in Example 10-4, the analysis had shown that the yard would not be able to get $2.0 million additional revenue under the current system of setting dry-docking fees, and the maximum additional profit is expected to be no more than $1.75 million, this creates a need to find the expected rate of return and to analyze its acceptability under the new conditions. The same basic formula is used except that the interest is unknown:

$$\text{PW of costs} = \text{PW of benefits}$$

$$\text{where PW of costs} = 7{,}200 + 12\ (P/A, i, 20) + 120\ (P/A, i, 10) + 1{,}500 (P/F, i, 10) + 540\ (P/A, i, 9)\ (P/F, i, 11)$$

$$\text{PW of benefits} = B\ (P/A, i, 20) = 1.75\ (P/A, i, 20)$$

In order to find the rate of return, it is necessary to calculate a number of interest rates and to interpolate for the result. The calculations are presented in the table.

Interest Factors & PWs	*Interest Rates* 16	18	20	22	24
P/A, i, 20	5.9288	5.3527	4.8696	4.4603	4.1103
P/A, i, 10	4.8332	4.4941	4.1925	3.9232	3.6819

Interest Factors & PWs	*Interest Rates* 16	18	20	22	24
P/F, i, 10	0.2267	0.1911	0.1615	0.1369	0.1164
P/A, i, 9	4.6065	4.3030	4.0310	3.7863	3.5655
P/F, i, 11	0.1954	0.1619	0.1346	0.1122	0.0938
PW benefits	10,375	9,367	8,522	7,806	7,193
PW costs	8,677	8,466	8,297	8,159	8,046
PW benefits – PW cost	1,698	901	225	–353	–853

Interpolation or a plot of the PW of benefits – PW of costs versus the rate of return will yield a result of 20.7 percent as shown in Figure 10-1.

Increase in capacity of the shipyard facilities very often appears the main objective of the capital investment. Economic analysis related to the subject of capacity is illustrated by the following example.

Example 10-7

A sliding caisson for closing the entrance of the dry dock has deteriorated and should be either reconstructed or replaced. Three alternatives are under consideration:

1. Overhaul of the present sliding type caisson
2. Building two regular single-keel floating caissons (second one to be used as a spare when repairs are made to the first one)
3. Building a new caisson with two keels to allow for caisson repairs to be carried out in place

The following data are used to choose the best alternative:

Item	*Alternatives* 1	2	3
Cost of initial overhaul, $ million	3.6	—	—
Downtime for initial overhaul or caisson installation, hours	2,700	200	200
Cost of midlife overhaul (10 years), $ million	2.8	0.4	0.6
Downtime for midlife overhaul, hours	2,000	—	400
Construction cost, $ million	—	5.0	7.2
Average time for one docking and undocking, hours	0.08	0.5	0.5
Docking crew size, workers	1	9	9
Cleaning and preparation time between dockings, hours	8	8	8
Cleaning/preparation crew size, workers	10	10	10

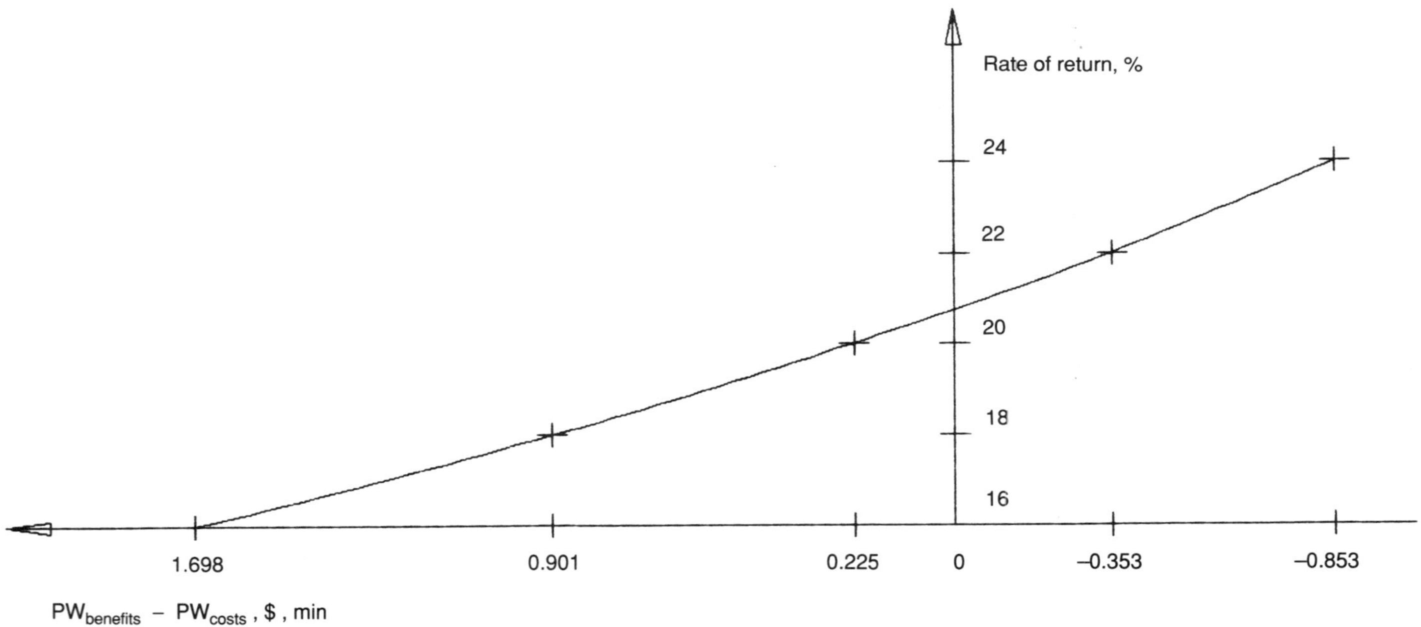

Figure 10-1. Net present worth versus rate of return

Item	*Alternatives* 1	2	3
Average labor rate, $/labor-hour	15	15	15
Labor related burden rate, %	45	45	45
Total available time for operation per year, hours	6,000	6,000	6,000
Average time in dock, hours	200	200	200
Downtime for repairs per year, hours	150	10	30
Additional maintenance per year, $000	200	125	150
Price rate per docking, $000	90	90	90
Remaining service life, years	20	20	20

Differences in capacity between the alternatives are reflected in the number of dockings per year and subsequently in the annual revenues from operation. Number of dockings per year is found using the formula:

$$N_{dock} = \frac{T_{year} - t_{down}}{t_{oper}} = \frac{T_{year} - t_{down}}{t_{rep} + 2\,t_{dock} + t_{cln}}$$

where T_{year} = available time for drydocking per year, hours
t_{down} = downtime for repairs per year, hours
t_{oper} = average time per one drydocking operation, hours
t_{rep} = average time in dock under repairs, hours
t_{dock} = average time for one docking and undocking, hours
t_{cln} = cleaning and preparation time per one docking operation, hours

In order to calculate annual operating cost A_{oper} the following formula is used:

$$A_{oper} = (t_{dock}\,W_{dock} + t_{cln}\,W_{cln})\,r_{lab}\,(1 + \frac{b}{100})\,N_{dock} + A_{maint}$$

where W_{dock} = docking crew size, workers
W_{cln} = cleaning/preparation team size, workers
r_{lab} = average labor rate, $/labor-hour
b = labor related burden rate, %
A_{maint} = additional maintenance per year, $000

The results of calculations are presented in the table:

Item	*Alternatives* 1	2	3
Average time per one operation, hours	208.16	209	209
Number of dockings per year, each	28.10	28.66	28.56
Number of dockings first year, each	15.13	27.7	27.6
Number of dockings 10th year, each	18.50	28.66	26.65
Operational cost per year, $ million	0.249	0.18	0.205

Item	Alternatives 1	2	3
Operational cost first year, \$ million	0.226	0.179	0.204
Operational cost 10th year, \$ million	0.232	0.18	0.202
Revenues per year, \$ million	2.529	2.580	2.570
Revenues first year, \$ million	1.362	2.493	2.484
Revenues 10th year, \$ million	1.665	2.580	2.394
PW costs, \$ million	6.337	11.478	8.928
PW benefits, \$ million	17.511	19.194	19.064

Present worth of additional costs and revenues shown in the table have been calculated assuming 12 percent required rate of return as follows:

$$\text{PW cost}_1 = 3.6 + 2.8\ (P/F, .12, 10) + 0.226\ (P/F, .12, 1) + 0.249\ (P/A, .12, 19)\ (P/F, .12, 1) - (0.249 - 0.232)\ (P/F, .12, 10)$$

$$= 3.6 + 2.8\ (0.3220) + 0.226\ (0.8929) + 0.249\ (7.3658)(0.8929) - 0.017\ (0.3220)$$

$$= \$6.337 \text{ million}$$

$$\text{PW cost}_2 = 5.0\ (2) + 0.4\ (0.3220) + 0.179\ (0.8929) + 0.18\ (7.3658)(0.8929) -0\ (0.3320)$$

$$= \$11{,}478 \text{ million}$$

$$\text{PW cost}_3 = 7.2 + 0.6\ (0.3320) + 0.204\ (0.8929) + 0.205\ (7.3658)(0.8929) -0.003\ (0.3220)$$

$$= \$8.928 \text{ million}$$

$$\text{PW benefits}_1 = 1.362\ (P/F, .12, 1) + 2.529\ (P/A, .12, 19)\ (P/F, .12, 1) - (2.529 - 1.665)\ (P/F, .12, 10)$$

$$= 1.362\ (0.8929) + 2.529\ (7.3658)\ (0.8929) - 0.864\ (0.3220)$$

$$= \$17.571 \text{ million}$$

$$\text{PW benefits}_2 = 2.493\ (0.8929) + 2.580\ (7.3658)\ (0.8929) - 0\ (0.3220)$$

$$= \$19.194 \text{ million}$$

$$\text{PW benefits}_3 = 2.484\ (0.8929) + 2.570\ (7.3658)\ (0.8929) - 0.176\ (0.3220)$$

$$= \$19.064 \text{ million}$$

The best alternative should have a maximum of:

$$\Delta = \text{PW benefits} - \text{PW costs}$$

Alternative 1: Δ = \$11.234 million

Alternative 2: Δ = \$7.716 million

Alternative 3: Δ = \$10.136 million

Alternative 1 under the current circumstances is the best choice.

SHIPYARD EQUIPMENT DEVELOPMENT

The major objective of this type of economic decision is a change in labor consumption. As a rule, certain changes in overhead might occur, most likely energy use or equipment maintenance and repair expenses. On the other hand, some labor-related items, or labor burden, included in the overhead are proportionate to the direct labor: benefits (vacations, insurance, pension fund, holidays), unemployment compensation, and FICA taxes.

In the majority of economic decisions related to equipment development, the following are the economic variables to be considered:

Initial Cost:
1. Cost of added or modernized equipment
2. Research and development costs
3. Salvage cost of equipment which is to be replaced or eliminated

Operational Expenses:
1. Basic and auxiliary materials
2. Direct labor and labor-related costs (burden)
3. Direct operation expenses
4. Indirect expenses (energy, equipment maintenance and repairs, etc.)

Example 10-8

Two alternative plans are being considered for a machine shop modernization lasting five years. Plan A calls for an initial investment of \$350,000 and \$80,000 annually thereafter. Plan B calls for an initial investment of \$450,000 the first year and \$210,000 the second year. The table below presents the variable cost items, assuming constant annual output.

Item	*Plan A*	*Plan B*
Direct labor, \$ million	1.350	1.260
Energy, \$ million	0.075	0.087
Maintenance & repairs, \$ million	0.113	0.126
Labor-related overhead, %	45	45

Assuming that the interest rate is 8 percent, the present worth for both plans is calculated as follows:

$$\text{PW plan A} = 0.35 + 0.08\,(P/A, .08, 5) + 1.35\,(1 + 45/100) + 0.075 + 0.113$$

$$= 0.35 + 0.08\,(3.993) + 1.35(1.45) + 0.075 + 0.113$$

$$= \$2.815 \text{ million}$$

$$\text{PW plan B} = 0.45 + 0.21\,(P/F, .08, 1) + 1.26\,(1.45) + 0.087 + 0.126$$

$$= 0.45 + 0.21\,(0.9259) + 1.827 + 0.087 + 0.126$$

$$= \$2.684 \text{ million}$$

Based on the results of the calculations, plan B is more attractive. However, the final decision regarding implementation of plan B should be taken after comparing it with the existing costs of the "do nothing" alternative and making an incremental analysis if a MARR exists.

Service or economic life of shipyard equipment is directly reflected in calculations as the number of years for interest compounding. When the service life varies among the alternatives or when during the same service life the payment or operation expense schedules are different, the service life might have a substantial impact on the results of the calculations. If possible, service life for the alternatives should be equal.

Example 10-9

Two semiautomatic painting machines are being considered. Machine A costs \$12,000, and has a useful life of five years, a salvage value of \$1,000, and an annual operating cost of \$7,000. Machine B costs \$30,000, and has a useful life of 10 years, a salvage value of \$5,000, and an annual operating cost of \$4,500. Compare their annual costs if the minimum attractive rate of return is 12 percent.

$$\text{EUAC plan A} = 12{,}000\,(A/P, .12, 5) - 1{,}000\,(A/F, .12, 5) + 7{,}000$$

$$= 12{,}000\,(0.2774) - 1{,}000\,(0.1574) + 7{,}000$$

$$= \$10{,}171$$

$$\text{EUAC plan B} = 30{,}000\,(A/P, .12, 10) - 5{,}000\,(A/F, .12, 10) + 4{,}500$$

$$= 30{,}000(0.1770) - 5{,}000(0.0570) + 4{,}500$$

$$= \$9{,}525$$

While evaluating many projects related to facilities development, it is important to consider the compatibility of the alternatives in terms of capacity and productivity of machines or processes compared and quality of manufactured products and their service life. If, for instance, the alternatives include machines with different capacities or life, adjustments must be made to assure rational comparisons using equal life and equal capacities.

The following detailed example illustrates methods of economic analysis of technology development when the capability of equipment changes. Some data from the publication [6] have been used.

Example 10-10

An improvement of the widely employed automatic submerged arc welding process by using metal powder additions is evaluated. Although there are a variety of welding seams used, a weighted average seam has been considered in the example. The data for calculation of the change in the machine capacity is presented in the table:

Symbol	*Item*	*Technology* Old	*Technology* New
A_s	Average seam section area, sq. in.	1.69	1.69
r_w	Specific weight of weld metal, lb./cu. in.	0.283	0.283
m_r	Weld metal deposition rate, lb./hr.	19	28
Top	Total operational time, hrs./yr.	4,000	4,000
ε_{op}	Average machine utilization rate, %	85%	85%
t_m	Manual time (installation, loading), hr. per ft. of weld	0.15	0.18

Average time per foot of weld (t) is calculated as:

$$t = t_a + t_m = \frac{12\ A_s\ r_w}{m_r} + t_m$$

where t_a = automatic machine time per foot of weld
t_m = manual time (installation, loading, etc.) per foot of weld
A_s = average seam section area, sq. in.
r_w = specific weight of weld metal, lbs./cu. in.
m_r = weld metal deposition rate, lbs./hr.

Based on the data from the table the average time per foot of weld is found as:

$$t_{old} = \frac{12\ (1.69)\ (.283)}{19} + .15 = .452 \text{ hr./ft.}$$

and

$$t_{new} = \frac{12\ (1.69)\ (.283)}{28} + .18 = .385 \text{ hr./ft.}$$

Calculation of operational costs is carried out per foot of weld based on the data in the table:

Cost Item	Technology Old	Technology New
Cost of consumables per 1 ft. of weld, $/ft.	7.860	8.150
Direct labor per 1 ft. of weld		
Manual labor, hr./ft.	0.150	0.180
Machine service (50 percent machine time), hr./ft.	0.151	0.103
Direct labor cost (@ $17 per labor-hour), $/ft.	5.117	4.811
Labor burden (47 percent), $/ft.	2.405	2.261
Total material and labor, $/ft.	15.382	15.222

The annual capacity of the welding machine or amount of work per year *Q*, measured in total length of weld, is found using a simple equation:

$$Q = \frac{T_{op}\ \varepsilon_{op}}{100\,t}$$

Old technology allows for:

$$Q_{old} = \frac{4{,}000\ (.85)}{.452} = 7{,}522 \text{ ft.}$$

and the new technology increases the capacity to:

$$Q_{new} = \frac{4{,}000\ (.85)}{.385} = 8{,}831 \text{ ft.}$$

Assume that the increased annual workload calls for 26,000 feet of weld using the submerged automatic welding process. The yard would be forced to buy another welding machine using old technology or to add three metal powder dispensers to the existing three machines in order to employ the new technology.

The cost calculation for the annual program of 26,000 feet of weld is carried out below:

Cost Item	Technology Old	Technology New
Initial cost		
Additional welding machine, $	70,000	—
Additional three powder dispensers, $	—	45,000
Labor and materials, $	399,932	395,772
Additional maintenance and repairs, $	3,900	2,500
Additional energy, $	—	5,000
Misc. additional expenses, $	—	1,500
Total additional annual expenses, $	403,832	404,772

Using the annual cash flow method under the assumptions of 12 percent interest and 10 years' service life, the results of calculations using the two alternatives are:

$$EUAC_{old} = 70{,}000 + 403{,}832\ (A/P, .12, 10) = 70{,}000 + 71{,}478 = \$141{,}478$$

$$EUAC_{new} = 45{,}000 + 404{,}772\ (0.177) = \$116{,}645$$

The results support the introduction of the new technology only if the required volume of work increases. It might be useful to evaluate the maximum cost of three dispensers which will still keep the proposal viable. The maximum cost is found when the alternatives become equivalent or

$$EUAC_{old} = EUAC_{new}$$

Substituting the assumed cost of dispensers of \$45,000 in the above equations, we find:

$$EUAC = 3D + 404{,}772\ (0.177)$$

$$3D = 141{,}478 - 71{,}645 = \$69{,}833$$

$$D = \$23{,}278$$

PRODUCT QUALITY IMPROVEMENT

The subject of product quality deserves a special discussion. Quality relates to newly built or repaired ships as products of shipyards, and to various ship components as products of manufacturers of pumps, engines, electronics, and other ship equipment and spares.

There are two aspects of quality. One reflects the adherence of production departments to established standards or requirements of specifications and drawings. This aspect of product quality is measured as a share of defect-free parts and final products in the entire output. For example, a low level of quality causes increased material and labor consumption due to rework, leads to delays in completion of contracts, draws additional overhead expenses, overloads shipyard capacities, and reduces total output. Improvement of quality requires replacement or modernization of some production equipment to achieve tighter tolerances, and improvement of quality control procedures and equipment. The Zero Defects System, Total Quality Management, and other modern approaches require substantial investment.

Example 10-11

A valve manufacturer is evaluating a program intended to reduce the level of defective final products and spares. The program considers replacement and modernization of production and control equipment and also improvement of the manufacturing processes. Total labor cost per unit increases, but due to reduction of defects, the amount of rework is reduced, which lowers the cost of the products. The principal data are presented in the following table.

Item	*Technology* Old	New
New equipment, $ million	—	0.46
Overhaul & modernization		
During the first year, $ million	0.16	0.41
During the second year, $ million	0.05	0.15
Average equipment service life, years	15	15
Salvage cost of replaced equipment, $ million	—	0.015
Direct labor per unit, $	215	219
Additional controller's labor cost, $	—	1.25
Labor burden, %	50	50
Additional annual energy, $ million	—	0.052
Additional maintenance & repair, $ million		0.065
Total labor losses for rework, $ million	0.157	0.010
Total material losses for rework, $ million	0.171	0.014
Average profit per unit, $	105	105
Total accepted annual output, units	3,000	3,100

Using the annual equivalent cost method and a 10 percent minimum required rate of return, the following calculation procedure could be applied:

$$\text{EUAC}_{\text{old}} = 0.16\,(P/F, .10, 1)(A/P, .10, 15) + 0.05\,(P/F, .10, 2)(A/P, .10, 15) + [3{,}000\,(0.000215) + 0.157]\,(1 + 50/100) + 0.171 - 3{,}000\,(0.000105)$$

$$= \$1.005 \text{ million}$$

$$\text{EUAC}_{\text{new}} = [0.46 + 0.41\,(P/F, .10, 1) + 0.15\,(P/F, .10, 2) - 0.015](A/P, .10, 15) + 3{,}100\,(0.000219 + 0.00000125)\,(1 + 50/100) + 0.052 + 0.065 + 0.010\,(1 + 50/100) + 0.014 - 3{,}100\,(0.000105)$$

$$= \$0.9611 \text{ million}$$

Another direction to aim for in product quality management is improvement of product characteristics and parameters such as capacity, fuel consumption, pressure, and service life. For example, higher-efficiency propellers, pumps, and motors could be considered an upgrade in product quality. As a rule, the economic impact on production is the increased cost of products and additional investment in manufacturing methods, quality control procedures, and production equipment. The manufacturer would be interested in this if either the price goes up or the market expands.

Example 10-12

A pump part manufacturer is planning to improve its production technology in order to substantially increase service life of circulating pump impellers. The existing impellers are sold for \$3,000 each and operational expenses amount to \$2,700. Total initial expenses for R&D, new equipment and control devices are \$700,000 and additional annual operational expenses (labor cost, materials, energy, and fixed overhead) are \$400 per impeller. Assuming a 12 percent minimum attractive rate of return and 10 years' useful life of the project, what should be the minimum price for a new impeller?

Total annual sales amount to 1,000 units. Using the rate of return procedures, the required level of additional annual profit ΔB_a is found as follows:

$$\text{PW of costs} = \text{PW of benefits}$$

$$\text{PW of costs} = \$700{,}000$$

$$\text{PW of benefits} = (\Delta B_a)(P/A, .12, 10) = (\Delta B_a)(5.481)$$

$$\Delta B_a = \frac{\$700{,}000}{5.481} = \$127{,}714$$

Additional profit per improved impeller should be \$127.7, and total profit \$300 + \$127.7 = \$427.7. New production cost is \$2,700 + \$400 = \$3,100, and the price should be not less than \$3,100 + \$427.7 = \$3,527.7.

DEVELOPMENT OF NONPRODUCTION FACILITIES

Normally, this type of project results in the saving of operating expenses related to nonproduction facilities. These expenses are reflected in overhead: indirect labor, cost of management and administration personnel, facilities maintenance, and energy costs. There might also be an impact on the amount of financial resources tied up in production, for example, the cost of materials and equipment in warehouses and storage areas. Some improvements of nonproduction facilities might even affect the duration of the production process, as is the case when lack of storage capacity causes delays in production.

Example 10-13

A complex program of computer-based automation of warehouses is considered. Additional initial costs are \$32,000 for computer hardware, \$85,000 for automated loading/unloading system, and \$25,000 for consulting services.

Cost benefits include auxiliary labor savings (with labor-related benefits) of \$35,000 a year, and elimination of a practice of short-term loans with an average annual interest payments of \$17,000 to cover material shortages. The new system allows for a substantial reduction of material assortment in storage. However, certain increases in other operational expenses would result: maintenance cost would increase by \$7,000 and energy consumption by \$5,700.

Assuming a seven-year useful life, the expected rate of return is found using the following procedure:

$$\text{PW of benefits} - \text{PW of costs} = 0$$

$$\$32{,}000 + \$85{,}000 + \$25{,}000 = (\$35{,}000 + \$17{,}000 - \$7{,}000 - \$5{,}700)\,(P/A, i, 7)$$

From this equation

$$(P/A, i, 7) = \frac{142{,}000}{39{,}300} = 3.613$$

From the compound interest tables the value of i where (P/A, i, 7) = 3.613 is found by interpolation between i = 20% (PW factor = 3.605) and 18% (PW factor = 3.812). The resultant value is 19.9 percent which is much higher than the minimum attractive rate of 12 percent for this yard. Therefore, the proposal would definitely be recommended. The maximum allowed value of initial expenses at 12 percent rate of return is found as follows:

$$\text{PW of costs maximum} = 39{,}300\,(P/A, 12, 7) = 39{,}300\,(4.564) = \$179{,}300$$

JUSTIFICATION OF A NEW SHIPYARD

As in any case of investment in a business, justification of a new ship construction or repair enterprise requires the investor to consider alternative uses of the money, for instance, to buy a baseball team or to build condos on the land where the shipyard is intended to be built. Or, as economists would say, the relevant cost is always the sacrificed alternative, i.e., opportunity cost.

Even before any economic evaluation is started, a detailed market analysis should be carried out. This analysis should concentrate not only on the feasibility of a yard and the type and size of ships to be built or repaired, but even more importantly on the projected yard capacity, or the average number of ships per year. Based on the capacity, the number, sizes, and types of building ways, floating and graving docks, outfitting piers, and production equipment would be determined.

Development of a new shipyard might be one of the alternatives to expansion or modernization of an existing shipyard. Experience shows that in many cases it appears preferable to build a new facility at another location rather than to rebuild an existing yard.

Example 10-14

A shipyard is evaluating a proposal to develop waterfront land adjacent to the existing property. It will serve as the building site for a large series of river-sea container-carrying barges. The land cost is $450,000, construction of buildings and a building way will cost $1.2 million, equipment costs are $400,000, and $200,000 are needed to start production. Total project price for 14 ships is $14 million, and it is planned that two ships will be built per year.

Profit per ship included in the price is $200,000. After the project is completed in seven years, the land with the buildings will be sold to a residential developer for a price of $600,000 and the equipment's salvage value is expected to be $50,000. Assuming a minimum attractive rate of return of 15 percent, determine if the proposal should be accepted.

Initial cost = $2,250,000

Profit = $400,000 per year

Salvage value = $1,080,000

PW of costs = $2,250,000 – $650,000 (P/F, .15, 7) = $2,250,000 – $650,000 (.3759)

= $2,006,000

The required annual profit to repay the initial cost is found as $2,060,000 = B(P/A,15,7), and B = $2,060,000/4.1604 = $482,165, which is more than the projected amount. The proposal would not be accepted unless either annual benefit is increased by reducing the operating expenses, or the minimum attractive rate of return is lowered to the following level: (P/A, i, 7) = $2,006,000 / $400,000 = 5.015 and i = 8.7%.

SHIPYARD EQUIPMENT MAINTENANCE AND REPAIRS

Any project involving maintenance and repair of shipyard equipment has an impact on the overhead costs: repair and maintenance personnel, material, energy, etc. It might also affect the production time and production output if the project reduces the machine down-time.

Example 10-15

A shipyard is considering turning the currently unused portion of the machine shop building into a specialized department for maintenance and repair of the shipyard equipment. The following initial cost would be incurred:

Design = $80,000
Procurement of the additional equipment = $150,000
Construction and modernization work = $70,000
Equipment installation = $9,000
Salvage cost of the replaced equipment = $6,000
Useful life of the new department = 20 years

The workforce of the department will include the mechanics and repair personnel already employed in various shipyard departments. Additional expenses related to operation of new equipment (energy, maintenance, and repairs) are estimated to be $12,000. Total reduction of repair costs due to elimination of certain service contracts with equipment manufacturers, transportation expenses, excessive costs of spare parts, etc., is estimated to be $68,000 per year.

The rate of return is calculated as follows:

$$\text{PW of costs} = 80{,}000 + 150{,}000 + 70{,}000 + 9{,}000 - 6{,}000$$

$$= \$303{,}000$$

$$\text{PW of benefits} = (68{,}000 - 12{,}000)\ (P/A, i, 20)$$

$$(P/A, i, 20) = \frac{303{,}000}{56{,}000} = 5.41$$

The rate of return can be calculated by interpolating between the values of the compound interest factor (P/A, i, 20) for 17 percent and 18 percent as follows:

$$i = 17\% + (5.6278 - 5.41) / (5.6278 - 5.3527) = 17.792\%$$

REFERENCES AND BIBLIOGRAPHY

Frankel, E. "The Path to U.S. Shipbuilding Excellence—Remaking the U.S. into a World-Class Competitive Shipbuilding Nation." *Ship Production,* 1992.

Hassani, J. "Shipyard Facilities—New and Old Closures for Dry Docks." *Ship Production,* Vol. 3, No. 3 (August 1987).

PD-337: Commercial Cargo Ship for Sealift in the Year 2000. Washington: Maritime Administration, 1992.

Relative Cost of Shipbuilding: A Report to the Congress. Washington: Maritime Administration, 1990.

Report on Survey of U.S Shipbuilding and Repair Facilities. Washington: Maritime Administration, 1992.

Thomas, Phillip D. "Automatic Submerged Arc Welding with Metal Powder Additions to Increase Productivity and Maintain Quality." Ship Production Symposium, 1986.

QUESTIONS AND PROJECTS

1. Explain the functional classification of shipyards.
2. Discuss the role of manufacturers of marine equipment and components in the shipyard system.
3. How would you describe the shipyard's resources and their role in the production process?
4. What are the specific features of shipyard facilities that make them different from regular manufacturing enterprises?
5. What are the principal categories of a shipyard labor force? Describe the specifics of allocation of labor cost.
6. Examine the labor productivity problems and the productivity measurement methods.
7. Analyze interrelations between labor cost and productivity.
8. What are the three groups of shipyard materials? Explain the methods of allocation of material cost.
9. Review the shipyard bidding system: request for proposal, bid preparation and estimation, and contract award.
10. Discuss the shipyard cost components and the cost formula.
11. Explain the components of labor cost: labor-hours and labor rates.
12. Give examples of shipyard production cost estimation.
13. Analyze types of shipyard contracts.
14. What are the principal directions of development of the existing shipyards?
15. Review the cost components and procedures of the economic evaluation of shipyard facility development.
16. How would you approach the analysis of the profitability of operating a shipyard with one shift, two shifts, or three shifts?
17. Are there any disadvantages to choosing the maximum output per labor-hour as the measure of shipbuilding productivity? What are they?
18. **Project.** Compare the following alternatives on an annual cost basis and estimate the initial cost level that makes the do-nothing alternative the most attractive.

    A pipe shop of a shipyard is evaluating three alternative proposals related to an old pipe-bending machine: (1) to carry a major overhaul with modernization for $17,000, (2) to buy a new computerized machine for $75,000, or (3) to do nothing but a regular overhaul for $3,000. The remaining relevant data are provided in the table (disregard salvage value):

Item	*Plan A*	*Plan B*	*Plan C*
Service life, years	20	25	5
Number of bends per year, ea.	40,000	100,000	10,000

Direct labor cost per bend, $	2	1.2	8
Labor related overhead, %	50	50	50
Annual energy cost, $	30,000	20,000	40,000
Maintenance and repair cost, $	3,000	8,000	2,000
Misc. operational expenses, $	7,000	25,000	4,000
Engineering cost, $	2,000	7,000	1,000

PROBLEMS

1. Compare the annual cost of two painting semiautomatic machines. Machine *A* costs $26,000, has a useful life of 10 years, a salvage value of $1,500 and an annual operating cost (maintenance, energy, and repairs) of $8,000. Machine *B* costs $40,000, has a useful life of 20 years, a salvage value of $2,000, and an annual operating cost of $10,000. The minimum attractive rate of return is 12 percent.
2. Test pump *A* in the pipe shop of a shipyard has an installed cost of $4,500, no salvage value, an expected life of 5 years, and will cost $900 annually to operate. Pump *B* has an installed cost of $6,500, no salvage value, an expected life of 10 years, and will cost $700 annually to operate. Compare the cost of both pumps on a present worth basis using a 10 percent rate of interest.
3. Two alternative plans are being considered for a machine shop modernization. Plan *A* calls for an initial investment of $300,000 and $50,000 annually thereafter. Plan *B* calls for an initial investment of $250,000 and $55,000 annually thereafter. Assuming that both plans are perpetual and that the interest rate is 8 percent, compare the plans on the present worth basis.
4. Two different dry dock gates are being considered for perpetual service. Type *A* gate has an installed cost of $1 million and an annual maintenance cost of $120,000, and will require major rehabilitation costing $750,000 every 20 years. Type *B* gate has an installed cost of $1.2 million and an annual maintenance cost of $110,000, and will require major rehabilitation costing $800,000 every 30 years. Using a 10 percent interest rate and the capitalized cost method of analysis, compare the present worth of each type of gate.
5. A shipyard forwarding department is purchasing a truck to be rented out to the owners of ships repaired at the yard. Income is expected to be $2,500 annually for the 10 years of its estimated life. Expenses are estimated to be $250 annually. If the purchase price is $20,000 and no salvage value is anticipated, what is the prospective rate of return before taxes?
6. A skid-mounted refrigeration unit with an initial cost of $11,000 has an estimated service life of six years and an estimated salvage value of $1,000. Operation and maintenance expenses are expected to be

$300 per year, and income from rental of the unit to ships under repairs is expected to be $3,000 annually. What is the expected rate of return?

7. An in-port underwater hull-cleaning project requires an initial investment of $78,000 and will have no salvage value at the end of its six years of estimated life. The expected income from the project is $100,000 annually. Expenses for the first year are estimated to be $80,000 with a $2,000 increase every year until they reach $90,000 at the sixth year. Calculate the expected rate of return.
8. Using straight-line depreciation, calculate the annual depreciation of a $200,000 dry dock crane having a service life of eight years and a salvage value of $20,000.
9. Compare two types of insulation for a 300-foot main condensate line. Type *A* insulation costs $1.20 per foot and the annual heat loss is estimated to be $0.33 per foot. Type *B* insulation costs $2.7 per foot and the annual heat loss is estimated to be $0.25 per foot. Assume 12 percent interest rate and 10 years' useful life; use the annual cost method.
10. Compare the annual cost of two types of tank-cleaning machines. Machine *A* costs $15,000, has a useful life of four years, a salvage value of $1,000, and an annual operating cost of $9,000. Machine *B* costs $20,000, has a useful life of six years, a salvage value of $1,800, and an annual operating cost of $8,500. The minimum attractive rate of return is 12 percent.
11. Automatic welding machine *A* has an installed cost of $35,000, a service life of 10 years, and an estimated salvage value of $5,000. Annual operating expenses are expected to be $9,000. Machine *B* has an installed cost of $45,000, a service life of 20 years, and an estimated salvage value of $7,000. Annual expenses are expected to be $8,500; a major overhaul at a cost of $12,000 will be required at the end of 10 years. Using an interest rate of 15 percent, compare the present worth cost of the machines.
12. Determine, using a 10 percent rate and the present worth method, whether it is more economical for an oil terminal to use a 10-inch or 12-inch pipe size for a 1,000-foot heating oil line from a pumping station to a holding tank. The 10-inch line will cost $95,000 to install and the pumping cost will be $22,000 annually. The 12-inch line will cost $110,000 to install and the pumping cost will be $18,000 annually. The life of both lines is estimated to be 20 years.
13. Determine, on a present worth basis, whether it is more economical for a shipyard to lease or purchase a mobile crane. The cost of purchasing the crane is $95,000; it has an expected life of six years, annual operating and maintenance cost of $20,000, and a salvage value of $12,000. The crane can be leased for $28,000 annually.

Operating expenses are expected to be $1,100. Use an interest rate of 12 percent.

14. Assuming a 10 percent interest rate, find the present cost of a concrete slab on the shipyard building way that will be used for perpetual service. The construction cost of the slab is $220,000; it has a service life of 50 years, a replacement cost of $190,000, and annual maintenance cost of $10,000.
15. A contractor is purchasing a machine for installation of nonskid deck coating for $35,000. Net annual income of $6,500 is expected during the 10-year service life. Calculate the rate of return.
16. In the above example, find the level of net annual income required to satisfy the 10 percent minimum attractive rate of return for this contractor.
17. A shop for servicing and repairing marine diesels in port is under development with an initial investment of $375,000. It is estimated that during the 20-year life of the facility, the annual income will reach $150,000 and total average annual expenses will be $105,000. The salvage value of the shop is expected to be $70,000. Find the expected rate of return of the project.
18. A marine compressor manufacturer has improved the design and manufacturing technology related to piston rings of their popular type of air compressor. The annual production of piston rings is 4,000 pieces per year, mainly to sell as spare parts. The production cost data: labor, $1.40 per piece; overhead, 140 percent; material, $1.15. Cost of R&D and additional equipment is $53,000.

    a. Calculate the production cost per one piston ring.
    b. Determine the annual cost of the project.
    c. Estimate the required profit per one piston ring that will make the project attractive at 15 percent interest rate.

19. A marine equipment manufacturer is planning to introduce a total quality management program and expects to substantially reduce losses due to production defects. Total cost of the program including consulting services, R&D, and additional equipment is estimated to be $255,000. The annual reduction of operational expenses is expected as follows: direct labor, $18,000; overhead, $13,000; materials, $9,000; energy, $1,200; and additional cost for quality control and new equipment operation, $8,000. Additional output will produce $12,000 of extra profit.

    a. Determine if the program is economically sound, assuming a 10-year useful life and 15 percent minimum interest rate.
    b. Estimate the rate of return of the project.

20. An improvement to the arc welding technology is considered at a shipyard. New equipment and R&D expenses together will cost

$240,000. Material cost will be reduced by $18,300; direct labor by $21,400; labor related overhead by $9,500. Additional annual cost will be incurred in energy consumption ($4,000) and maintenance and repairs of new equipment ($2,400). Assume the life of the project is 10 years.

a. Determine if the project is justified at 12 percent interest rate.
b. Find the rate of return of the project.

Chapter 11

Ship Construction

For a yard to thrive it is essential that reasonable profit should be made for each contract. This can only be achieved by careful tendering and sound judgment in the first instance and constant check and control of expenditure throughout the construction of the vessel.

—*G.H. Houlden,* N.E.C.I.E.S. Transactions, Vol. 71

SHIPBUILDING IMPROVEMENT OPPORTUNITIES

The National Shipbuilding Research Program, a combined effort of the shipbuilding industry and the U.S. Government, was organized in the 1970s with an objective to improve production organization, management, and technology at American shipyards. The program has since created an array of practical ideas intended to help the shipyards regain their competitive edge by substantially increasing productivity.

A modern ship production system requires continuing investments in new production facilities and equipment, in advanced organization and management systems, and in up-to-date design procedures. The results of these capital expenditures are reduction of ship construction cost and shortened production cycle. A gradual and continuing improvement of the production system, which is being implemented by many American shipyards, appears to be a more pragmatic approach compared to the alternative of a major shipyard rebuilding program.

The general methodology of economic analysis and justification of capital projects, which has been discussed in detail in the previous chapters, is fully applicable to the case of ship construction. However, the individual projects differ in the specific procedures to be applied and in the required data. The variety of projects, both completed and planned, might be categorized based on their objectives and the resultant impact on the economics of production. The specific procedures of economic analysis are more or less uniform for the projects that belong to the same classification. The following is a simplified list of the classifications which incorporate most of the capital projects intended for improvement of the principal components of a shipbuilding system:

1. *Design quality.* Improvement of the design procedures and the resultant technical documentation in order to reduce errors and inconsistencies causing rework, alterations, and work changes. Quality of design also affects the production organization and manufacturing methods, and, as a result, production cost and the production cycle.
2. *Production technology.* Development of efficient manufacturing processes, equipment, and tooling intended to enhance labor productivity, reduce production cost, and optimize the production cycle.
3. *Service and support systems.* Improvement of those shipyard manufacturing activities which serve to support and service the main production process, i.e., material handling and warehousing, maintenance and repairs of the production equipment and facilities, and management of energy. As a rule, projects of this type result in reduction of the corresponding items of the overhead expenses.
4. *Labor quality.* Improvement of qualification of the workforce, creation of labor participatory concepts, development of incentives, and improvement of worker satisfaction. The primary result of these projects should be productivity growth and improvement of product quality.
5. *Product quality.* Improvement of workmanship and other quality parameters of the shipyard products, i.e. new-built or converted ships. The primary result of these efforts is the reduction of ship operation cost and ship operation delays. It may also result in reduced expenses for correction of warranty defects.
6. *Production quality.* Stabilization of the production process in terms of reduction of the percentage of defective parts, subproducts, and products, and the adequacy of production and quality control equipment to satisfy the quality requirements of the final product.
7. *Materials and ship components.* Improvement of the procurement system of material and ship components, including quality, selection process, purchasing specifications, standardization, and incoming quality control.
8. *Production organization and management.* Development of an efficient production organizing, planning, scheduling, monitoring, and reporting system with an objective to improve labor productivity, to reduce production cost, and to enhance the production cycle.
9. *Production cycle reduction.* Most of the projects intended to improve technology, organization, and management methods affect the production cycle. This same result is achieved by improvement in material and component delivery procedures, by subcontracting for fabrication of hull modules and equipment assemblies, etc.

TABLE 11-1. Impact of Projects Intended to Improve Ship Construction on the Elements of Production Cost.

Principal Areas of Improvement Projects / *Elements of Production Cost*	Design Quality	Production Technology	Aux. Prod. Technology	Labor Quality	Product Quality	Production Quality	Material & Components	Organization & Mgmt.	Production Cycle
1. Direct production labor	−	−		±	+	−		−	±
2. Direct material expenses	−	−	−	−	+	−	−		
3. Indirect expenses									
a. Labor burden	−	−		±	+	−		−	±
b. Auxiliary labor		±	±			+	−	+	±
c. Supervisory labor		−		−				+	−
d. Maint. & Repairs		+			+	+			
e. Energy & Utilities		+	±	−	+	+	±		±
f. General & Admin.			±					+	−
g. Engineering labor	±	+			+	+	±	+	+
4. Production cycle	−	−	−	−	+	−	−	−	−

Table 11-1 is a guide to the possible production cost and cycle reductions available from the improvement projects listed under the above categories. Although the table covers the majority of possible cases and conditions, it is not intended as a replacement for a detailed analysis of a specific project.

IMPROVEMENT OF DESIGN QUALITY

The following is a partial list of areas where the efforts to improve design quality are useful:

1. Introduction of modern computer-aided design procedures
2. Improvement of accuracy of the contract documentation
3. Development and utilization of standards in design practices
4. Integration of design and production or "design for production"
5. Zone-oriented design

As a rule, these efforts increase the cost of design and may prolong the design stage. The benefits in the production area are direct labor and material savings and reduction of the production cycle time. The design portion of modern zone-oriented production technology is a good example of such a project. The following is a simplified algorithm of a corresponding economic analysis based on mathematical expressions rather than actual numerical data.

Example 11-1

A shipyard has been awarded a contract requiring the yard to complete the design and construct the ship in three years. The shipyard considers introducing modern assembly and outfitting procedures using the zone approach versus the traditional system approach. This would be based on the modernization of the yard, which is in progress with substantial capital investments. The yard expects to build many other ships using this new technology. The portion of total capital investment in production facilities, equipment, and tooling which is allocated to this contract is distributed as follows:

K_0 = immediate investments
K_1 = investments of the first year
K_2 = investments of the second year
K_3 = investments of the third year

An additional design stage would be required to convert the system-based contract drawings provided by the shipowner into zone-oriented detail documentation. Additional design cost is D, and although the design will last longer than in the traditional case, it will be done by the end of the first year. Additional cost L_{eng} would be incurred during the first year by the shipyard engineering staff.

The primary expected benefit of the undertaking is direct labor saving L_2 during the second year and L_3 during the third year, and a corresponding reduction of the labor-related portion l_o of the overhead. More thorough routing of piping and cables will result in material savings M_2 during the second year and M_3 the third year. On the other hand, additional equipment will require additional maintenance and repair expenses R_2 and R_3 and energy costs E_2 and E_3.

Another primary benefit is a substantial reduction of the production cycle by t days, a portion of which is offset by the increased engineering time. Resultant economic benefits are as follows:

1. Additional profit provided by the contract in the amount of b per each day of delivery ahead of the contract date
2. Saving of management and administration labor of l_a per day
3. Reduction of total fees for use of an outfitting pier, P
4. Increase of drydocking fees due to lengthening of the assembly period before launching DD
5. Saving in the payments of interest on the money borrowed for the performance bond, to cover shortage of resources due to the differ-

ence in the shipyard's spending schedule and the owner's progress payment schedule *q* per each day of reduction of the production cycle.

The present worth method is the most convenient for this type of economic analysis; however, rate of return calculations might also be useful. The criterion for a positive decision on the proposal is $PW_{benefits} - PW_{costs} \geq 0$, assuming *i* as the minimum acceptable rate of return at the shipyard.

$$PW_{benefits} = \left[L_2 \left(1 + \frac{l_o}{100}\right) + M_2 \right] (P/F, i, 2) + \left[L_3 \left(1 + \frac{l_o}{100}\right) + M_3 + P - DD + (b + l_a + q)\, t \right] (P/F, i, 3)$$

$$PW_{costs} = K_0 + \left[K_1 + D + L_{eng} \left(1 + \frac{l_o}{100}\right) \right] (P/F, i, 1) + K_2 (P/F, i, 2) + K_3 (P/F, i, 3)$$

PRODUCTION TECHNOLOGY

Improvement of production technology is the cornerstone of any effort intended to raise productivity and quality and to shorten the production cycle in shipbuilding. The potential projects range from purchasing a more efficient machine tool to completely rebuilding a shipyard for introduction of modern methods of ship construction. A partial list of the principal areas of improvement of the production technology includes:

1. Introduction of modern computer-aided manufacturing processes (lofting, hull and pipe fabrication, hull fitting, joiner work, cable pulling, and ship equipment installation)
2. Development and installation of high efficiency production equipment and tooling, including automatic machines and robots
3. Introduction of modern ship assembly processes, i.e., modular construction with pre-outfitting, zone-oriented pre-assembly, etc.
4. Installation of heavy-weight handling equipment, modern ship lifting and launching facilities
5. Introduction of elements of serial and mass production, e.g., process lanes and group technology
6. Standardization of manufacturing processes, equipment, and tooling

In the majority of the above projects the primary benefit is the direct production labor savings with corresponding reduction of related expenses (benefits, insurance, training). However, there are additional capital expenses for modernization of facilities and installation of new production equipment and tooling, as well as increased costs of energy and equipment maintenance and repairs. As an additional benefit, many projects might result in reduction of the production cycle and/or higher quality of the shipyard products.

Example 11-2

A shipyard is considering upgrading procedures for painting small parts. Three systems with different levels of automation are evaluated and compared with the existing work arrangement: system *A* is the existing, primarily manual operation using airless or air spray painting apparatus and forklifts for material transportation; system *B* employs manual blast cleaning and painting and conveyor drying and material handling; system *C* combines manual and semiautomatic blasting, semiautomatic coating, forced oven drying, and automated material handling; and system *D* includes an automatic airless blast cleaning machine, automatic spray coating, forced drying, and automatic material handling.

The total service life is assumed at 10 years for all alternatives. The basic data for the analysis is presented in the table below:

Items	*Symbol*	*Value by Systems, $* A	B	C	D
Annual production cost	A				
Direct labor cost with related overhead costs	L	600,000	410,000	165,000	110,000
Quality control labor	Q	20,000	20,000	10,000	8,000
Production energy	E	12,000	20,000	40,000	65,000
Repairs & maintenance	R	2,000	6,000	13,000	20,000
Rework	W	10,000	7,000	3,000	1,000
Initial cost	I	—	140,000	270,000	400,000
Equipment salvage value	S	—	10,000	20,000	30,000

Using the annual cash flow method and assuming the minimum acceptable interest rate of 8 percent, the calculations are carried out in accordance with the following formulae and the results are presented in the table below:

$$A = (L + Q + E + R + W)$$

$$EUAC = I\,(A/P, .08, 10) + A - S\,(A/F, .08, 10)$$

Items	*Symbol*	*Value by Systems, $* A	B	C	D
Annual production cost	A	644,000	463,000	231,000	204,000
Compound factor	A/P,.08,15	0.149	0.149	0.149	0.149
Compound factor	A/F,.08,15	0.069	0.069	0.069	0.069
Annual cash flow	EUAC	644,000	483,170	269,850	261,530

Based on the available data, system D is much more effective than the current arrangement, and it is preferable compared to other systems considered. System D is expected to provide savings of $644,000 – $204,000 = $440,000 in annual production costs. A payback period, which is calculated as a ratio of initial cost to the annual benefits, appears to be less than one year.

When evaluating the alternatives of technology improvement projects for the purpose of choosing the best, it is necessary to compare the data for the same physical output. However, measuring the output of a shipyard in physical units is a difficult task. Therefore, it is a common practice to create interim units of output such as the number of parts with weighted dimensions (e.g., a ton of metal installed, a foot of weld, 1,000 feet of cables pulled, etc.) and compare alternatives based on the costs and benefits per these units.

Another important point is the necessity to include all pertinent information in an analysis. The following example will illustrate the sensitivity of the results of economic analysis to the completeness of the data.

Example 11-3

Two modern methods of CNC machine cutting of steel plates into brackets are compared. The limited data, based on production of a certain assortment of metal brackets using one ton of steel plates, is presented in the table:

Method	*Labor Cost, $*	*Material Cost, $*	*Total Direct Cost, $*
Flame cutting	20.93	618.82	639.75
Plasma cutting	9.08	618.82	627.90

This preliminary comparison suggests an advantage for plasma cutting. However, such a conclusion might be misleading if the following factors are not taken into consideration:

1. Repair and maintenance expenses
2. Cost of energy per annual output
3. Machine initial and salvage cost
4. Annual capacity of machine
5. Required annual output
6. Machine effective use (actual operation time)

If only the first two factors are considered, the result will be as follows:

Method	*Direct Cost, $*	*Energy Cost, $*	*Maintenance Cost, $*	*Total Operation Cost, $*
Flame cutting	639.75	8.0	6.0	653.75
Plasma cutting	627.90	20.0	10.0	657.90

Now, the flame cutter appears to be a better choice. The below table presents the remaining relevant data and interim calculations:

Item	*Symbol*	*Flame*	*Plasma*
Weighted cutting time, min./ton	t_c	240	48
Weighted length of cut per 1 ton of plates, ft./ton	l_t	300	300
Average cutting speed, in./min.	V	15	75
Non-cutting time (plate handling, marking, etc.), min./ton	t_n	60	60
Weighted operational time, min./ton	t_o	300	108
Annual available machine time, hrs./year	T_a	1,805	1,710
Total annual machine time, hrs./year	T_y	1,900	1,900
Down time, %	D	5	10
Total machine capacity, tons/year	C	361	950
Service life, years	n	10	10
Initial cost of machine, $	I	80,000	125,000
Salvage cost, $	S	3,000	4,000
Required annual output, tons/year	Q	700	700
Number of required machines	N	2	1
Minimum interest rate, %	i	8	8
Operational expenses, $/ton	a	653.75	657.90

The formulae used in the interim calculations are as follows:

$$t_c = 12 \times \frac{l_t}{V}$$

$$t_o = t_c + t_n$$

$$T_a = T_y \left(1 - \frac{D}{100}\right)$$

$$C = 60 \times \frac{T_a}{t_o}$$

$$N = \frac{Q}{C} \text{ (round to the nearest integer)}$$

Annual cost for each of the alternatives is found using the formula:

$$EUAC = I \times N\,(A/P, .08, 10) + a \times Q - S \times N\,(A/F, .08, 10)$$

$$EUAC_{flame} = 80{,}000 \times 2 \times 0.149 + 653.75 \times 700 - 3{,}000 \times 2 \times 0.069$$

$$= 23{,}840 + 457{,}625 - 414 = \$481{,}051$$

$$\text{EUAC}_{\text{plasma}} = 125{,}000 \times 1 \times 0.149 + 657.90 \times 700 - 4{,}000 \times 1 \times 0.069$$

$$= \$478{,}879$$

And now, finally, the plasma method is proved to be preferable.

Example 11-4

The table below contains closing data for a technology development project aimed at labor savings and production quality improvement:

Item	*Symbol*	*Value*
Initial cost of additional production equipment, $	I_p	135,000
Initial cost of additional control equipment, $	I_c	42,000
Cost of worker training, $	C_t	35,000
Direct labor reduction, labor-hours	H_d	1,100
Hourly production labor rate, $/hr	l_p	15.00
Labor burden rate, %	b	45.00
Control labor reduction, labor-hours	H_c	150
Hourly control labor rate, $/hr	l_c	18.00
Rework reduction—labor, labor-hours	H_r	170
Rework reduction—materials, $	M_r	7,500
Service life, years	n	10
Minimum accepted interest rate, %	i	10

The following formulae are used for the calculations as follows:

$$\text{Labor savings} = (H_d \times l_p + H_c \times l_c + H_p \times l_p) \times \left(1 + \frac{b}{100}\right) = \$31{,}538$$

The resultant calculations:

$$PW_{\text{costs}} = PW_{\text{benefits}}\ ({}^{P}\!/\!_{A}, i, 10)$$

$$PW_{\text{costs}} = 135{,}000 + 42{,}000 + 35{,}000 = \$212{,}000$$

$$PW_{\text{benefits}} = \$31{,}358 + \$7{,}500 = \$39{,}038$$

$${}^{P}\!/\!_{A}, .10, 10 = {}^{212{,}000}\!/\!_{39{,}038} = 5.4306$$

The value of the compound factor 5.4306 corresponds to the rate of return 13.045 percent, which is more than the minimum acceptable rate.

PRODUCT QUALITY

Product quality improvement appears to be one of the most important areas where additional investments might affect the competitiveness of shipyards and related industrial facilities. Among the popular programs, the following ones have the high priority: improvement of workmanship and ship's appearance; improvement of the ship trial memoranda and procedures in order to reduce the warranty problems after ship's delivery; improvement of quality parameters of the ship and its components: service life, periods between repairs, energy consumption, capacity, reliability, repairability, etc., and meeting ISO 9000 standards.

Practically, in order to improve the product quality, the manufacturer is forced to carry additional design and production expenses. Besides enhancement of ship design, principal investments in the production area include, but are not limited to, development of manufacturing processes and quality control procedures, material and component improvement, and standardization. As a rule, the economic savings are realized in the product operation, i.e., by the customer. Practically any improvement of product quality creates an economic and marketing problem for manufacturers, including shipyards. In order to reimburse the additional expenses and also to make the efforts worthwhile, the manufacturer has to raise the price, which is normally not a very popular move. Therefore, economic evaluation of the product quality improvement is done in two stages:

1. The first stage is a determination of the limit price which makes it equally profitable for the manufacturer to produce either the improved quality product or the basic product: i.e., the manufacturer has no incentive to improve the quality of the product. On the other side, the limit price reimburses all additional expenses, therefore the manufacturer does not lose money if it carries out the program.
2. The objective of the second stage is to find out if the market can bear an increase in price that will provide the manufacturer with an incentive, and, at the same time, give the customer a chance to realize additional savings in operation.

A very important feature of the economic analyses involving improvement in the quality of a durable product such as a ship or a ship component is a whole life cycle approach. It generally means that the entire service life is considered when the operating expenses and revenues are analyzed. Another specific feature is a dual-area calculation: separately for the production and for the operation.

Example 11-5

A shipyard is working on improving its fishing boats in order to increase the market. As a result of substantial design and production efforts, the new boats will have increased service life, longer periods between shipyard repairs, lower fuel consumption, and reduction in maintenance expenses. The improved boat is expected to last for a period of at least 15 years, with an average annual time in operation of 250 days, or 3,750 days for its entire service life. The present worth (the first year of production of the improved boats) of the additional design efforts is estimated at $12,000 per boat, and the present worth of the additional production equipment and construction is expected to come to $20,000 per boat. Assuming a 10 percent minimum acceptable interest rate, identify the following: (1) whole life cycle savings in operation of the improved boat, and (2) suggested retail price increase to make profitable the production of the improved boat, as well as its acquisition and operation.

Total increase of production cost and other pertinent data is presented in the table:

Items	*Boat in Production*	*Improved Boat*
Construction cost, $/boat	145,000	162,000
Retail price, $	165,000	?
Days in operation (See table below)	—	—
Operating cost:		
Fuel and lubricants, $/day	275	263
Auxiliary materials, $/day	67	67
Subtotal (See table below)	—	—
Labor, $/year	120,000	120,000
Insurance, fees, etc., $/yr.	78,000	78,000
Maintenance (1 yr.), $/yr.	10,050	9,500
Maintenance growth, %/yr.	10	10
Repairs (See table below)	—	—
Revenue in operation, $/day	1,680	1,680
Annual revenue (See table below)	—	—

Annual revenues, annual operating costs (maintenance and repair, fuel and auxiliary materials) per whole life cycle, and number of days in operation are presented in the table below:

	Repair Cost per Year, $		*Days in Operation*		*Fuel/Auxiliary Materials, $*		*Revenues per Year, $*	
Year	*Old*	*New*	*Old*	*New*	*Old*	*New*	*Old*	*New*
1	1,000	500	250	250	85,500	82,500	420,000	420,000
2	2,500	550	249	250	85,158	82,500	418,320	420,000
3	13,800	8,950	248	249	84,816	82,170	416,640	418,320

Year	*Repair Cost per Year, $* Old	New	*Days in Operation* Old	New	*Fuel/Auxiliary Materials, $* Old	New	*Revenues per Year, $* Old	New
4	3,300	1,800	247	249	84,474	82,170	414,960	418,320
5	35,000	23,000	246	248	84,132	81,840	413,280	416,640
6	4,500	3,000	245	247	83,790	81,510	411,600	414,960
7	4,700	3,200	244	247	83,448	81,510	409,920	414,960
8	22,500	13,500	243	245	83,106	80,850	408,240	411,600
9	5,850	4,350	242	245	82,764	80,850	406,560	411,600
10	55,000	40,000	241	244	82,422	80,520	404,880	409,920
11	7,150	5,550	243	246	83,106	81,180	408,240	413,280
12	7,500	5,850	243	246	83,106	81,180	408,240	413,280
13	48,000	30,000	240	245	82,080	80,850	403,200	411,600
14	8,800	6,100	242	245	82,764	80,850	406,560	411,600
15	9,300	6,500	240	243	82,080	80,190	403,200	408,240

Present worth of whole life cycle operating expenses $PW_{\Sigma-operations}$ is calculated as follows:

$$PW_{\Sigma-operations} = O_{maintenance} + O_{fuel} + O_{labor} + O_{insurance} + O_{repairs}$$

Present worth of maintenance expenses is found by using the geometric gradient formula:

$$O_{maintenance} = A_1\ (P/A, g, i, n)$$

Where $O_{maintenance}$ = present worth of whole service life maintenance expenses

A_1 = maintenance expenses of the first year

Old boat: $O_{maintenance} = 1{,}050\ (P/A, .10, .10, 15)\ (A/P, .10, 15)$

$= 1{,}050 \times 13.6363$

$= \$14{,}318$

New boat: $O_{maintenance} = 950 \times 13.6363$

$= \$12{,}955$

Present worth of repair expenses, and also of fuel and auxiliary materials, is found by applying present worth factor to expenses of every year in operation and adding the expenses for the whole life cycle. The same principle is used when the present worth of revenues is identified. The results of the calculations are presented in the table:

Expenses & Revenues	*Annual Expenses, $*		*Present Worth Factor*	*Present Worth Expenses , $*	
	Old	*New*		*Old*	*New*
Maintenance	—	—	—	143,180	129,550
Fuel/lube	—	—	—	638,352	620,963
Labor	135,000	135,000	7.606	1,026,810	1,026,810
Insurance, etc.	78,000	78,000	7.606	593,268	593,268
Repairs	—	—	—	99,805	65,686
Total expenses	—	—	—	2,501,415	2,436,277
Revenues	—	—	—	3,135,766	3,161,268

Calculation of the present worth of production costs at the shipyard brings the following results:

$$\text{Old technology: PW costs} = \$145{,}000$$

$$\text{New technology: PW costs} = 152{,}000 + 5{,}000 + 9{,}000 = \$166{,}000$$

The present worth of benefits per old technology is equal to the retail price, \$165,000, and the profit is equal to \$20,000. The limit price concept assumes the same ratio of profit to the cost as for the old technology, i.e., 20,000/145,000 = .138. The new technology profit should be \$166,000 × .138 = \$22,900, or the limit price should be \$166,000 + \$22,900 = \$188,900.

Calculation of the whole life cycle costs and benefits in the operation provides the following results:

Old technology:

$$\begin{aligned} \text{PW costs} &= 165{,}000 + 2{,}501{,}415 \\ &= \$2{,}666{,}415 \\ \text{PW benefits} &= \$3{,}135{,}766 \\ \text{PW benefits} - \text{PW costs} &= \$3{,}135{,}766 - \$2{,}666{,}415 \\ &= \$469{,}351, \text{ or } 17.6\% \text{ profit} \end{aligned}$$

New technology, using the limit price:

$$\begin{aligned} \text{PW costs} &= 188{,}900 + 2{,}436{,}277 \\ &= \$2{,}625{,}177 \\ \text{PW benefits} &= \$3{,}161{,}268 \\ \text{PW benefits} - \text{PW costs} &= \$536{,}091, \text{ or } 20.4\% \text{ profit} \end{aligned}$$

New technology brings total savings of \$536,091 – \$469,351 = \$66,740. This amount could be split between the manufacturer and the boat operator by adding to the limit price a certain portion of the additional savings, for instance, \$10,000. This will still keep the profit for the operator on a level higher than the old technology: \$536,091 – \$10,000 = \$526,091/ \$2,625,177 = 20.04%. It will create an incentive for the shipbuilder by increasing the profit to (22,900 + 10,000)/166,00 = 19.8%.

PRODUCTION QUALITY

Principal areas of improvement of the production quality are introduction of modern methods of quality control; modernization and replacement of inadequate equipment and tooling; introduction of TQM and similar quality management systems; installation of modern efficient quality control equipment; and standardization of quality control equipment and methods.

A widely publicized area of improvement of ship production organization is standardization of production processes, management and organizational procedures, product quality requirements, etc. As a result, a certain saving of direct labor might be achieved due to reduction of rework and duplications. An improvement of product quality would make another substantial contribution. Additional engineering effort would be required at the stage of standard development, as would additional or overhauled equipment.

Example 11-6

A company owning several shipyards is considering a program to create standards for materials, production methods, management procedures, and product quality requirements.

The research and development cost of a three-year program is distributed as follows: first year, \$100,000; second year, \$150,000; and third year, \$150,000. These costs include outside engineering assistance and the yard workforce. In addition, the overhaul and modernization of equipment will be done with costs of \$50,000 during the second year and \$100,000 during the third year. New equipment will be bought during the third year for \$220,000. The following benefits are expected: annual labor savings (with labor burden) per year based on this year production program: \$100,000; and annual savings due to reduction of rework (labor and materials): \$30,000.

Using an annual cash flow method at 10 percent MARR and 15 years' economic life, the following calculations apply:

$$EUAC_{costs} = [100{,}000(P/F, .10, 1) + (150{,}000 + 50{,}000)\ (P/F, .10, 2) + (150{,}000 + 100{,}000 + 220{,}000)\ (P/F, .10, 3)\,]\ (A/P, .10, 15)$$

$$= [100{,}000\ (.9091) + 200{,}000\ (.8264) + 470{,}000\ (.7513)]\ (.1315)$$

$$= \$80{,}123$$

$$EUAB_{benefits} = 100{,}000 + 30{,}000$$

$$= \$130{,}000$$

The benefits are larger, therefore the proposal should be adopted.

MATERIALS AND SHIP COMPONENTS

The principal directions of the potentially beneficial projects are:

1. Improvement of the material selection system and an increase in size of the vendor network
2. Improvement of purchasing specifications
3. Standardization of materials and ship components
4. Introduction of incoming quality control of materials and ship components
5. Analysis and improvement of assortment, dimensions, and principal parameters of materials

PRODUCTION ORGANIZATION AND MANAGEMENT

The following are some areas of improvement of the ship construction management and organizational system:

1. Development of computerized production planning, management, and project monitoring procedures
2. Development of product-oriented work breakdown and work packaging
3. Development of computerized database and management information system

A global task of improving organization and management should be addressed to the entire shipyard. However, for this discussion we are interested in the introduction of some elements of modern principles related to ship construction which are the subject of economic evaluation. Generally, this type of a project is undertaken to reduce production time, delays, and disruptions. These efforts will require purchasing of computers and software, conducting studies, training personnel, etc. The following example is an economic evaluation of this type of project.

Example 11-7

A small shipyard has suffered contractual penalties ($17,760 a year) for production delays. It was decided to install a computerized scheduling and monitoring system to handle all projects. Increased productivity (due to elimination of delays and congestion) and corresponding direct labor saving are anticipated. Reduced production cycle will also provide reduction of indirect labor (cranes, firewatch, cleaning, etc.) and also energy saving of $12,000 a year. The relevant data are as follows:

Item	*Value*
Initial cost	
PC computers ($3.750 × 5), $	18,750
Software, $	5,000
Training of personnel, $	3,000
Consultant services, $	15,000
Additional operating expenses	
Salary (one scheduler), $	30,000
Labor burden (40%), $	12,000
Computer maintenance: First year, $	1,500
Annual increase, $	185
Annual benefits	
Direct labor saving, labor-hrs.	2,800
Direct labor rate, $/labor-hr.	13.00
Direct labor cost saving, $	36,400
Indirect labor saving, labor-hrs.	800
Indirect labor rate, $/labor-hr.	10.00
Indirect labor cost saving, $	8,000

Assuming a 10 percent MARR and 5 years economic life, the annual cash flow method will give the following results:

$$\text{Total initial costs} = 18{,}750 + 5{,}000 + 3{,}000 + 15{,}000$$

$$= \$41{,}750$$

$$\text{Additional operating expenses} = 30{,}000 + 12{,}000 + 1{,}500 + 185\,(A/G, .10, 5)$$

$$= 43{,}500 + 185\,(1.810)$$

$$= \$43{,}836$$

$$\text{EUAB} = \text{Annual benefits} = 36{,}400 + 8{,}000 + 17{,}760 + 12{,}000$$

$$= \$74{,}160$$

$$\text{EUAC} = 41{,}750\,(A/P, .10, 5) + 43{,}836$$

$$= 41{,}750\,(.2638) + 43{,}836$$

$$= \$54{,}850$$

$$\text{EUAB} - \text{EUAC} = 74{,}160 - 54{,}850 = \$19{,}310.$$

The proposal might be accepted because it provides a positive EUAB – EUAC.

PRODUCTION CYCLE REDUCTION

Most of the production technology improvement methods affect the production cycle, especially modern assembly methods with zone outfitting, automation, heavy-weight handling equipment, etc. Similar results attend the development of production organization and management. Among other major factors leading to the reduction of the production cycle are:

1. Improvement in material and component delivery procedures, which reduces delays and correspondingly the production cycle
2. Cooperation with other shipyards in ship construction by subcontracting for fabrication of hull modules and equipment assemblies
3. Introduction of modern computerized project scheduling and monitoring systems

The methodological principles related to the above projects have been discussed earlier in this chapter.

QUESTIONS AND PROJECTS

1. What are the specific features of economic justification of projects related to improvement of ship construction?
2. Discuss the classifications which incorporate the majority of the capital projects intended for improvement of the principal components of a shipbuilding system. Provide examples.
3. What are the principal areas of improvement of the quality of ship design?
4. Examine the impact of improvement of ship design quality on the components of the construction cost and on the length of the production cycle.
5. Technology improvement normally leads to a reduction of certain components of the production cost and to an increase in the cost of other components. Identify both types of production cost components, and provide examples.

6. Identify the elements of the additional capital expenses related to the improvement of technology in ship construction.
7. Analyze example 11-2 and suggest the conditions under which system *C* might become the more efficient alternative.
8. Analyze example 11-3 and discuss the impact on the final result of the number of factors being considered.
9. What is product quality? How does it differ from production quality? What is the economic impact of product quality improvement on the financial results of the shipyard?
10. Discuss the significance of creating incentives for the improvement of the quality of products. What is the role of price in creating the incentives? If the price for a higher quality product has been raised, what is the rationale for it?
11. What are the principal components of the improvement of production quality? What are the economic results for a shipyard?
12. How would an improvement of the material supply system affect the economic results of a shipyard operation? Give examples.
13. Discuss the principal areas of further development of the ship construction management and organizational system.
14. Analyze example 11-6 and identify the margins of the changes of the labor cost when the proposal continues to be profitable. How much more investment might be borne?
15. What improvements in a shipyard would affect the duration of the production cycle? How would a reduction of the production cycle impact the construction cost? Discuss the major components of the production cost savings and additional benefits.

PROBLEMS

1. A shipyard is analyzing a proposal to install a CNC machine for the sheet metal shop to replace a manual operation being performed by highly qualified cutters/welders. They estimate that the equipment will cost $320,000 and will last 10 years with a salvage value of $50,000. The annual cost of manual labor to be replaced is $80,000, and the related indirect cost is 70 percent. Other components of annual saving are $7,000 of energy cost and elimination of production losses due to defects in products consisting of $20,000 of material costs and $12,000 of total labor costs or cost of rework. Additional annual production expenses are as follows: maintenance and repairs of new equipment: $11,000, increasing by $1,000 annually; and computer programming: $8,000. The shipyard's marginal tax rate is 35 percent over the period, and minimal interest rate is 10 percent. Estimate the net after-tax annual cash flow, and evaluate the project using the annualized cost method.

2. Two alternative milling machines are being considered for the shipyard's machine shop. Machine A costs \$90,000 with a salvage value of \$25,000 at the end of 10 years of estimated service life. Total operating cost allocated to the machine is \$7,500 per year. Machine B costs \$50,000 with a salvage cost after 10 years of operation of \$5,000 and an annual operating cost of \$12,000. Compare the alternatives, assuming the minimum rate of return is 18 percent.
3. A shipyard is considering submitting a bid for a ship design and construction contract with total cost of \$80 million. The shipowner proposes that the shipyard gets 10 percent of the total contract cost at the contract signing, another 10 percent upon the ship's delivery, and 5 percent a year after delivery when the warranty period expires. The remaining 75 percent would be paid in equal amounts as progress payments for every 10 percent of the physical completion of the ship (see Figure 11-1, the curve of the expected progress of ship completion).

   The total design and construction time is four years. During the first year of the contract the shipyard estimates it will have the following expenses.

Cost of design by the outside designer	\$6 million
Partial modernization of the building ways and tooling	\$3 million
Steel and piping procurement	\$10 million

   The remaining \$50 million are distributed per Figure 11-1. Compute the expected annual cash flow and the return on investment for the entire project. Assuming a minimum interest rate of 12 percent, should the yard submit the bid?
4. A small shipyard is evaluating a project to improve management of the information flow related to delivery, storage, and distribution of materials. The capital cost is estimated as follows:

10 computer stations	\$120,000
Hardware for a local network	\$20,000
Software development	\$25,000
System installation	\$6,000

   The service life of the system is estimated to be seven years, with no salvage value at the end of the period. The system would eliminate an estimated \$70,000 of the cost of clerical labor, including the related indirect expenses. The additional annual operational expenses include \$2,000 for software maintenance, \$10,000 for hardware maintenance and repair, and \$1,000 for additional energy cost. The yard's before-taxes rate of return is 20 percent. Determine if the project is feasible.

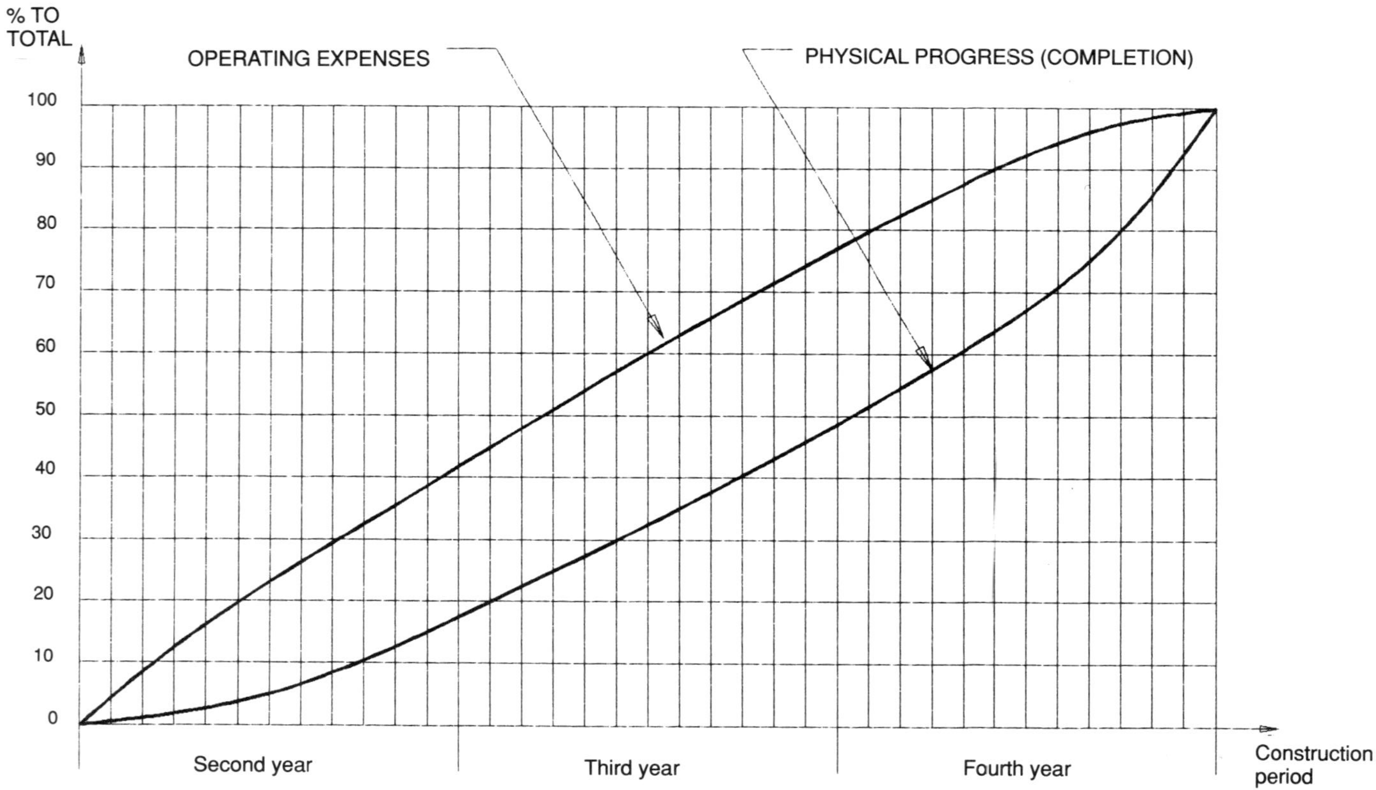

Fig. 11-1. Progress of ship construction and expenses

5. At the request of the shipowner, the shipyard is evaluating a comprehensive change for a ship which is currently under construction. The intention is to improve the operating efficiency of the ship by reducing fuel costs and the maintenance and repair costs. The principal improvements include main engine substitution, change of shell plating material, rearrangement and material change of piping, and improvement of the hull painting schedule with use of new durable paint. The shipyard would have to incur substantial expenses. During the first year of the project, additional design cost of $65,000; equipment and tooling costs of $85,000; materials at $73,000; labor cost, including related indirect expences, of $26,000; and energy and equipment maintenance cost of $6,500. During the second year, there would be an additional labor cost of $12,000; materials cost of $24,000; additional fees for the use of the drydock and the outfitting pier due to a longer production cycle, $27,000; and other production expenses in the amount of $7,000. The ship would be delivered at the end of the second year. The following are the components of the expected savings in operation of the ship: fuel cost saving, $33,000 per year; reduced maintenance cost, $21,000 per year; and reduced shipyard repair cost (average per year), $15,000.

   Determine the whole life cycle saving and estimate the minimum required increase in the ship construction cost. Assume a 15-year ship service life and 20 percent average before-taxes rate of return at the yard.
6. A shipyard is deciding whether to automate certain manufacturing and quality control operations in the pipefitting shop. It is estimated that the cost of equipment would be $230,000, and the installation would cost $80,000. Training of operators would require another $15,000. New technology would provide an estimated saving as follows: saved production labor cost (with labor burden), $35,000; saved labor cost for quality control, $28,000; reduced losses due to defects in products, $36,000. Additional energy and equipment service cost is estimated to be $12,000. Based on a 10-year life of the project with a salvage value of equipment of $15,000, determine if the proposal is efficient. Assume before-taxes rate of return of 17 percent.
7. A shipyard is modernizing its sewage treatment plant. There are three alternatives under consideration, each of them allowing it to process 30 tons of sewage per day. The following data have been included in evaluation:

Cost Item	*Plan 1*	*Plan 2*	*Plan 3*
Equipment cost	$400,000	$250,000	$200,000
Installation cost	$125,000	$100,000	$100,000

Annual operation cost	$ 30,000	$ 45,000	$ 50,000
Service life	20 years	15 years	12 years
Salvage value	$ 25,000	$ 17,000	$ 12,000

Assume the minimum rate of return at the yard is 17 percent. Determine processing cost per ton of sewage for each plan and the most efficient plan.

Chapter 12

Ship Repair and Overhaul

It cost much labor and many days before all things were brought to perfection.

—*Daniel Defoe,* Robinson Crusoe

REPAIRS AS AN ELEMENT OF SHIP OPERATIONS

As is any machine in use, a ship is subjected to constant wear and tear. Unlike most ordinary machines, due to its relative complexity and harsh operational conditions, a ship is exposed to wear that has a profound effect on operation. The effect is twofold: physical and economic. Physical deterioration gradually reduces the ship's capacity, speed, and other performance parameters, increases nonproduction delays, and decreases available time in operation due to malfunction and failures. The economic effect reduces the market value of the ship and causes financial losses due to delays and nonproduction time.

In order to continue normal ship operation and compensate for both physical and economic effect of ship's wear, additional expenses for maintenance and repairs must be incurred. It makes maintenance and repairs a vital and necessary element of ship operations, together with physical movement of cargoes and passengers, with loading and unloading in ports, or with other activity for which the ship has been designed.

Maintenance of ship's components, including the hull, machinery, and equipment, consists of routine procedures involving inspection, lubrication, painting, gland repacking, adjustment, etc. Specific maintenance procedures are normally prescribed by the shipbuilder, the classification society, and the manufacturers of ship equipment. It is generally understood that ship's maintenance is carried out while the ship is underway or in port, i.e., without taking the ship out of operation. Therefore, ship maintenance is discussed as a ship operation subject in Chapter 13.

Ship repairs are carried out in order to replace worn-out elements of hull, ship systems, machinery, and equipment, and the refurbishment or replacement of whole machinery and equipment units. They are classified as voyage, emergency, and planned periodic shipyard repairs. Major voyage and turnaround repairs are normally done during a voyage or while in port by a riding team of repair workers or professional contractors, with or without participation of the ship's crew (a topic that will be discussed in Chapter 13). Emergency repairs are caused by accidents (collisions, fires, groundings, etc.) or by breakdowns or malfunctions of the equipment.

Planned shipyard repairs are performed on a periodic basis with preplanned intervals between repairs, and the scope and the content of the repair is also preplanned. During the 1980s, a series of ship management initiatives including planned on-board maintenance, contractor-provided riding maintenance crews, adoption of continuous surveys, use of inwater equipments and divers, and improved coatings began an ongoing reduction of the work undertaken in shipyards. However, shipyard repairs still remain an important element of ship operation and a principal procedure for periodic upgrading of the ship's condition to satisfy the operational and classification requirements, and/or to correct damages caused by accidents.

Planned periodic shipyard repairs are executed in accordance with repair specifications prepared in advance. If a company operates more than one ship of a certain design, standard repair specifications are used. These specifications are based on the project documentation and the data on previous repairs of ships of the same class, and contain a standardized list of repair items typical for the given type of repairs. The list includes replacement of the deteriorated elements and certain preventative operations. Obviously this list contains more work items than will be eventually assigned to a particular ship. The principal data sources for the individual repair specifications are:

1. Analysis of wear of machinery components based on hours in operation and on statistical data,
2. Results of inspections carried out by the crew and based on the continuous machinery and hull survey procedures which have been set by the classification society,
3. Outstanding items per classification survey reports,
4. Any defects found by the port engineer and the crew during survey.

The scope and the goal of a planned shipyard repair differ, depending on the age of the ship and special considerations of the shipowner. More comprehensive repairs are carried out every four to five years together with a special classification survey. Every two or two-and-a-half years an intermediate survey is performed and the corresponding repairs are reduced in scope. At the ship's midlife a major overhaul might be scheduled if the shipowner decides that a substantial expenditure is warranted. Another type of comprehensive shipyard work is life extension repairs which allow continued ship operations for a certain period of time. The most extensive repair work involves ship conversion, with or without overhaul. This type of work is carried out when the owner wishes to change the assignment of the ship.

Shipyard repairs provide a twofold impact on ship operation: (1) by affecting the operating cost and (2) by increasing the nonproduction time of the ship, and, as a result, operating losses in the amount of cargo moved or services rendered. Therefore, the shipowner is vitally interested in both reduction of duration of repair and its cost. On the other hand, the shipyard is interested in increasing repair cost and duration if the contract price is based on actual expenditures, and in reducing both if the contract is fixed-price.

Relative to the shipyard repairs, the following are the principal engineering economic analyses required:

1. By the shipowner:
   a. How much work should be done during a certain repair?
   b. What is more efficient: to use a riding team, or to give the work to a shipyard? How much work should be given to a riding team?
   c. Should a certain unit of machinery or equipment be repaired or replaced?
   d. How long should the ship be used? How much should be spent on repairs? When should repairs be done?
   e. What is the preferred action at a certain period in ship's life: to overhaul or to replace?
   f. Should the shipowner finance extension of the ship's life or scrap the ship?
   g. How much work should be done during an emergency repair and how much should be left for the next planned shipyard repair?
   h. Should the ship be modernized or scrapped and replaced?
2. By the shipyard:
   a. How effective would a capital expenditure project be that is intended to improve the shipyard's organization, management, and/or technology?
   b. What should the proposed price be for a contract bid?
   c. Should the shipyard production machinery unit be repaired, modernized, replaced, or scrapped?

REPAIR COSTS AND PRICES

Ship repair prices are based on the specifications and on the type of the contract. Fixed contract prices are based on the standard specification and are normally applied to certain complex work which is repetitive and clearly identified. An example of such work might be a modernization item or standardized scope of repairs prescribed by a manufacturer or by a classification or statutory institution.

The "fixed plus" contracts assume a fixed price for the core of repairs as per the repair specification plus a negotiated price for the changes and additional work, i.e., work changes. Emergency and any other unscheduled repairs are priced on a "labor and materials" basis which means actual expenses and profit. The structure of the shipyard prices has been discussed in Chapter 10.

Relative uncertainty regarding the work scope of shipyard repairs, especially unscheduled repairs, puts an additional burden on the shipyard cost estimator. After receiving the repair specifications and other materials from the shipowner, the shipyard estimator uses the following cost rates to estimate the cost of the contract:

1. Accounting data, cost returns in labor hours, and estimates on the similar jobs carried out during the previous repairs of the comparable ships;
2. Average unit rates in labor-hours per major standard operations, for example, per one ton or 100 pounds of steel, per running foot of pipe, per 100 square feet of the hull to be painted, etc.;
3. Material requirements and rates based on the drawings, accounting records, and data from the previous repairs;
4. Labor rate which was quoted to the shipowner in the competitive bid or upon a special request. This rate is created as a composite figure containing an average flat labor rate which includes all crafts, labor-related expenses (insurance, social security payments, vacations, etc.), overhead, and profit. This labor rate is a billing rate which will be used later for estimating additional work or work changes, in the same way as the unit rates will be used. Material expenses are included in the price estimate as per invoices, plus the shipyard's mark-up.

SHIP REPAIR COMPETITIVE BIDDING

It is generally recognized that shipowners will repair ships in the yard which has the lowest total cost for the repairs. This includes the owner's expense to bring the ship to the yard and other subjective and quantifiable factors which should be considered in establishing total cost.

There are several approaches used in ship repair yards to establish prices and solicit business from owners. Competitive bidding appears to be the most widely accepted procedure. As shipowners recognize the benefits of requesting competitive bids based on preparing detailed repair specifications, the sealed bid marketplace for ship repair and conversion will grow.

There are a few special situations in which the competitive bidding is either not advantageous or simply cannot be arranged. For example, when there is only one contractor available who is qualified to perform the specific repairs, no competitive condition exists and the contract price is set during negotiations. Another nonbidding situation occurs when time is of more value than the repair cost only, as in the case of emergency repairs when the similar replacement tonnage is not readily available. An even more important consideration against competitive bidding is the case wherein a unique job must be done, and there is a shipyard which is well known by past experience, whose technical competence related to this job outweighs the possible price difference that might result from competitive bidding.

In order to develop a competitive bidding environment, the following conditions should exist:

1. A list of qualified contractors who are reliable and capable of performing the repair work is created.
2. Detailed repair specifications and appropriate drawings and procedures have been prepared.
3. Preliminary cost estimates of the planned repairs have been made by the shipowner's engineering department.
4. Instructions and procedures for preparation and submission of competitive bids by the prospective contractors have been developed.

After the competitive bids have been received by the shipowner, they are thoroughly studied for adequacy to the procedures which have been announced in advance, and classified and ranked by the cost quoted. The lowest bids should be analyzed in their details for errors and possible pricing manipulations and underbidding.

The Competitive Bidding Marketplace

When a true competitive bidding marketplace exists, the astute shipyard has an opportunity to collect data, perform analyses, and prepare optimum bids for future requests from shipowners. The bids will be optimum in the sense that the shipyard will have the highest possible expected income for a given price.

The true competitive bidding marketplace exists when the following characteristics are present:

1. There are a relatively large number of opportunities to bid competitively each year, i.e., more than 30.
2. The same shipyards submit bids for most of the bid requests.
3. All yards bid to the same detailed specifications prepared by the owner.
4. All bid results are made public.
5. Only qualified bidders are allowed to bid.

6. There is no collusion among bidders.
7. The low bidder is awarded the contract for the repair work.
8. The shipyard interested in an analysis must prepare an out-of-pocket cost estimate for each bid opportunity included in the analysis.

A Ship Repairer's Approach to Optimum Competitive Bidding

A general model for optimum competitive bidding has been suggested by Churchman, *et al.*, and is refined here based on the authors' successful experience. Let C equal the estimated out-of-pocket cost for fulfilling a ship repair contract: that is,

$$C = L_d(1 + Br) + M + L_p$$

where: L_d = cost of direct labor
B_r = direct benefits cost as a fraction of direct labor
M = direct material cost
L_p = labor premiums, such as overtime and holiday pay

For a given competitive bid, let $r_1, r_2, \ldots r_n$ equal the ratios of the competitors' bid prices to the analyst's out-of-pocket cost for each bid. If, over a period of time, these ratios are collected for the published bid situations, it is possible to determine $f(r_1), f(r_2), \ldots f(r_n)$, the probability density functions for each of the competitor ratios. These density functions encode the historical bidding patterns for each of the potential competitors. Figure 12-1 illustrates three typical competitor ratio distributions.

Let x equal the amount bid for a specific contract. If x is the lowest price quoted, the contribution margin will be $x - C$. If $f(r_1), f(r_2), \ldots f(r_n)$ are the

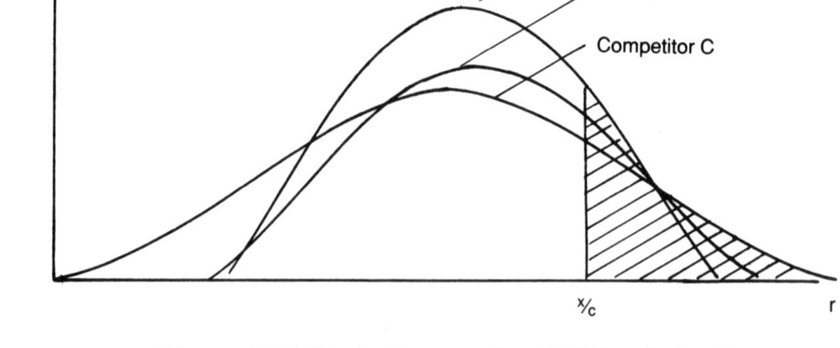

Figure 12-1. Typical competitor bidding behavior

probability densities for the specific bidders' price to the analyst's out-of-pocket cost, then the probability of the analyst's bid being equal to or lower than the competitor's bid is

$$\int_{\frac{x}{c}}^{\infty} f(r)\, dr$$

Case I

If it is known beforehand exactly who will bid in a competitive situation, then the probability of x being the low price is

$$P(x) = \prod_{a=1}^{n} \int_{\frac{x}{c}}^{\infty} f(r_a)\, dr_a$$

and the expected contribution margin C_m is

$$C_m = (x - C)\, P(x)$$

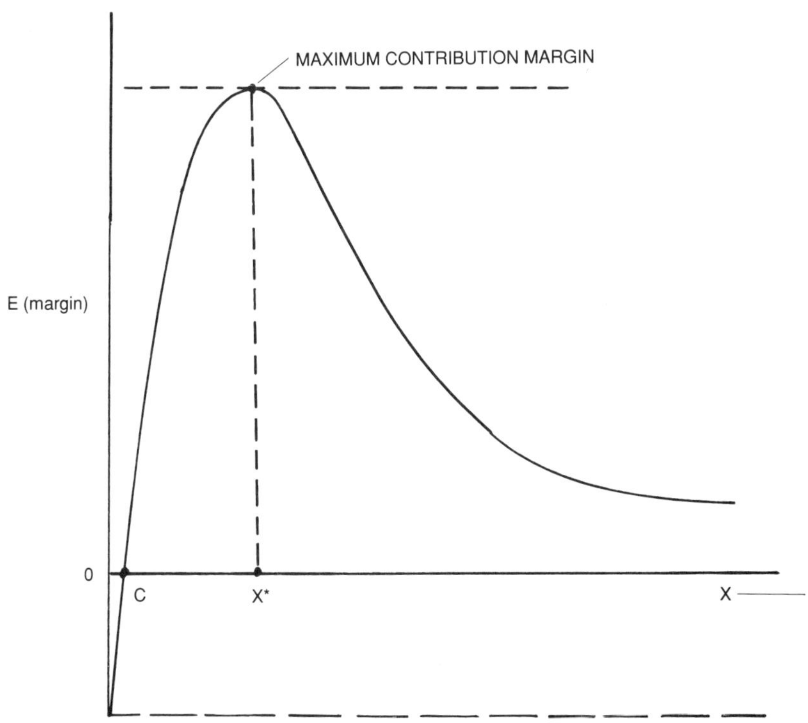

Figure 12-2. Optimum bid price x occurs at maximum contribution margin*

which can be calculated for increasing values of x. Figure 12-2 is a plot of a typical expected contribution margin versus the bid price x.

Case II

If the specific bidders are unknown, but the numbers of bidders n is known, then all the information contained in the bidder-specific r ratios may be combined into a single average ratio to determine $f(\bar{r})$, the probability density function of the average bidder's bid to the analyst's out-of-pocket cost. Then the probability of x being the low bid price is

$$P(x) = \left[\int_{\frac{x}{c}}^{\infty} f(\bar{r})\, d\bar{r}\right]^{n}$$

and the expected value of x for each bid price considered is the same as Case I.

Case III

It is most likely that neither the specific bidders nor the number of bidders will be known to the analyst. In this case, the probability of a given x being equal to or lower than k average bidders is

$$P(x) = \left[\int_{\frac{x}{c}}^{\infty} f(\bar{r})\, d\bar{r}\right]^{k}$$

where k is a random variable. If the probability density function of k is $g(k)$ then

$$P(x) = \sum_{k=0}^{\infty} \left[g(k) \left(\int_{\frac{x}{c}}^{\infty} f(\bar{r})\, d\bar{r}\right)^{k}\right]$$

In all cases, $f(r_a)$ is found by fitting the collected ratio data for specific competitor-published bid prices to an appropriate two-parameter distribution where $0 < x/c < \alpha$ and $f(\bar{r})$ includes the data for all bidders. Churchman, *et al.*, recommends that r or x/c be fitted to a gamma distribution and that k be considered a Poisson distribution.

Therefore,

$$f(r) = \frac{\alpha^{(\beta+1)}}{\beta!}\, r^{\beta}\, e^{-\alpha r}$$

and

$$g(k) = \frac{\lambda^{k} e^{-\lambda}}{k!}$$

Therefore,

$$P(x) = e^{-\lambda} \sum_{k=0}^{\infty} \left\{ \frac{1}{k!} \left[\lambda \left(\int_{\frac{x}{c}}^{\infty} \frac{\alpha^{(\beta+1)}}{\beta!} r^{\beta} e^{-\alpha r} dr \right) \right]^{k} \right\}$$

$$P(x) = e^{-\lambda} \exp\left[\lambda \left(\int_{\frac{x}{c}}^{\infty} \frac{\alpha^{(\beta+1)}}{\beta!} r^{\beta} e^{-\alpha r} dr \right) \right]$$

$$P(x) = \exp\left[-\lambda \left(1 - \sum_{i=0}^{\beta} \frac{1}{i!} \left(\frac{\alpha x}{c} \right)^{i} e^{\frac{-\alpha x}{c}} \right]$$

where $\alpha > 0$ and $\beta > 0$ are gamma parameters and λ is the Poisson parameter.

Since the derivative of P(x) (x-c) does not exist in a closed form, the optimum bid may be determined by plotting the expected contribution margin versus a range of the analyst's bid prices to select the highest expected contribution margin by inspection as shown in Figure 12-2.

The mean number of bidders *k* can be calculated from the historical bidding database and used as the estimated parameter for *g(k)*. The mean bidders *k* may be refined by the potential observation that the estimated *k* is a function of the size of the contract: that is, large contracts attract more bidders. Regression analysis could be used to get a better estimate of *k* if a functional relationship exists.

The authors have found that fitting recent historical data (i.e., the past six months) for the ratio *r* to a Weibull distribution provides better results than the gamma distribution for the sealed-bid ship repair marketplace. Regardless of the distribution selected for *f (r)*, a statistical goodness-of-fit test should be made to assess the adequacy of the fit.

If the two-parameter Weibull distribution is used for *f (r)* or $f(\bar{r})$, then the analysis proceeds as follows:

$$f(r) = \left(\frac{\beta}{\alpha}\right) \left(\frac{r}{\alpha}\right)^{\beta-1} e^{-\left(\frac{r}{\alpha}\right)^{\beta}}$$

$$\text{where } F(r) = 1 - e^{-\left(\frac{r}{\alpha}\right)^{\beta}}$$

$$\text{mean}(r) = \alpha\, \Gamma\left(\frac{1}{\beta} + 1\right)$$

$$\text{variance}(r) = \alpha^{2} \left\{ \Gamma\left(\frac{2}{\beta} + 1\right) - \left[\Gamma\left(\frac{1}{\beta} + 1\right) \right]^{2} \right\}$$

and *g(k)* is the same as above. Then

$$P(x) = e^{-\lambda} \sum_{k=0}^{\infty} \left\{ \frac{1}{k!} \left[\lambda \left(\int_{\frac{x}{c}}^{\infty} \left(\frac{\beta}{\alpha}\right) \left(\frac{r}{\alpha}\right)^{\beta-1} e^{-\left(\frac{r}{\alpha}\right)^{\beta}} dr \right) \right] \right\}$$

$$P(x) = e^{-\lambda} \sum_{k=0}^{\infty} \left\{ \frac{1}{k!} \left[\lambda\, e^{-\left(\frac{r}{\alpha}\right)^{\beta}} \right]^{k} \right\}$$

$$P(x) = e^{-\lambda}\, e^{\lambda e^{-\left(\frac{r}{\alpha}\right)^{\beta}}}$$

where α is the Weibull scale parameter, β is the Weibull shape parameter, k is the number of bidders, and the optimum bid x_{opt} is the ratio r which provides the highest expected value of $x - c$, the contribution margin,

$$\max_i \; P(x_i)\; (x_i - c)$$

Two computer programs should be written for the regular use of this method of finding an optimum bid price. The first program should be designed to fit the r or x/c ratios for each competitor and for all competitors combined to the Weibull distribution function using an appropriate least squares algorithm. For each $f(r)$ and $f(\bar{r})$, the program would output the following information:

α = parameter of $f(r)$

β = parameter of $f(r)$

$\bar{\mu}$ = estimated mean of $f(r)$

$\overline{\sigma^2}$ = estimated variance of $f(r)$

λ = parameter of $g(k)$

This program would be run after each bid situation to update the historical data by adding new bid results and removing old bid results as appropriate.

Ranking competitors by the mean of their r ratio density function provides a comparative measure of the competitors' pricing aggressiveness. The lowest mean value of r would identify the most aggressive bidder.

A second computer program would be designed to calculate the optimum bid price for the analyst in a given competitive situation. The input to such a program would be:

c = out–of–pocket cost estimate

r = range and increment of r ratios to investigate

α = Weibull scale parameter

β = Weibull shape parameter

λ = Poisson parameter

If profit is to be calculated, the information to determine total cost must be included.

The output calculated for each ratio r in the range of interest might be:

$f(\bar{r})$ = distribution function

$F(\bar{r})$ = left tail cumulative probability of r

$1 - f(\bar{r})$ = right tail cumulative probability of r

$P(x)$ = probability that x is the low bid price

x = bid prices in the range of interest

$x - c$ = contribution margin

c_t = total cost

$P(x)(x - c)$ = expected contribution margin

$P(x)(x - c_t)$ = expected profit

Typical functions of $f(\bar{r})$ and $F(\bar{r})$ are shown in Figure 12-3.

Example 12-1

Assume that the competitive data have been collected for sealed bids over a period of six months, that more than 100 ratios r have been calculated, and that all of the potential competitors are included in the sample. The information has been fitted to the Weibull distribution and the two parameters are found to have the following values:

$$\beta = 4.57358$$

$$\alpha = 1.47931$$

The 100+ results occurred in approximately 20 competitive bid situations so that the Poisson parameter may be estimated as $\lambda = 5.2222$.

For the ratio $r = 1.18$, calculate the expected contribution margin for an out-of-pocket cost determined as follows:

Direct labor and benefits	$1,960,314
Direct material	$2,134,904
Labor premiums	$6,700
Out-of-pocket cost	$4,101,918

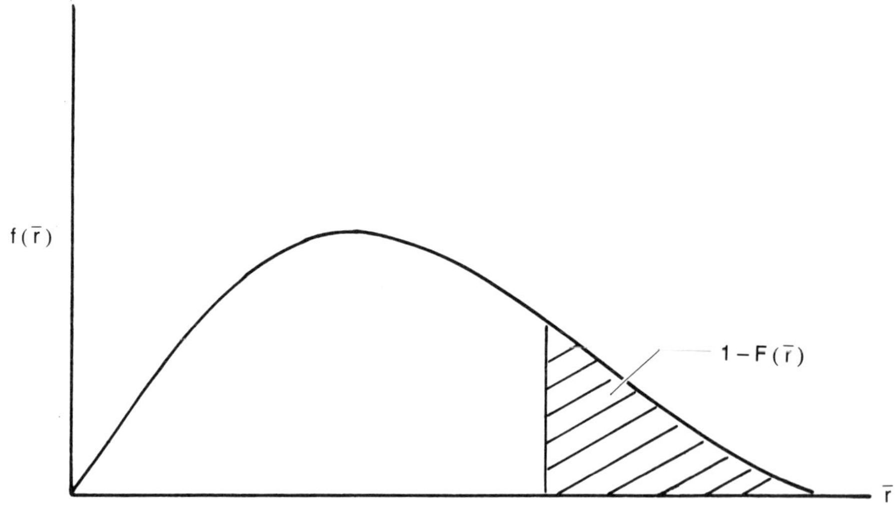

(a) density function curve

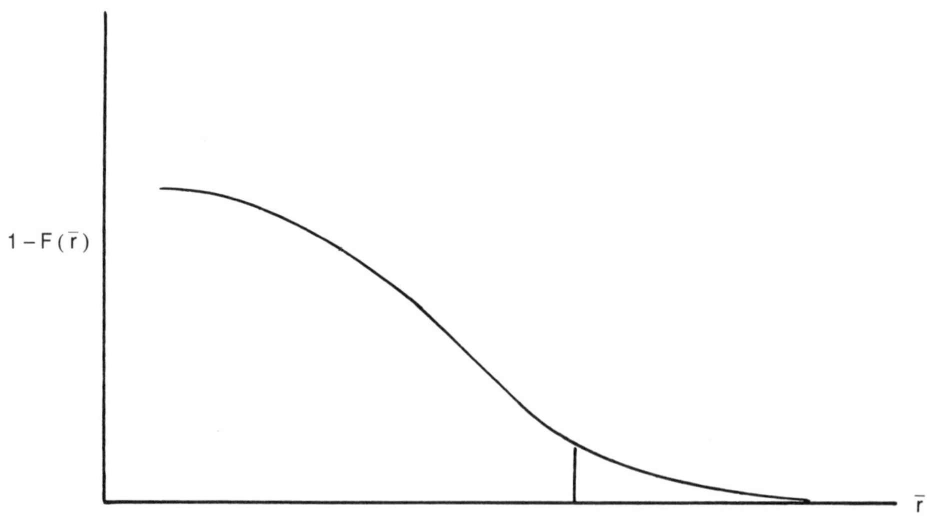

(b) probability curve

Figure 12-3. Typical density function and right tail cumulative probability curves

Bid price at this out-of-pocket cost is:

$$x = c\,(x/c) = 4{,}101{,}918\ (1.18) = \$4{,}840{,}263$$

$$1 - F(\bar{r}) = e^{-\left(\frac{r}{\alpha}\right)^{\beta}} = e^{-\left(\frac{1.18}{1.4793}\right)^{4.57358}} = .70074$$

$$P(x) = e^{-\lambda}\, e^{\lambda(1-F(\bar{r}))} = e^{-5.222}\, e^{5.227\,(.70074)}$$

$$= (.005396)\ (38.838) = .20959$$

$$\text{Expected contribution margin} = P(x)\ (x - c)$$

$$= (.20959)\ (4{,}840{,}263 - 4{,}101{,}918) = \$154{,}749$$

If the analyst's shipyard total cost is $5,177,133, the expected profit at r = 1.18 is a loss of $70,604. To find the optimum bid price, the expected contribution margin for increasing values of *r* must be calculated until the maximum margin is located. Table 12-1 illustrates the determination of optimum price to out-of-pocket cost for ship repair bid.

TABLE 12-1

Determination of Expected Margin in a Ship Repair Competitive Bid Scenario for Four Values of *r*

r	1.18	1.19	1.20	1.25
f(r)	.96778	.98351	.99881	1.06782
F(r)	.30003	.30979	.31970	.37142
1 - F(r)	.69997	.69021	.68030	.62858
P(x)	.20871	.19834	.18834	.14376
price	4,848,263	4,881,282	4,922,301	5,127,397
margin	738,345	779,364	820,383	1,025,479
profit	–336,869	–295,850	–254,831	–49,735
E(margin)	154,749	154,581	154,508	147,420
E(profit)	–70,604	–58,679	–47,994	–7,149

The optimum expected contribution margin occurs at about r = 1.195, which is located by plotting bid price versus expected margin as shown in Figure 12-2. Figure 12-4 is a computer program output for the optimum bid in a competitive situation. The optimum bid is indicated by a change in slope of the expected margin marked by the asterisk in the output tabulation.

SHIP REPAIRS AND SERVICE LIFE

To repair or to replace? The need to make this decision follows the owner of any machine during the entire life of the unit, especially at the time of

ratio	f(r)	F(r)	1-F(r)	P(x)	price	margin	profit	E(margin)	E(profit)
0 90000	0 47361	0 09818	0 90182	0 59885	3691725 8	-410191 8	-1485407 2	-245645 3	-889543 1
0 91000	0 49005	0 10300	0 89700	0 58397	3732745 0	-369172 6	-1444388 0	-215587 5	-843486 3
0 92000	0 50674	0 10799	0 89201	0 56897	3773764 2	-328153 4	-1403368 8	-186710 2	-798478 0
0 93000	0 52366	0 11314	0 88686	0 55387	3814783 3	-287134 3	-1362349 7	-159034 7	-754563 1
0 94000	0 54079	0 11846	0 88154	0 53869	3855802 5	-246115 1	-1321330 5	-132579 0	-711783 8
0 95000	0 55814	0 12395	0 87605	0 52345	3896821 7	-205095 9	-1280311 3	-107357 5	-670179 1
0 96000	0 57567	0 12962	0 87038	0 50818	3937840 8	-164076 8	-1239292 2	-83380 7	-629784 8
0 97000	0 59337	0 13547	0 86453	0 49290	3978860 0	-123057 6	-1198273 0	-60655 5	-590633 1
0 98000	0 61124	0 14149	0 85851	0 47764	4019879 2	-82038 4	-1157253 8	-39185 0	-552752 6
0 99000	0 62924	0 14769	0 85231	0 46242	4060898 3	-41019 3	-1116234 7	-18968 1	-516167 9
1 00000	0 64737	0 15408	0 84592	0 44726	4101917 5	-0 1	-1075215 5	-0 0	-480899 9
1 01000	0 66560	0 16064	0 83936	0 43219	4142936 7	41019 1	-1034196 3	17727 9	-446965 0
1 02000	0 68391	0 16739	0 83261	0 41722	4183955 9	82038 3	-993177 1	34228 2	-414375 6
1 03000	0 70227	0 17432	0 82568	0 40239	4224975 1	123057 5	-952157 9	49517 2	-383139 9
1 04000	0 72068	0 18143	0 81857	0 38771	4265994 2	164076 7	-911138 7	63614 9	-353261 9
1 05000	0 73910	0 18873	0 81127	0 37321	4307013 4	205095 9	-870119 6	76544 8	-324741 3
1 06000	0 75752	0 19622	0 80378	0 35891	4348032 7	246115 1	-829100 3	88333 5	-297573 4
1 07000	0 77589	0 20388	0 79612	0 34482	4389051 9	287134 3	-788081 1	99011 0	-271749 8
1 08000	0 79421	0 21173	0 78827	0 33097	4430071 1	328153 5	-747061 9	108610 1	-247257 8
1 09000	0 81244	0 21977	0 78023	0 31738	4471090 2	369172 7	-706042 7	117166 5	-224081 0
1 10000	0 83055	0 22798	0 77202	0 30405	4512109 4	410191 9	-665023 6	124718 1	-202199 2
1 11000	0 84852	0 23638	0 76362	0 29101	4553128 6	451211 1	-624004 4	131305 0	-181588 9
1 12000	0 86632	0 24495	0 75505	0 27826	4594147 9	492230 3	-582985 1	136969 4	-162223 1
1 13000	0 88392	0 25370	0 74630	0 26583	4635167 1	533249 5	-541965 9	141754 8	-144071 9
1 14000	0 90128	0 26263	0 73737	0 25372	4676186 2	574268 6	-500946 7	145706 2	-127102 6
1 15000	0 91838	0 27173	0 72827	0 24195	4717205 4	615287 8	-459927 6	148869 6	-111280 0
1 16000	0 93518	0 28100	0 71900	0 23052	4758224 6	656307 0	-418908 4	151291 7	-96566 6
1 17000	0 95166	0 29043	0 70957	0 21944	4799243 8	697326 2	-377889 2	153019 5	-82923 0
1 18000	0 96778	0 30003	0 69997	0 20871	4840263 1	738345 4	-336869 9	154100 3	-70308 2
1 19000	0 98351	0 30979	0 69021	0 19834	4881282 2	779364 6	-295850 7	154581 1	-58679 8
1 20000	0 99881	0 31970	0 68030	0 18834	4922301 4	820383 8	-254831 6	154508 6**	-47994 2
1 25000	1 06782	0 37142	0 62858	0 14376	5127397 3	1025479 7	-49735 7	147420 9**	-7149 9
1 30000	1 12137	0 42622	0 57378	0 10798	5332493 2	1230575 6	155360 2	132877 1**	16775 7
1 35000	1 15577	0 48324	0 51676	0 08017	5537589 1	1435671 4	360456 1	115103 3**	28899 1
1 40000	1 16796	0 54143	0 45857	0 05916	5742684 9	1640767 3	565551 9	97074 4**	33460 3
1 45000	1 15596	0 59963	0 40037	0 04366	5947780 7	1845863 2	770647 7	80585 7**	33644 5
1 50000	1 11914	0 65661	0 34339	0 03242	6152876 6	2050959 0	975743 6	66494 5**	31634 7
1 55000	1 05844	0 71114	0 28886	0 02439	6357972 5	2256054 9	1180839 5	55016 7**	28796 2
1 60000	0 97647	0 76210	0 23790	0 01869	6563068 4	2461150 8	1385935 4	45996 1**	25901 6
1 65000	0 87735	0 80850	0 19150	0 01467	6768164 2	2666246 7	1591031 2	39105 2**	23335 3
1 70000	0 76646	0 84964	0 15036	0 01183	6973260 1	2871342 5	1796127 1	33972 9**	21251 3
1 75000	0 64987	0 88506	0 11494	0 00983	7178356 0	3076438 4	2001223 0	30252 9**	19679 5
1 80000	0 53377	0 91463	0 08537	0 00843	7383451 8	3281534 2	2206318 8	27651 4**	18591 2
1 85000	0 42384	0 93854	0 06146	0 00744	7588547 7	3486630 1	2411414 7	25931 8**	17934 9
1 90000	0 32468	0 95720	0 04280	0 00675	7793643 6	3691726 0	2616510 6	24907 8**	17653 4
1 95000	0 23940	0 97124	0 02876	0 00627	7998739 4	3896821 8	2821606 4	24432 8**	17691 3
2 00000	0 16952	0 98139	0 01861	0 00595	8203835 3	4101917 7	3026702 3	24390 0**	17996 8

LAMBDA=5 222200 BETA=4 573580 ALPHA=1 478310 E(r)=1 350338 V(r)=0 112544

TOTAL COST= 5177133 0 DIRECT COST= 4101917 6 LABOR COST= 1473920 0 MATERIAL COST= 2134904 0

LABOR PREMIUM COST= 6700 0

Figure 12-4. Computer program output for competitive bid model

major repairs or overhaul. This question appears even more important for ship operators due to very high costs of ship overhauls. In essence, the shipowner should make a decision regarding the "actual life" of the ship.

As with any other machine, the term "life" can mean several things: physical, depreciation, actual, and economic. Physical life is normally understood as the period until a machine becomes worn out or deteriorated as a result of normal operation, to the level at which it can no longer be safely operated. Relative to a ship, the term physical life does not make a lot of sense: most ships can be kept in normal operating condition for an almost indefinite period of time by carrying out necessary maintenance operations, repairs, and overhauls.

There will be only one problem: eventually, the cost of all these repairs becomes unacceptable. Obviously, the economic life is substantially less than the duration of the physical availability of the ship. The concept of economic life assumes the optimal duration of service life while the ship, or any other machine, will require the lowest total expenses per unit of work performed.

The depreciation life, or accounting life, as described in chapter 5, is the period of time required to depreciate the cost of a ship based on tax regulations or other accounting procedures. For instance, the IRS publication No. 456, *Depreciation Guidelines and Rules,* suggests the following depreciation lives for the assets of maritime-related industries:

Water transportation equipment	18 years
Ship- and boat-building equipment	12 years
Wharves and docks	20 years

Determination of the economic life of a ship should be a vital portion of the economic analysis at the design stage when the whole life costs are analyzed in order to choose the best design alternative. These life costs should be used as the reference data at the later stage when the ship is in operation.

The information necessary for the analysis includes the following principal components:

1. The initial cost of the ship, i.e., combined cost of design, construction, delivery, original set of spares, initial fees for registration, classification, etc.
2. Annual cost in operation per each year of the ship's life.
3. Residual value at the time of the ship's retirement (market sale price or scrap costs).

From the standpoint of their dependency on the ship's service life, the annual operating costs might be classified as relatively independent (costs of crew, fuel, fees and dues, general and administrative expenses) and variable (cost of stores and auxiliary materials, maintenance, shipyard repairs, insurance, etc.). Efficient shipyard repairs during the first 10 to 15 years of operation are capable of bringing consumption of fuel and other materials very close to the levels of a new ship. However, later in the ship's life, the increase of the fuel and lubricant consumption becomes quite substantial due to the aging of the ship's propulsion plant and auxiliary machinery and deteriorating hull condition.

As a general rule, all annual operating expenses should be considered when an economic life is evaluated. However, in practice the costs which are dependent exclusively on the annual amount of work are excluded, such as port and canal charges and fees, cost of loading and unloading, cleaning, container charges, etc.

The example below illustrates the principal methodology of estimating the economic life of a ship.

Example 12-2

A large commercial cargo ship is planned to be built at a foreign shipyard for an American shipowner. The initial (construction) cost of the ship is $60 million. Table 12-2 contains the principal data for analysis by the years of ship operation. The annual expenses reflect the latest improvements in ship design: for instance, use of modern machinery which reduces fuel cost, engine room automation which allows a sharp reduction in the crew size, etc. On the other side, the cost of repairs and maintenance might go up due to the crew reduction and the increased complexity of machinery. The following assumptions have been accepted in the example: crew cost includes wages and salaries, insurance, medical and training costs, leave pay and travel, food and crew related supplies, etc.; maintenance cost includes the overtime expenses and also the cost of technical services, additional temporary crew members and riding teams, emergency and voyage repairs, and also stores and supplies cost; fuel cost includes also lubricating oil and other materials; other expenses embrace hull and machinery insurance, and administration costs.

TABLE 12-2

Elements of Annual Operating Expenses in $ million (per years of service life)

Year of Service Life	*Maintenance & Auxiliary Material* A_{maint}	*Shipyard Repairs* A_{rep}	*Crew Cost* A_{crew}	*Fuel Cost* A_{fue}	*Other Annual Costs* A_{oth}	*Annual Operating Expenses* A_{an}
1	0.75		1.60	1.95	1.20	5.50
2	0.75	0.35	1.60	1.95	1.15	5.80
3	0.75		1.60	1.98	1.22	5.55
4	0.78		1.60	2.01	1.22	5.61
5	0.75	0.70	1.60	1.95	1.15	6.15
6	0.80		1.60	1.99	1.24	5.63
7	0.77	0.50	1.60	1.95	1.20	6.02
8	0.85		1.60	1.99	1.24	5.68
9	0.93		1.60	2.04	1.24	5.81
10	0.82	2.30	1.60	1.95	1.18	7.85
11	0.95		1.60	2.00	1.25	5.80
12	0.85	1.30	1.60	1.97	1.20	6.92
13	1.10		1.60	2.08	1.28	6.06
14	0.80		1.60	1.95	1.29	5.64
15	1.20	3.50	1.60	2.00	1.22	9.52
16	1.00		1.60	2.15	1.33	6.08
17	1.40	2.40	1.60	2.10	1.23	8.73
18	1.00		1.60	2.35	1.33	6.28
19	1.85	3.10	1.60	2.15	1.31	10.01
20	1.30		1.60	2.45	1.37	6.72
21	2.05	3.90	1.60	2.15	1.30	11.00
22	1.90		1.60	2.50	1.38	7.38
23	2.30	3.20	1.60	2.35	1.35	10.80
24	2.70		1.60	2.60	1.41	8.31

Table 12-3 presents the results of calculations of the annualized capital recovery costs C_N for a ship's service life of N years, using the following formula:

$$C_N = (A/P, i, n)\ (C_{sh} - S_n) + S_n i$$

where: C_{sh} = initial cost of ship

S_n = salvage (residual) ship's value of n year

TABLE 12-3

Calculation of Capital Costs, $ million

Service Life, N Years	*Residual Value S_n*	*Lost Value $(C_{sh} - S_n) A/P$*	*Invested Capital $S_n i$*	*Annualized Capital Recovery Cost C_N*
1	56	4.48	6.72	11.20
2	52	4.73	6.24	10.97
3	48	5.00	5.76	10.76
4	45	4.94	5.40	10.34
5	42	4.99	5.04	10.03
6	39	5.11	4.68	9.79
7	36	5.26	4.32	9.58
8	33	5.44	3.96	9.40
9	30	5.63	3.60	9.23
10	27	5.84	3.24	9.08
11	24	6.06	2.88	8.94
12	21	6.29	2.52	8.81
13	18	6.54	2.16	8.70
14	15	6.79	1.80	8.59
15	12	7.05	1.44	8.49
16	9	7.31	1.08	8.39
17	7	7.45	0.84	8.29
18	5	7.57	0.61	8.18
19	3	7.74	0.36	8.10
20	2	7.77	0.24	8.01
21	2	7.67	0.24	7.91
22	2	7.59	0.24	7.83
23	2	7.52	0.24	7.76
24	2	7.45	0.24	7.69

Annualized operating costs, shown in Table 12-4, represent summary annual costs which have been adjusted by time factor coefficients A/P and P/F for every option N of ship's service life. The calculations are done using the following formula:

$$A_N = (A/P, .12, n) \sum_{n=1}^{N} A_n (F/P, .12, n)$$

where A_N = annualized operating costs for N years' service life

A_n = operating expenses during nth year of service life

TABLE 12-4

Calculation of Annualized Operating Costs, $ million

Year	*Annual Operating Expenses,* A_n	*Present Worth Factor,* $P/F, .12, n$	*PW of Annual Operating Costs of n-year*	*Summary PW of Operating Expenses for N-year Service life*	*Capital Recovery Factor,* $A/P, .12, n$	*Annualized Operating Expenses,* A_N
1	5.50	0.8929	4.91	4.91	1.1200	5.50
2	5.80	0.7972	4.62	9.53	0.5917	5.64
3	5.55	0.7118	3.95	13.48	0.4163	5.61
4	5.61	0.6355	3.57	17.05	0.3292	5.61
5	6.15	0.5674	3.49	20.54	0.2774	5.70
6	5.63	0.5066	2.85	23.39	0.2432	5.69
7	6.02	0.4523	2.72	26.12	0.2191	5.72
8	5.68	0.4039	2.29	28.41	0.2013	5.72
9	5.81	0.3606	2.10	30.50	0.1877	5.73
10	7.85	0.3220	2.53	33.03	0.1770	5.85
11	5.80	0.2875	1.67	34.70	0.1684	5.84
12	6.92	0.2567	1.78	36.48	0.1614	5.89
13	6.06	0.2292	1.39	37.86	0.1557	5.89
14	5.64	0.2046	1.15	39.02	0.1509	5.89
15	9.52	0.1827	1.74	40.76	0.1468	5.98
16	6.08	0.1631	0.99	41.75	0.1434	5.99
17	8.73	0.1456	1.27	43.02	0.1405	6.04
18	6.28	0.1300	0.82	43.84	0.1379	6.05
19	10.01	0.1161	1.16	45.00	0.1358	6.11
20	6.72	0.1037	0.70	45.70	0.1339	6.12
21	11.00	0.0926	1.02	46.71	0.1322	6.18
22	7.38	0.0826	0.61	47.32	0.1308	6.19
23	10.80	0.0738	0.80	48.12	0.1296	6.23
24	8.31	0.0659	0.55	48.67	0.1285	6.25

The calculation of combined annualized expenses as a sum of recovery cost C_N and operating cost A_N is presented in Table 12-5. Another important component should be considered in these calculations—annual reduction of amount of work performed by the ship. While the ship is aging, the duration of planned repairs increases and more emergency repairs occur with the corresponding increase in days out of operation. Annual loss of work, for all practical purposes, is proportionate to reduction of days in operation.

This is reflected in economic evaluations by using a special equivalency coefficient ε_n which is calculated as a ratio of a number of standard or typical voyages for the *n*th year to that of the first year in operation. It might be calculated also as a ratio of the *n*th year operational time to that

TABLE 12-5

Calculation of Summary Annual Cost (per Options of Ship's Service Life)

Service Life, Years	*Annualized Operating Expenses* A_N, *\$ million*	*Annualized Capital Recovery Cost,* C_N, *\$ million*	*Annual Equivalency Coefficient,* ε_n	*Average Equivalency Coefficient for N-year Service Life,* ε_N	*Equivalent Summary Annual Costs,* $\frac{A_N + C_N}{\varepsilon_N}$, *\$ million*
1	5.50	11.20	1.0000	1.0000	16.70
2	4.56	10.97	0.9836	0.9918	16.75
3	4.48	10.76	1.0000	0.9945	16.46
4	4.47	10.34	0.9973	0.9952	16.03
5	4.59	10.03	0.9781	0.9918	15.86
6	4.57	9.79	0.9973	0.9927	15.59
7	4.61	9.58	0.9808	0.9910	15.44
8	4.60	9.40	0.9973	0.9918	15.24
9	4.60	9.23	0.9945	0.9921	15.07
10	4.77	9.08	0.9397	0.9868	15.13
11	4.77	8.94	0.9808	0.9863	14.99
12	4.83	8.81	0.9589	0.9840	14.94
13	4.84	8.70	0.9699	0.9829	14.85
14	4.88	8.59	0.9671	0.9818	14.74
15	4.99	8.49	0.8493	0.9730	14.87
16	5.01	8.39	0.9397	0.9709	14.81
17	5.08	8.29	0.8959	0.9665	14.82
18	5.11	8.18	0.9370	0.9648	14.75
19	5.19	8.10	0.8027	0.9563	14.86
20	5.22	8.01	0.8685	0.9519	14.84
21	5.29	7.91	0.7589	0.9427	14.94
22	5.33	7.83	0.8301	0.9376	14.95
23	5.39	7.76	0.7808	0.9308	15.03
24	5.42	7.69	0.8082	0.9257	15.06

of the first year. For instance, if the total time in operation for the first year is 365 days, and the prognosis for the tenth year calls for 85 days to be spent for planned shipyard repairs and various emergency repairs, the equivalency coefficient for the tenth year is

$$\varepsilon_{10} = \frac{365 - 85}{365} = .767$$

The ratio of number of voyages gives the same result. If, for instance, a typical voyage takes 36.5 days, the first year allows for

$$\frac{365}{36.5} = 10 \text{ voyages}$$

and the tenth year for

$$\frac{365 - 85}{36.5} = 7.67 \text{ voyages}$$

And the equivalency coefficient is

$$\varepsilon_{10} = \frac{7.67}{10} = .767$$

The equivalency coefficient varies by years of service life. In order to analyze summary costs per any option of service life N, a simple arithmetic average value of the coefficient for N years is used:

$$\varepsilon_N = \frac{\sum_{n=1}^{N} \varepsilon_n}{n}$$

The summary annualized costs adjusted by the equivalency coefficient are called *equivalent summary annual costs (ESAC)*. The following formula is used

$$ESAC_N = \frac{A_N + C_N}{\varepsilon_N}$$

The above formula might be interpreted as follows: equivalent summary annual costs represent prorated summary annualized costs per every service life option N years assuming that the annual amount of work stays the same throughout the entire service life as it is during the first year in operation.

Figure 12-5 presents a plot of the equivalent annual costs and the variable data for capital recovery and operating cost. The curve shows a gradual fall down of the annual costs with the rise of the ship's service life. Based on Table 12-5 and Figure 12-5 two almost equal choices of the economic life option can be made—14 and 18 years. The fact that the pace of the fall slows down after the service life of 12 years and almost flattens between 13 and 20 years service life signals the necessity to bring forward additional information in order to make an intelligent decision. For example, actual change of fuel consumption might appear more significant than is assumed; the change of insurance costs for an aging ship might significantly affect the curve. Reduction of the annual amount of work might be quite substantial for the 14- to 20-year-old ship due to a sharp increase of failures and corresponding loss of operation time. Maintenance and repair cost might grow even higher, for instance, due to structural failures as in the case of very large tankers and bulk cargo carriers.

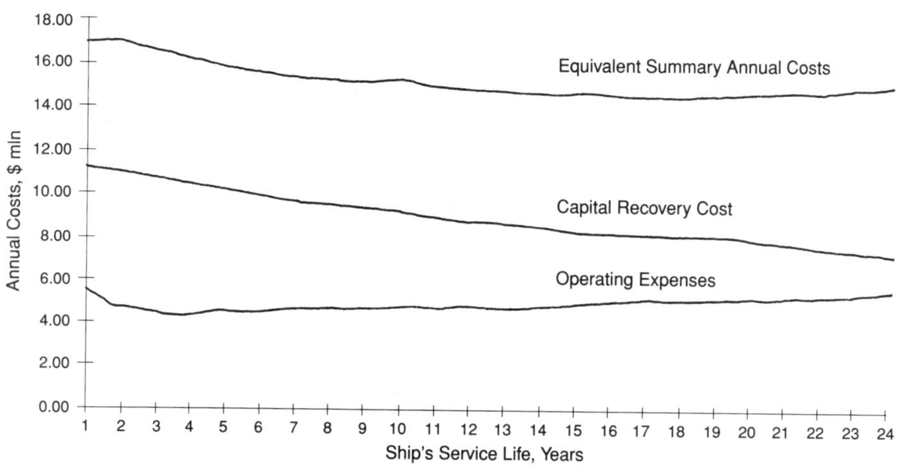

Figure 12-5. Capital, operating and summary annual costs for options of ship's service life

Another important factor to be taken into consideration is depreciation and corporate income taxes, which is normally the case of the after-tax economic analysis. Depreciation has a twofold impact on the economic analysis. On one hand, it reduces the initial cost of the ship by a fixed annual amount, depending on the accepted method of depreciation. The remaining cost of the ship is called "book value" and it might differ from the salvage or market value. On the other hand, income taxes are reduced by a fixed percent, currently a maximum of 34 percent in the United States, of the annual amount of depreciation. Table 12-6 reflects the impact of the 15-year straight-line depreciation schedule and the calculation of annual expenses reduced by the tax benefits.

TABLE 12-6

Calculation of Summary Annual Costs with Depreciation

Service Live, Years	*Depre- ciation, %*	*Depre- ciation Amount, $ million*	*Tax Benefit @ 34%, $ million*	*PW Tax Benefit, $ million*	*Annual Tax Benefit, $ million*	*Summary Annual Expenses, $ million*
1	5.0	3.80	1.02	0.91	1.02	15.68
2	9.5	5.70	1.94	1.55	1.46	15.29
3	8.55	5.13	1.74	1.24	1.54	14.92
4	7.7	4.62	1.57	1.00	1.55	14.42
5	6.93	4.16	1.41	0.80	1.53	14.33
6	6.225	3.74	1.27	0.64	1.49	14.10
7	5.905	3.54	1.20	0.54	1.46	13.98
8	5.905	3.54	1.20	0.48	1.44	13.80
9	5.905	3.54	1.20	0.43	1.42	13.65
10	5.905	3.54	1.20	0.39	1.41	13.72
11	5.905	3.54	1.20	0.35	1.40	13.59
12	5.905	3.54	1.20	0.31	1.39	13.55
13	5.905	3.54	1.20	0.28	1.39	13.46
14	5.905	3.54	1.20	0.25	1.38	13.36
15	5.905	3.54	1.20	0.22	1.38	13.49
16	2.95	1.77	0.60	0.10	1.36	13.45
17	—	—	—	—	1.33	13.49
18	—	—	—	—	1.31	13.44
19	—	—	—	—	1.29	13.57
20	—	—	—	—	1.27	13.57
21	—	—	—	—	1.25	13.69
22	—	—	—	—	1.24	13.71
23	—	—	—	—	1.23	13.80
24	—	—	—	—	1.22	13.84

In a case where the non-fully depreciated ship is retired, i.e., when the salvage value is less than the book value, the law allows a reduction in the income tax by the same percent to the financial loss. Furthermore, the loss of revenues is reflected by decreased income tax. The resultant curve (Figure 12-6) differs slightly from the previous data, and the minimum cost/maximum benefit criteria identifies the economic life as fourteen years.

REPLACEMENT VS. OVERHAUL

The economic life of a ship, which was set at the design stage, might be used only as guidance when the decision of remaining life is made. The actual life of a ship might be defined as the actual service time until the

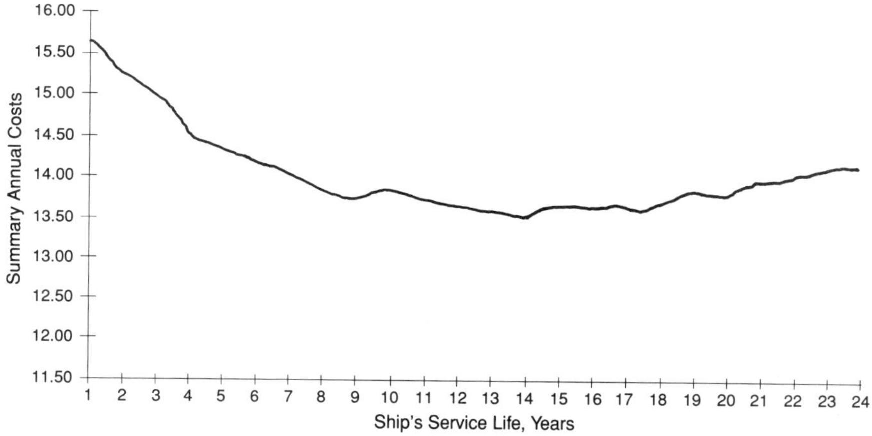

Figure 12-6. Annual expenses reduced for depreciation tax benefit

ship is sold for scrap or lost to an accident. The decision to retire the ship is based on a thorough economic analysis, normally when a major overhaul or even a middle-size shipyard repair is planned. Some additional criteria should be included in the analysis: first of all, the characteristics of a ship to replace the one under consideration. It is widely accepted in the economic literature to define the aging ship as a "defender," and the newer one which is intended as a substitute as a "challenger."

The pace of technological changes in ship design and the fluctuation of economic conditions affect the actual life of ships. For instance, in the early eighties the reopening of the Suez Canal after the Arab-Israeli war made many super-tankers obsolete in spite of the fact that most of them were 10 to 12 years old. Another example is the introduction of specialized container carriers and RO/RO ships, which ended the life of many general cargo ships. Therefore, when shipyard repairs are planned, especially in the case of middle-aged or older ships, a thorough consideration of a potential challenger should be made.

When the possibility of a replacement for the ship is discussed, for all practical purposes, it is not the necessity of replacement which is being questioned; it is, *a priori,* accepted that the ship has to be retired eventually, and a new one purchased. The only uncertainty regarding the aging ship is: When should it be retired?

Based on the definition of the economic life given above, the following are the principal technical analyses to be carried out when the decision of an overhaul or replacement is taken:

1. Characteristics and operational parameters of newer ships that the market offers to perform the same functions as the ship under consideration, i.e., the challenger.
2. Current condition of the defender in terms of its ability to perform the task compared to a possible replacement ship, e.g., speed, fuel consumption, cargo capacity, crew size and prospectus of its reduction, loading/unloading time, etc.
3. Expected condition of the defender after the overhaul compared to that of the possible replacement ship.

The entire overhaul-or-replacement analysis is normally performed in two steps. In the first stage, the remaining economic life of the defender is evaluated using the methodology described above. However, one principal difference should be noted: it is not the initial cost of the ship that is considered, but the current market value. In the second stage, the annual expenses of the defender are compared with those of the challenger, and the final decision is taken.

Example 12-3

A shipping company is planning to overhaul a 15-year-old cargo carrier with original cost of $35 million; the overhaul is supposed to provide for ten more years of the ship's service life. The cost of the overhaul (A_{ov}) is estimated to be $4.15 million. The following is the evaluation of the remaining service life of the defender. The current market value of the ship *(S_0)* is $6 million; it will grow to $8.0 million after the overhaul and gradually fall down until it reaches scrap value of $1.5 million seven years after the overhaul. The ship is fully depreciated.

It is required to determine the remaining economic life of the ship if the overhaul is carried out. The Table 12.7 presents the annual expenses for each year of the remaining service life. High maintenance expenses are the result of substantial structural problems and main engine design defects. Due to reduced reliability of the old ship the company will incur substantial profit losses due to breakdowns and corresponding repair costs. It is reflected in low values of the equivalency coefficient ε_n.

TABLE 12-7

Elements of Annual Operating Cost, $ Million

Remaining Service Life, Years n	*Maintenance Cost,* A_{maint}	*Shipyard Repair Costs,* A_{rep}	*Fuel Cost,* A_{fuel}	*Crew Cost,* A_{crew}	*Other Operating Costs,* A_{oth}	*Annual Operating Expenses,* A_n
1	2.15	4.15	2.35	3.20	2.15	14.00

(continued on next page)

TABLE 12-7 *(continued)*

2	2.45	—	2.90	3.20	1.55	10.10
3	2.26	2.55	2.55	3.20	1.95	12.51
4	3.25	—	3.15	3.20	1.75	11.35
5	1.85	3.45	2.50	3.20	2.35	13.35
6	3.45	—	3.25	3.20	2.10	12.00
7	2.50	2.65	2.65	3.20	2.55	13.55
8	3.40	—	3.30	3.20	2.20	12.10
9	2.75	3.10	2.60	3.20	2.95	14.60
10	3.55	—	3.50	3.20	2.65	12.90

The annualized operating expenses (Table 12-7) and capital recovery cost (Table 12-8) have been calculated based on the 12% interest rate using the following formulae:

$$C_N = (A/P, .12, n)\ (A_{ov} + S_0 - S_n) + S_n i$$

$$A_N = (A/P, .12, n) \sum_{n=1}^{N} A_{op}\ (P/F, .12, n)$$

TABLE 12-8

Calculation of Capital Recovery Cost, $ million

Remaining Service Life, Years	*Residual Value, S_n*	*Lost Value, $(A_{ov} + S_o - S_n)$ $(A/P, .12, n)$*	*Invested Capital $S_n i$*	*Annual Capital Cost C_N*
1	8.0	2.41	0.96	3.37
2	7.0	1.86	0.84	2.70
3	6.0	1.73	0.72	2.45
4	5.0	1.70	0.60	2.30
5	4.0	1.71	0.48	2.19
6	3.0	1.74	0.36	2.10
7	2.0	1.79	0.24	2.03
8	1.5	1.74	0.18	1.92
9	1.5	1.62	0.18	1.80
10	1.5	1.53	0.18	1.71

Table 12-9 presents the results of the calculations of the equitable summary annual cost as a sum of operating and capital costs adjusted by the average equivalency coefficient.

TABLE 12-9
Calculation of Equivalent Summary Annual Cost

Remaining Service Life, Years	*Total Operating Expenses, A_N $ million*	*Annual Capital Cost, C_N $ million*	*Summary Annual Costs, $A_N + C_N$ $ million*	*Equivalency Coefficient ε_n*	*Average Equivalency Coefficient ε_N $ million*	*Equivalent Summary Annual Costs, A_Σ $ million*
1	14.00	3.37	17.37	0.85	0.850	20.43
2	12.16	2.70	14.86	0.92	0.885	16.80
3	12.26	2.45	14.71	0.90	0.890	16.53
4	12.07	2.30	14.37	0.94	0.903	15.92
5	12.27	2.19	14.46	0.81	0.884	16.36
6	12.24	2.10	14.34	0.87	0.882	16.26
7	12.37	2.03	14.39	0.78	0.867	16.60
8	12.35	1.92	14.27	0.84	0.864	16.52
9	12.50	1.80	14.31	0.79	0.856	16.72
10	12.52	1.71	14.24	0.82	0.852	16.71

The above data identifies that the minimal expenses occur at the fourth year after the overhaul, four years being the economic life of the ship.

The analysis of the summary annual costs for the five- to six-year service life options shows that they are very close to the four-year life. Therefore, when the decision to replace the aging ship is made and other practical reasons are considered (availability of a replacement tonnage, financial situation, etc.), it might be economically reasonable to extend further the remaining ship's life.

The fact that the ship is fully depreciated makes its book value equal to the cost of scrap, or $1.5 million. As a result, if the ship is sold now, before the overhaul, a capital gain of $4.5 million would be realized and the tax consequences would add $1.53 million to the annual expenses. The ship's sale in the later years would decrease this burden, but would not eliminate it until the market value exceeds the book value.

The cargo carrier from Example 12-2 is intended eventually to replace this aging ship. It is assumed that the annual performance of the defender is comparable with the challenger. Simple comparison of the annual expenses of the ships reveals the economic preference of replacing the ship immediately, before the overhaul ($14.74 million for the challenger with a 14-year service life ahead vs. $15.92 million for the best service life option for the defender).

When the overhaul-or-replace analysis is made, the decision to carry out the planned overhaul and to continue ship operation in accordance with the preplanned service life is only one of several alternative scenarios to be considered. Another option is to sell the ship for scrap or for a further use, and to buy a new one with a similar assignment. The third possible alternative is to commence an extensive and expensive ship overhaul with modernization which will allow the shipowner to extend the ship's service life for another 12 to 15 years. One more scenario is to perform repairs on a limited scale in order to support the ship operation until the next shipyard repairs when another round of analyses would be made. In this case the annualized expenses for the first two years would differ from those calculated in the above example. For instance, if the extended overhaul in the above example is reduced to a scope of $2.0 million minimally necessary for a two-year operation of the ship, the calculations would provide the following results: the annual expenses would become $15.51 million vs. $14.74 million, which allows the owner to consider at least two more years for the old ship until the final decision is made.

If the defender and the challenger differ in the amount of work they are capable of performing per year, the comparison should involve the cost of a unit of work (a ton-mile, a passenger, an hour of service, etc.) or freight rates. An even more practical result is achieved if the annualized benefits and/or profits are compared.

The whole life cost approach described above should also be utilized when the periodic shipyard repairs are planned, especially when the ship reached the age of 10 years. This type of economic analysis becomes a necessity in the cases of new domestic and international regulations, substantial changes of operational and market conditions, sharp fluctuations in ship construction costs, and introduction of new shipping technology.

ECONOMIC ANALYSIS OF SHIPYARD REPAIRS

In the process of preparation for repairs and during the performance of repairs, various engineering economic analyses are carried out by both the shipowner and the shipyard.

Economic Analyses by Shipowner

After the decision to carry the planned or emergency repairs has been made, the shipowner considers approaches which will yield the shortest repair period and the lowest cost.

After the ship has been delivered to the shipyard, the primary concern of the owner's representative is the constant changing of the work scope which requires him or her to make immediate decisions. The economic analysis in this case is normally very limited and involves a comparison of the cost of the work estimated by the owner's representative and by the shipyard. However, the field for the economic justification should be much wider.

First of all, the necessity of the change should be economically approved by comparing losses and benefits for two options: the original work scope and the change. Obviously, the technical necessity is not discussed here: correction of the design defects, adherence to the statutory or classification requirements, safety considerations, etc., in most cases do not require a detailed economic study unless there are several alternatives for a repair job. On the other hand, the economics of carrying out a certain work option during this repair should be compared with either performing it during the next repair or having it performed by riding teams without taking the ship out of operation. (The onboard activities are discussed in Chapter 13).

Another important subject for consideration by the shipowner is the quality of repairs. The shipowner does not have much leverage over the shipyard when it concerns quality of repairs. However, the increased presence of qualified owner representatives at the yard is very effective in controlling work quality. The fact that more monitoring is done allows the shipowner to uncover shipyard defects on time and to prevent substantial financial losses. Increased presence at the yard means additional cost, but compared to benefits in the operation it is advantageous. Principal elements of the corresponding economic analysis are additional expenses for keeping an office at the shipyard (salaries of extra personnel, travel cost, equipment, communication cost, etc.) and benefits in operation such as reduction of maintenance and repairs, increase of revenues as a result of decreased out-of-service time, etc. It is important to keep in mind that the benefits might accrue until the next shipyard repairs, or even longer.

The shipowner might affect the quality of repairs also by supplying certain materials and equipment. It is a common practice for the shipyard to acquire most of the material resources needed for repairs. The reason is that, this way, many possible shipyard's claims to the owner for cost adjustments due to alleged owner-caused delay might be eliminated. However, when the economic analysis indicates that better materials, or painting schedule, or equipment will bring substantial financial benefit in operations, the shipowner might request the yard either to supply them for an additional cost, or to accept the owner's supply. The components of this economic analysis are the increased expenses for acquisition with possible designer's cost and very likely savings on the shipyard's mark-up compared to the benefits in ship operation.

Another economic analysis performed in the shipping company deals with choosing the best way of carrying out certain work: (1) to include it in a repair specification to be done at the shipyard or (2) to use its own or contracted labor as a riding team without taking the ship out of operation.

Economic Analyses by Shipyard

Any shipyard's decisions which are intended to reduce production cost, improve quality, and/or reduce the production cycle during ship repairs should be evaluated using the principles discussed in the previous chapters. However, the specifics of ship repairs demand certain adjustments to the methods used, for instance, in the case of ship construction. It concerns, first of all, a much higher degree of uncertainty regarding the future use of the evaluated technology. For example, for any projects intended to introduce more efficient production equipment there are no clear data on the volume of production employing this new technology. Obviously, ship repair uses more universal equipment and tooling than the ship construction does.

Another specific feature of ship repair is a very short production cycle. As a result, the impact of projects intended to reduce the duration of repairs has normally a higher value than the cost reduction. Therefore, improvement of project scheduling, material and equipment delivery, drydocking operations, etc., appear economically quite efficient. Moreover, these projects are less affected by the fluctuation of the ship repair market.

REFERENCES AND BIBLIOGRAPHY

Churchman, C. West, et al. *Introduction to Operations Research.* New York: John Wiley & Sons, 1957.

QUESTIONS AND PROJECTS

1. Discuss the place of ship repairs in the ship operation process.
2. What is the difference between ship repairs and ship maintenance?
3. What are the advantages of the planned repairs? Give examples of preventative features of planned repairs.
4. What are the engineering economic analyses related to ship repairs? Provide examples of economic analyses carried out by the shipowner and by the shipyard.
5. What is the preferred type of contract for a planned shipyard repair? For an emergency repair? Give examples of the cost rates employed by shipyards while estimating repair specifications.
6. Is a competitive bidding model valid if the shipowner negotiates with the two lowest bidders to achieve the most favorable price?
7. Why are detailed work specifications essential to a successful competitive bidding model?
8. Suggest ways to use the $f(r)$ and $1 - F(r)$ distribution curves to compare the bidding behavior of individual competitors.
9. What are the advantages of involving seagoing staff in the preparation of repair specifications?
10. In the United States the law prohibits competitors from working together in bid situations. This is not so in all countries. What impact will collusion among competitors have on a bidding model result?
11. Why would a ship repair yard quote a price which has a negative profit for the work quoted?
12. If a ship repairer has an unannounced policy of making rebates to selected shipowners or their representatives, what impact would

this have on the validity of a bidding model? What do you think of such practices?

13. Write a computer program to calculate the optimum bid price given the parameters of the $f(r)$ and $g(k)$ distributions.
14. Write a computer program to calculate the parameters of a Weibull distribution given any size historical data base for r.
15. What is economic life of a ship? Discuss different definitions of service life: physical, depreciation or accounting, actual, etc.
16. What is called "variable operation expenses"? List the principal components of variable operation expenses which are normally considered in determining the economic life of a ship.
17. Discuss what impact the market cost of the ship at various periods of its service life will have on the economic life.
18. In the economic analysis, how will the deterioration of the aging ship be reflected relative to evaluation of the economic life?
19. What are the shipowner's alternatives when the decision is being made regarding an overhaul of an aging ship?
20. Describe a step-by-step procedure of the overhaul vs. replacement economic analysis.

PROBLEMS

1. A shipbuilder is scheduled to bid for repair of a vessel in competition with the only two other qualified competitors. The Weibull parameters for the competitors' bid ratio distributions are

Competitor	*Alpha*	*Beta*
A	1.3185	5.8006
B	1.4766	6.0793

The out-of-pocket cost estimate is $246,000. What is the optimum bid for the shipbuilder?

2. A shipbuilder has maintained a databank of results for competitive ship repair bids for the past year. The data have been combined and fitted to a Weibull distribution $f(r)$ to provide the following parameters:

Alpha	1.7321
Beta	5.2244

The shipbuilder knows that only three competitors are in a position to quote. If the out-of-pocket cost is $496,000, what price should be quoted as the optimum bid?

3. A competitor's average bidding history is encoded in the following Weibull and Poisson distribution parameters:

Alpha	1.5605
Beta	5.3243
Lambda	4.6

The builder's out-of-pocket cost is $368,000. What price should be quoted for an optimum bid? Calculate the means and variances of $f(r)$ and $g(k)$.

4. In a recent sealed-bid situation, a competitor's price was $450,000. The analyst's costs were as follows:

Direct labor	$200,000
Labor benefits	.33 (labor)
Material	$70,000
Labor premium	$8,000

Calculate the value of r to be entered in the database for this competitor during this competition.

5. Given the following parameter values of $f(r)$, rank the competitors by the aggressiveness of their bidding behavior.

Competitor	*Alpha*	*Beta*
A	1.3185	5.8406
B	1.4766	6.0793
C	1.5239	4.3209
D	1.5222	4.8082
E	1.5605	5.3243
F	1.7321	5.2244

6. Determine the annual capital recovery cost of a newly built ship for options of the service life up to 25 years and plot the results. Assume the original cost of the ship was $50 million; the market (residual) value decreases by $3 million a year until it reaches a scrap value of $2 million. Interest rate is 10%.
7. Determine the remaining economic life of a ship based on an interest rate of 15 percent and the data provided in the table below. Assume equivalency coefficient fixed at 1.00. Carry out necessary calculations and plot the results on a chart.

Remaining Life, Years	*Residual Value, \$ million*	*Maintenance Cost, \$ million*	*Shipyard Repair, \$ million*	*Other Operating Costs, \$ million*
0	8	.5	0	1.6
1	7.4	.3	5.5	1.4

Remaining Life, Years	*Residual Value, $ million*	*Maintenance Cost, $ million*	*Shipyard Repair, $ million*	*Other Operating Costs, $ million*
2	6.8	.4	0	1.8
3	6.2	.6	1.5	1.6
4	5.6	.8	0	1.8
5	5.0	1.0	2.7	1.6
6	4.4	1.2	0	2.0
7	3.8	1.4	1.8	1.6
8	3.2	1.6	0	2.1
9	2.6	1.8	4.2	1.8
10	2	2.0	0	2.0
11	2	2.2	2.5	2.2
12	2	2.4	0	2.4

8. Calculate the annual operating cost and annual capital recovery cost by years of service life of a small coastal ship, plot the results, and determine the economic life. Original cost of the ship is $800,000 and no residual value is expected by the end of the life. The market value is decreasing by $50,000 a year. The shipyard repairs are carried out every two years and their cost goes up by $10,000 starting with $30,000—the cost of the first repair. Annual maintenance is $10,000 during the first year and goes up by $3,000 every year. Other operating expenses and annual ship performance might be assumed unchanged. The interest rate is 15 percent.
9. A shipping company is planning to overhaul a 19-year-old tanker with an original cost of $30 million; the overhaul is supposed to provide for eight more years of the ship's service life. The cost of the overhaul is estimated to be $4 million. By the time of the overhaul, the ship is fully depreciated. The ship's current market cost is $6 million, and it is decreasing by $500,000 a year, until it reaches the scrap value of $2 million.

   Cost of annual maintenance, shipyard repairs, fuel, other annual expenses, and also decline of the value of equivalency coefficient due to reduced reliability of the old ship are shown in the table below. Other annual operating expenses are equal to $3.5 million and assumed constant.

   The best of new ships of comparable capacity could be bought for $55 million and will have an economic service life of 20 years. Assume that the average total operating expenses are $6 million per year and there is no salvage value by the end of the service life.

It is required to determine the following:

1. Economic remaining life of the old ship.
2. Whether the overhaul should be performed.
3. When the replacement should happen if the planned overhaul is carried out.

The following data should be used for the calculations:

Remaining Life, Years	*Fuel Cost, $ million*	*Maintenance Cost, $ million*	*Shipyard Repair, $ million*	*Other Operating Costs, $ million*	*Equivalency Coefficient*
0	1.8	.4	4.0	1.0	.90
1	2.2	.7	0	1.0	.96
2	2.0	.6	1.5	1.0	.92
3	2.4	.9	0	1.15	.95
4	2.1	.8	2.4	0.8	.91
5	2.6	1.1	0	1.2	.94
6	2.3	1.0	2.1	1.0	.90
7	2.8	1.3	0	1.5	.92
8	2.5	1.2	3.3	1.2	.88
9	3.1	1.4	0	1.8	.90

Chapter 13

Ship Operations

There is need of a method of finding out the truth.
—René Descartes, Rules

COST OF SHIP OPERATIONS

Assignment, design type and specific features, ship's registry and flag—all these and other factors affect the operating cost of a ship. The crew cost (which includes salaries and wages, benefits, feeding and lodging expenses) differs based on the crew size and the salary level. Obviously, for open registry ships the crew cost is substantially lower than for American flag ships. The same is true for ships with automated engine rooms and correspondingly reduced crews. For example, the crew cost as a percentage of total cost for a large full-crew American flag general cargo carrier is about 40 percent, while the same ship carrying a foreign flag might have a crew cost as low as 20 percent of the total. In the case of a reduced-crew ship, this share might go down to 10 percent.

Although the annual cost of fuel and lube oil primarily depends upon the type and capacity of the propulsion plant and auxiliary machinery, its percentage of the total operating cost varies substantially, reaching the maximum for reduced-crew ships, especially in the case of extra large tankers and cargo carriers. Using the example of the American registry large cargo carrier, the fuel cost amounts to 25-30 percent and it increases to 45-55 percent for the open registry ship. Obviously, the actual cost of fuel is practically the same for both cases.

The third major component of the operating cost, maintenance and repair expenses, is very low, about 5-7 percent of the total cost, when a new ship enters into operation. However, it goes rapidly up to 15-20 percent as the ship ages. After the midlife has been reached, the maintenance and repair portion of operational expenses might reach 25-30 percent and even more. The actual financial value of this cost component for an aging ship is evident when the cost of repairs and maintenance is combined with the losses of revenue because the ship is taken out of operation for repairs (see the examples in Chapter 12).

The above discussion serves to support a simple fact: a thorough analysis of operating expenses should precede practical efforts aimed at improvement of economic performance of a shipping company or of a single ship. Because financial resources normally are limited, such analysis would help to prioritize the areas of capital investment intended to reduce operating cost. Comparison of operating expenses of similar ships or even shipping companies appears to be the first step of "economic troubleshooting" which should result in identifying the problem areas.

ANALYSIS OF VOYAGE DATA

Microcomputer-based ship management systems are available that provide a plethora of ship performance data for all aspects of ship operations. This information includes indices of machinery performance, hull performance, and crew performance. Performance deviations from norm are quickly uncovered, the cause identified, and action taken to correct the problems. For example, a database of fuel oil and lube oil chemical/physical analysis and consumption can be paired with appropriate software to predict maintenance problems. Speed/power statistical data can be employed to predict the break-even points for hull cleaning and recoating or as an indicator of propeller damage and fouling.

A vessel in a liner service where the same distance is traveled each voyage and the same ports visited has an opportunity to collect information, calculate statistics from randomly collected data, and perform tests of significance to determine if the data is different from previous voyages. The following voyage information would be useful for cost control decisions:

1. Weekly or daily overtime wage payments for licensed and unlicensed and all personnel.
2. Fuel consumption corrected for speed, displacement, and heating value.
3. Voyage consumption of stores, supplies, and spare parts.
4. Voyage analysis of personnel feeding costs.
5. Electrical power usage, at sea and in port.
6. Voyage maintenance and repair costs.
7. Voyage revenue, costs, and profit.

Speed-power Analysis for Cost Control

The torque coefficient K_q (see Comstock in the References) for a propeller in open water is defined as

$$K_q = \frac{3600\,Q}{\rho\, n^2 D^5} = g\left(\frac{Q}{\rho\, n^2 D^5}\right)$$

where Q = torque in ft-lbs
ρ = water density
n = propeller rpm
D = propeller diameter, ft

and the advance coefficient *J* is defined as

$$J = \frac{101.33\,V}{n\,D} = F\left(\frac{V}{n}\right)$$

for a given propeller

where V = ship speed, knots.

Figure 13-1 shows a typical propeller coefficient plot versus the speed of advance. In the range of J = .4 to J = 1.0, this plot approximates a straight line. Let $\dot{f}$ = fuel consumption rate in pounds per hour, then Qn = function of power = $g(\dot{f})$, and substituting into the Kq relationship

$$K_q = g\left(\frac{\dot{f}}{n^3}\right)$$

since *D* and ρ are constants for a given ship operating in salt water.

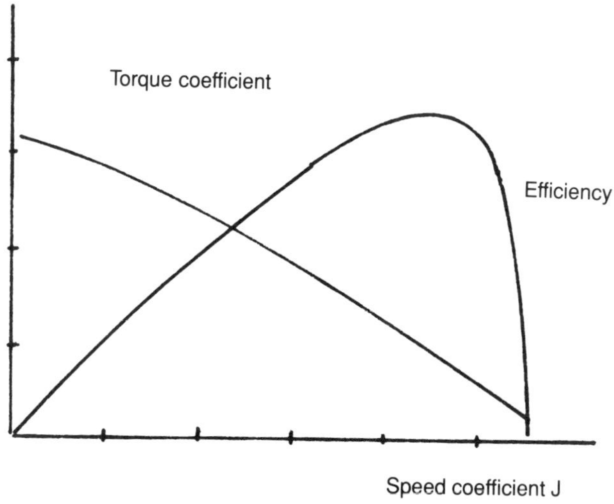

Figure 13-1. Typical propeller torque coefficient and efficiency curve

Therefore, for a given ship, the following linear relationship is true in the range of interest:

$$\frac{\dot{f}}{n^3} = a + b\left(\frac{V}{n}\right)$$

Figure 13-2 illustrates this relationship enclosed in statistical prediction intervals.

$$\text{Let } y = \frac{\dot{f}}{n^3} \text{ and } x = \frac{V}{n},$$

then the prediction interval for a future observation of y corresponding to x with a probability of $1 - \alpha/2$ is stated in all statistical texts, including Bowker (see References), as follows:

$$\hat{a} + \hat{b}x_o \pm t_{m-2;\frac{\alpha}{2}}\ \hat{\sigma}_e \sqrt{1 + \frac{1}{m} + \frac{(x_o - \bar{x})^2}{\sum_{i=1}^{m} (x_i - \bar{x})^2}}$$

where

$$\hat{b} = \frac{\sum_{i=1}^{m} (x - \bar{x})\ (y - \bar{y})}{\sum_{i=1}^{m} (x_i - \bar{x})^2}$$

is slope estimate,

$$t_{m-2;\frac{\alpha}{2}}$$

is Student's t distribution,

$$\hat{a} = \bar{y} - \hat{b}\,\bar{x}$$

is intercept parameter estimate,

$$\hat{\sigma}_e = \left[\frac{\sum_{i=1}^{m} (y_i - \hat{y}_i)^2}{n-2}\right]^{\frac{1}{2}}$$

is the estimate of the error in y,

$$\bar{x} = \frac{\sum_{i=1}^{m} x_i}{m}$$

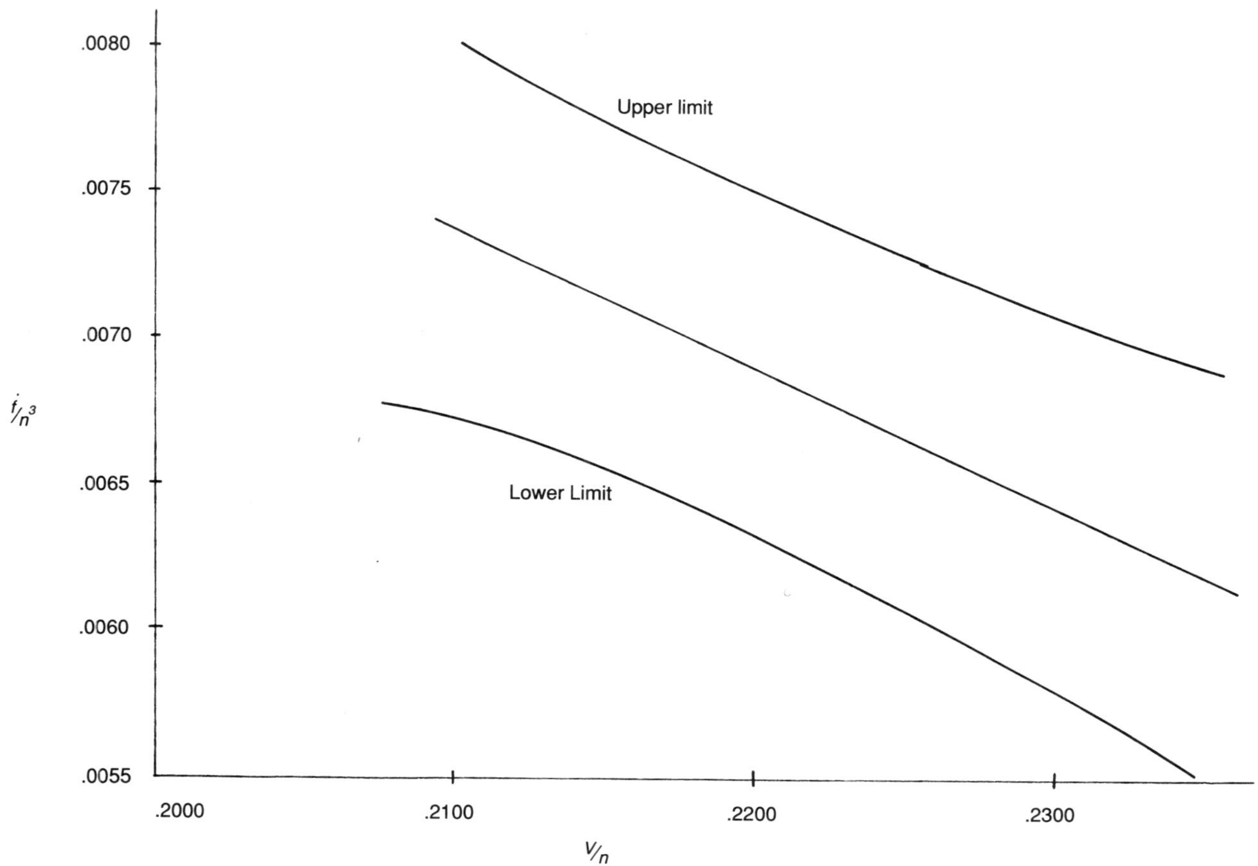

Figure 13-2. Regression analysis of v/n *versus* $\dot{f}/n^3$

is the sample mean of *x*, and

$$\bar{y} = \frac{\sum_{i=1}^{m} y_i}{m}$$

is the sample mean of *y*.

Subsequent points taken during normal commercial operation at about the same displacement, sea conditions, and steady state conditions should be randomly distributed about the line a + bx and within the interval. A lack of randomness within the interval or values outside the interval indicate that the overall system is not performing the same as trial performance. The ship personnel may then investigate to assign a cause and correct the problem. Possible causes include:

1. Deterioration of propulsion machinery
2. Hull fouling
3. Propeller damage or fouling
4. Change in heating value or specific gravity of fuel
5. Change in hull resistance due to waves

Example 13-1

Prior to being placed in service, the following data are collected during a standardization trial for a cargo liner.

TABLE 13-1
Sea Trial Data

Speed, knots	*PRPM*	*Shp*	*Fuel, lbs/hr*	$\dot{f}/n$	V/n
18.5	78.5	9,000	3,200	.00662	.2357
19.0	83	9,500	3,487	.00610	.2289
19.5	86	10,003	3,900	.00613	.2267
20.0	88	11,500	4,634	.00680	.2273
20.5	93	13,100	5,350	.00665	.2204
21.0	95	15,000	6,045	.00705	.2211
21.5	100	17,300	7,100	.00710	.2150
22.0	104	20,000	8,211	.00730	.2115
22.5	107	22,300	9,150	.00745	.2103
23.0	110	25,000	10,075	.00760	.2091

Find an approximate curve of $\frac{\dot{f}}{n^3}$ versus $\frac{V}{n}$ and the prediction interval of 90 percent confidence.

$$\bar{x} = \frac{\sum_{i=1}^{10} x_i}{10} = .2206$$

$$\bar{y} = \frac{\sum_{i=1}^{10} y_i}{10} = .006877$$

$$\hat{b} = \frac{\sum_{i=1}^{10} (x_i - .2206)\ (y_i - .006877)}{\sum_{i=10}^{10} (x_i - .2206)^2} = .04678816$$

$$\hat{a} = \bar{y} - \hat{b}\bar{x} = .006877 + .04678816(.2206) = .01719847$$

$$\hat{\sigma}_e = \left[\frac{\sum_{i=1}^{10} (y_i - \hat{y}_i)^2}{10 - 2}\right]^{\frac{1}{2}} = .0003062605$$

$$t_{10-2;\frac{10}{2}} = 1.86$$

$$.01719847 - .04678816\, x_0 \pm 1.86\,(.0003062605)\ \frac{\sqrt{1 + .1 + (x_0 - \bar{x})^2}}{\sum_{i=1}^{m} (x_i - \bar{x})^2}$$

Table 13-2 is the result of these calculations showing the estimated $\frac{\dot{f}}{n^3}$ for a given $\frac{V}{n}$ and the upper and lower limits of the intervals.

TABLE 13-2

Prediction Interval at 90 Percent Confidence

V/n	*$\dot{f}/n^3$*	*Lower Limit*	*Upper Limit*
.2357	.006171	.005491	.006881
.2289	.005489	.005866	.007112
.2267	.006592	.005980	.007204
.2273	.006564	.005949	.007179

(continued on next page)

TABLE 13-2 *(continued)*

.2204	.006886	.006289	.007483
.2211	.006854	.006252	.007452
.2150	.007139	.006530	.007748
.2115	.007303	.006674	.007932
.2103	.007359	.006722	.007996
.2091	.007415	.006769	.008062

A reduction of the confidence level on the interval from 90 percent to 80 percent will reduce the interval about the regression line with a slight decrease in the confidence of the prediction. An increase of the sample size above 10 will provide only a slight reduction in the confidence interval.

CAPITAL PROJECTS FOR SHIP OPERATION IMPROVEMENTS

When capital projects related to ship operations are analyzed, the problem of their compatibility from the standpoint of volume of work arises. A project can result in reduction of one or several elements of the operational cost without any impact on the ship's capability or performance. In this case only the relevant elements of cost or variable cost items are considered in the project justification. For example, when an energy-saving project is under consideration, the principal variable element of operating expenses is the cost of fuel, and possibly the maintenance cost.

Example 13-2

A riding team consisting of manufacturer's representatives and extra crew members are to carry out a retrofitting project intended to stabilize the fuel injection temperature on an older main diesel engine. The project includes installation of heating coils in the fuel line, in-line viscosity gages and controllers, etc., to provide for an automatic adjustment of the fuel injection temperature when engine load or supply fuel viscosity changes. It is estimated that fuel consumption will be reduced, which will result in an average annual saving of $18,000 during the six years of the remaining service life of the ship. Total cost of the project consists of the following components:

Manufacturer's representatives cost	$27,000
Extra crew cost	10,000
Fabrication cost	12,000
Installed equipment cost	23,000
Total project cost	$72,000

The only variable operating cost component except fuel is maintenance expenses which will go up by an average of $4,000 per year. Assuming 12 percent required rate of return, is the project justifiable?

Applying the rate-of-return method, the ratio of capital cost to annual benefit is found as follows:

$$(A/P, i, 6) = \frac{\$72{,}000}{(\$18{,}000 - \$4{,}000)} = 5.1429$$

which corresponds to a rate of return of 4.65 percent by interpolating between (A/P, .04, 6) = 5.2421 and (A/P, .05, 6) = 5.0757. Based on the company average rate of return of 12 percent, this project could not be justified.

An outcome of many capital projects is increase of volume of work performed by the ship during a certain period of time; it might be expected when the project expands the ship's cargo or passenger capacity, or the ship's speed. Similar results might be expected when the reliability of the ship is improved, and the ship loses less operational time because of elimination of failures and breakdowns. A similar effect provides an improvement of the ship's repairability, which means that the ship will spend less time under repairs; for instance, there might be a longer period between shipyard repairs of the underbody or the main propulsion plant when the total time at the shipyard is lessened.

Any of the above projects will most likely result in an overall increase of the amount of work performed during the period of time, say, *a* years. At the same time, the annual operational expenses would grow at a lower rate, and a saving of cost per unit of volume of work (a container, a passenger, a ton-mile of cargo carried, etc.) would be realized. In practice it means that those components of operational costs that have not been affected directly, like crew cost, administrative overhead, and sometimes maintenance and repairs, should also be included in the analysis, because a relative savings by each element per unit of work might be achieved.

Example 13-3

It has been proposed to refurbish two unused crew cabins into passenger accommodations, which might be done while the small cruise ship is underway. The estimate shows that the company will be able to increase the annual revenues from $4.55 million to $4.61 million for the remaining eight years of the ship's service life. Total annual operational expenses of $2.87 million will grow (due to additional costs related to the new passengers) to an estimated level of $2.895 million. The estimated cost of refurbishing is $110,000, but it would not affect the salvage value of the ship. It is required to make a decision regarding the proposal, assuming an interest rate of 15 percent.

Applying the annual cash flow method, the equivalent uniform annual benefit *EUAB* is represented by the revenues before and after refurbishing.

The equivalent uniform annual cost *EUAC* will be found using the formula

$$\text{EUAC} = \text{C}\,(\text{A/P}, \text{i}, \text{n}) + \text{A}$$

$$\text{EUAC}_1 = 0 + \$2{,}870{,}000 = \$2{,}870{,}000$$

$$\text{EUAC}_2 = \$110{,}000\,(\text{A/P}, .15, 8) + \$2{,}895{,}000 = \$2{,}919{,}500$$

The maximum profit criterion EUAB − EUAC will show the following results:

$$\text{Before refurbishing} = \$4{,}550{,}000 - \$2{,}870{,}000 = \$1{,}680{,}000$$

$$\text{After refurbishing} = \$4{,}610{,}000 - \$2{,}919{,}500 = \$1{,}690{,}500$$

The proposal appears to be sound.

Reduction of Fuel Consumption

The principal operational areas of the fuel savings are modernization and various improvements to the propulsion plant, modernization of the fuel system to allow use of heavier and less expensive grades of fuel, various additives to fuel, improvement of the combustion process, etc. As a rule, the cost of improvement projects includes expenses for design and engineering services, purchase of equipment and machinery, and installation by a shore team either while the ship is in port or underway. The principal components of the savings include reduction in fuel consumption and/or unit costs and possible reduction of lubricating oil consumption. The additional expenses include additional maintenance cost, and possible increase in the unit cost of fuel.

Example 13-4

An improvement in the boiler combustion control system has been made by a riding team of the automation equipment manufacturer. Total cost is $75,000. Resultant annual fuel savings is an average $26,000 during the 10 years of the remaining ship's life. Additional expenses for automation repairs are $7,000.

Using the net present worth criterion and 15 percent interest rate in order to consider the proposal effective, the formula *NPW* = *PW of benefits* − *PW of costs* should yield a positive result.

$$\text{NPW} = (\$26{,}000 - \$7{,}000)\,(\text{P/A}, .15, 10) - \$75{,}000 = \$20{,}361$$

Hull-Cleaning Economics

Ship hulls become rough over time, increasing ship resistance due to fouling by the growth of animal and plant life. This condition is aggravated by slow operating speed, operations in tropical waters, extended time in port, and the poor quality of the last hull coating. Depending on the propulsion system design, the ship operator adjusts for the increasing resistance by increasing power to maintain speed, which results in higher operating costs, or accepting a lower speed at a constant power with the consequent loss in revenue.

Normal dry docking for a tanker is usually scheduled at 2.5-year intervals. During the period between docking for hull recoating, hull resistance increases. Regular in-water cleaning of the hull will restore the hull smoothness and maintain the resistance at a level near the out-of-dock condition. Figure 13-3 shows that the 1991 cost for such cleaning ranges from $5,000 to $15,000 for typical tankers (see "Fouling-free Fortunes" in the References).

Shipowners who carefully monitor their ship speed deterioration in a given trade find that the speed reduction between in-water cleanings is .05 to .2 knots per month at constant power. Given the revenue per month loss for each knot reduction in speed, the economic interval for in-water hull cleaning may be determined by calculating the number of months required for the equivalent monthly cost of hull cleaning to equal the revenue loss for the speed reduction during the interval.

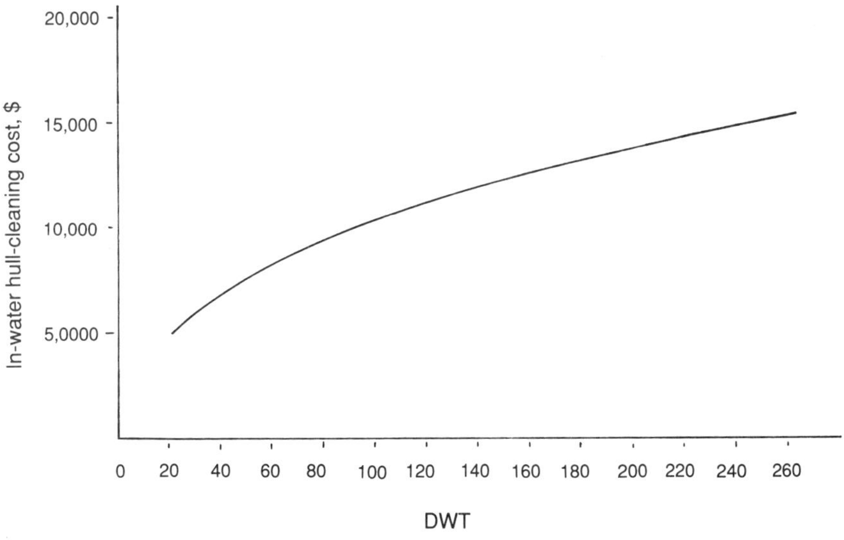

Figure 13-3. In-water hull-cleaning costs

$$R_d = H_c\left[\frac{i}{(1 + i)^n - 1}\right]$$

where R_d = monthly revenue loss per knot speed reduction
H_c = cost of in-water cleaning
i = monthly interest rate
n = months between cleaning

Example 13-5

A 42,000-dwt tanker is engaged in a coastwise trade between Texas and New Jersey. The average of the nominal speeds for loaded and ballast conditions is 16.375 knots. At this speed, the revenue per year is $19.96 million.

In-water hull cleaning is accomplished in Texas during the normal cargo-loading operation. At an average voyage speed of 15.375 knots, the annual revenue is $19,815,000 and at 14.375 knots it is $19,665,000. The shipowner's minimum attractive rate of return is 8 percent. The cost of in-water cleaning is shown in Figure 13-3. How often should the hull be cleaned if the owner's data indicate that speed at constant power decreases .1 knot per month between cleaning?

Average monthly revenue loss at operating speeds:

$$\text{at 14.375 to 15.375 knots} = \frac{19{,}815{,}000 - 19{,}665{,}000}{12} = \$12{,}500$$

$$\text{at 15.375 to 16.375 knots} = \frac{19{,}960{,}000 - 19{,}815{,}000}{12} = \$12{,}083$$

$$\text{Average revenue decrease per month} = \frac{12{,}500 + 12{,}083}{2} = 12{,}291.5 \text{ \$/knot}$$

Monthly interest rate

$$i_{eff} = (1 + i)^m - 1$$

$$.08 = (1 + i)^{12} - 1$$

$$i = .00643$$

From Figure 13-3, the cost of the hull cleaning is $5,200 with no schedule delay.

If the speed reduction is .1 knot per month, the revenue reduction is (.1 knot/month) (12,292 $/month) = $1,229.2 and

$$5{,}200 \left[\frac{.00643}{1.00643^n - 1} \right] = \$1{,}229.2$$

$$n = 4.19 \text{ months}$$

Figure 13-4 is a plot of the economic cleaning interval versus the speed deterioration for the same vessel with the same revenue and operating conditions.

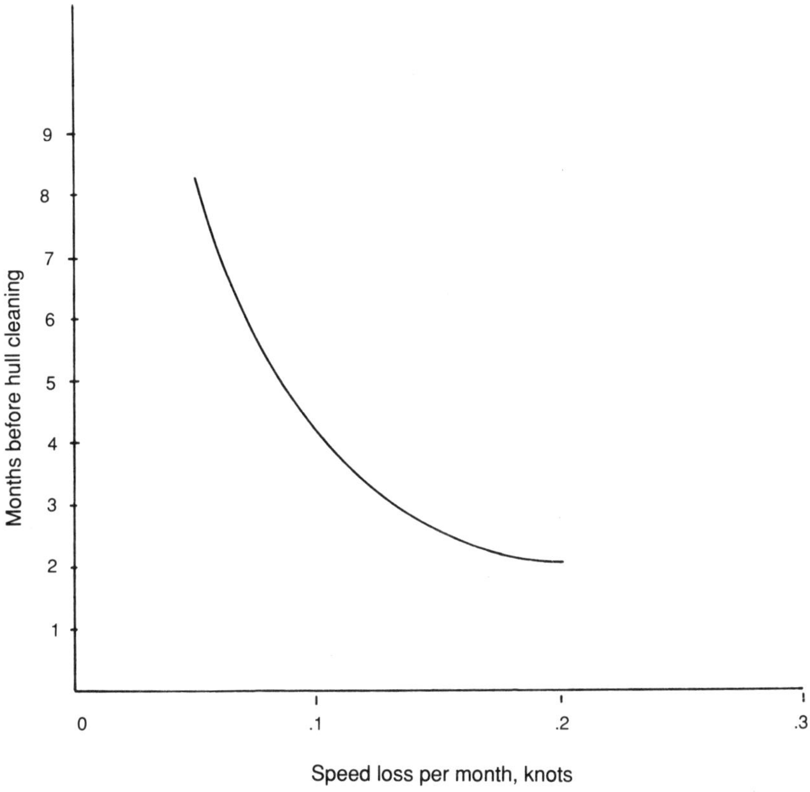

Figure 13-4. Economic cleaning interval versus speed of deterioration

Crew Reduction Efforts

Necessary conditions to carry out crew reduction on the ship in operation include engine room automation, provision for loading/unloading to be done by shore equipment and labor, deck activity automation and mechanization (automation in the pump room of a tanker, for instance), special crew training in carrying out simple engineering responsibilities, and a computerized preventive maintenance system. All the above represent additional expenses; some of the improvements might be done while the ship is underway. The savings come from reduction of the crew cost, including wages and salaries, overtime, food, etc. Additional operating expenses include increased shore engineering support personnel as well as increased maintenance and repair cost because of new automated equipment and decreased maintenance onboard. A probability factor might also be considered related to a possible increase of accidents and emergencies.

Reduction of Maintenance Cost

This might be achieved by replacement of aging machinery and equipment; introduction of modern computerized maintenance management systems; development of the continuous hull and machinery survey system; development of proper maintenance procedures and methods, equipment condition monitoring procedures, spare parts inventory management, etc. Cost of projects normally includes engineering and design cost, computer stations and software, training, and some equipment. Cost saving is achieved by reduction of overtime and consumption of spare parts, and also of stock of spares onboard. Another component of saving is reduction of future repair costs.

Example 13-6

A shipowner is considering enhancement of the shipboard maintenance system by setting up modern procedures for equipment condition monitoring and analyses of wear of components. The system will provide better information for shipyard repair and routine maintenance, and, as a result, is expected to significantly reduce equipment failures and related repair costs and ship delays. Total cost of the system consists of the following: engineering and R&D cost: $100,000; computers, condition monitoring, and measuring tools: $45,000. The estimated reduction of emergency delays is three days per year and the average annual cost of emergency repairs will go down from $65,000 to $49,000. The remaining ship's service life is eight years. Average per-voyage-day fuel expenses are $5,000 and port costs are $900. The introduction of the system would require increased overtime for equipment wear analyses and for additional maintenance operations: $12,000.

Total initial cost of the project = $100,000 + $45,000 = $145,000.

Additional annual operational expenses will consist of the extra cost of three voyage days [3 × ($5,000 + $900) = $17,700] and increased maintenance cost ($12,000), or a total of $29,700. Reduction of emergency repair cost is estimated to be $65,000 – $49,000 = $16,000. Thus, total increase of annual operational expenses is $29,700 – $16,000 = $13,700.

Additional revenues are expected to be $18,000 × 3 = $54,000.

Assuming a 12 percent rate of return, the present worth of total annual benefits of the project during the eight-year remaining life of the ship is found as follows: (\$54,000 − \$13,700) × (P/A,.12,8) = \$40,300 × 4.968 = \$200,210.

Because the benefits are substantially higher than the initial cost, the project should be considered for implementation.

It is important to note that the above example is actually a rough economic estimate of the project because average per-year cost and benefit values have been used. Actual annual expenses and revenues vary substantially from year to year depending on various factors, such as aging of the ship, shipyard repairs, etc. Therefore, if the rough estimate appears close to the borderline, a more precise analysis should be done of operating costs and revenues per every year of the remaining service life.

Reduction of Repair Expenditures

The costs related to repairs include the shipyard invoices and the support charges for technicians, designers, tugboats, cleaning stations, etc., as well as the revenue losses caused by taking the ship out of service. Reduction of repair expenditures may be achieved by improved operational practices, improved maintenance, and modernization and replacement of defective units of machinery and equipment. Engineering and consulting costs for development of the improved methods and procedures appear to be an important element of the total cost.

Example 13-7

The modernization of a large group of ships has been carried by a leading U.S. flag company without taking the ships out of operation. The primary reason was the apparently faulty design of the existing auxiliary diesels which caused excessive repair expenditures, high down-time, and higher than normal fuel costs. After comparing several alternatives, the company decided to replace all the generator engines with newer-design low-maintenance machines which, in addition, provided lower fuel cost by burning lower-grade fuel. The idea and certain data for this example have been adopted from Rynn (see References). The objective is to illustrate how the company rate of return and the estimated remaining service life affect the results of the economic analysis.

Total cost of re-engining is composed of the following components:

1. Engine cost
2. Installation cost
3. Rental and installation of a temporary replacement unit
4. Project management
5. Additional crew cost
6. Modifications to existing systems onboard
7. Rental of additional portable onboard accommodations

Per company estimates, the total savings during the remaining 26 years of each ship's service life will be \$800,000 due to increased time between overhauls and \$110,000 due to reduced fuel cost. Without taking into consideration other possible saving components like reduced revenue losses and safety claims, the following calculations are performed:

1. What would be the maximum allowed total cost of re-engining? Employing the rate of return method, PW of costs = PW of benefits = \$800,000 + \$110,000, i.e., if the cost of the re-engining is less than \$910,000, the project is sound at the assumed 15 percent interest rate.
2. What should be an average annual saving assuming 15 percent rate of return?

$$\text{PW of benefits} = \text{EUAB}\ (P/A, .15, 26) = \text{EUAB}\ (6.0541)$$

$$\text{EUAB} = \frac{\$910{,}000}{6.0541} = \$150{,}300$$

3. What would be the maximum allowed cost of re-engining if the remaining economic life is not 26 years, but 15 years?

$$\text{PW of costs} = \text{PW of benefits} = \text{EUAB}\ (P/A, .15, 15) = \$150{,}300\ (5.5279) = \$830{,}900$$

4. What should be the annual saving to satisfy the above if the rate of return is 10 percent?

$$\text{PW of benefits} = \text{EUAB}\ (P/A, .10, 15) = \text{EUAB}\ (7.3867)$$

$$\text{EUAB} = \frac{\$910{,}000}{7.3867} = \$123{,}200$$

Employing riding teams to carry out enhanced maintenance, repairs, and even modernization is one of the most promising efforts. The cost of riding teams might be higher than the same type of shipyard personnel; however, the mere fact that the ship is not taken out of service provides the necessary advantage in very many cases. In making a decision regarding a riding team, the shipowner should first determine the cost of shipyard work plus losses from out-of-service time, or simply from staying at the yard if the planned work is the principal reason for the shipyard repairs. In the next stage, the shipowner evaluates the cost of the riding team to do the same job, and compares these costs with the shipyard alternative.

Example 13-8

Installation of sacrificial anodes in the tanks of a large tanker prevents shell and tank bottom corrosion, weakening of material, and possible cracks in the hull stringers and stiffeners. Local repairs and welding of discovered cracks is normally done at the same time; touch-up painting is necessary upon completion of anode installations. It is estimated that a team of workers rides the ship for three months on the ballast legs of the voyages. Total cost of the riding team would be $60,000, which includes wages, insurance, accommodations, and food. Total cost of anodes, welding, and painting materials is $65,000. The competing alternative is to do the same work during the upcoming dry docking of the ship. Planned duration of the dry docking is seven days, but in order to carry out the repairs in the tanks the ship would be delayed, incurring a cost of $40,000 in lost profit. Total cost of tank repairs with installation of anodes, welding, and painting carried out at a shipyard is estimated to be $95,000. Additional time in the dock based on the per-diem drydocking fee costs an additional $13,000.

The cost of the work if performed by a shipyard ($95,000) is substantially lower than if it were done by a riding team ($60,000 + $65,000 = $120,000) due to the relatively low productivity of the riding team. However, reduction of shipyard time will bring additional benefits ($40,000 + $13,000 = $53,000) to the riding team alternative. The total count gives $28,000 = ($95,000 + $53,000) – $120,000 cost advantage to the riding team.

TIME FACTOR IN OPERATION

Time factor in operation should be looked at from two different perspectives: first, impact on the duration of the ship's service life or the period between shipyard repairs and overhauls, and, second, as an impact on a structure of the ship's operational and nonoperational time. The first subject has been addressed in the previous chapters. The subject of the ship's operational and nonoperational time comprises analyses of the ship's voyage time and its components, and the time that the ship is temporarily taken out of service either for repairs or because of absence of work. Any effective projects directed at better use of the time in operation or toward reduction of the time out of use should result in increased revenues and most likely in reduction of certain operational costs. The following are the principal areas and objectives of this type of cost improvement project:

1. Reduction of duration of loading and unloading operations, both in port and while lightering at sea. Examples of the projects: new equipment onboard and in ports, improvement of loading and unloading procedures, improvement of cargo inspections and analyses, increased capacities of lightering ships and/or elimination of lightering, etc.
2. Reduction of delays in ship departure due to surveys, inspections carried onboard, or late delivery of supplies, food, spares, etc. The principal projects in this area are dealing with improvement of

management and scheduling of the above activities, which can be achieved with general computerization of the ship management process.

3. Reduction of the sailing component of the voyage time by increasing ship's speed or by improvement of navigation practices and procedures. The increase in speed of the ship in operation might be achieved by various technical projects to be carried out during shipyard repairs or even while the ship is in operation (in port, at a temporary lay-up, etc.) Examples of such projects are a new or improved propeller, installation of a propeller duct, improved painting (providing less friction), increase of the main engine output by increasing the compression ratio or adding super-charging. Important components of the justification for any projects related to the reduction of the wasted time in operation have to involve daily cost of ship delay, or the lost daily profit per ship.
4. Reduction of the nonproduction delays of the ship, related to emergency repairs. Reduction of the failures in operation can be achieved by improving maintenance procedures and planning shipyard repairs.

Example 13-9

For a properly designed vessel, there is little that can be economically accomplished to improve performance above the design level once the vessel is placed in service. One of the few examples of productive efforts in this direction was developed by Schneekluth (see References). As per Bush and O'Leary, his device, called a wake-equalizing duct or a wake improvement duct, has become a popular retrofit device for propulsion enhancement.

The wake improvement duct consists of two half-rings with airfoil cross sections which are welded to the hull forward of the propeller. In place, the rings assist the flow in resisting separation from the hull to improve the propulsor performance. The fuel rate reduction for a 41,000 LT displacement parcel tanker with a 21.32-foot propeller at full load and design speed was measured at 7.2 percent.

The following data for a parcel tanker is used to determine the economic benefit from retrofitting a wake improvement duct. The ship operates at the same power in ballast at a higher speed. The interest rate is 8 percent, assuming no inflation.

Propulsion fuel rate	.3550 lbs/shp-hr
Days in operation	350 days per yr.
Bunker fuel cost	$100/short ton
Port days	20 days/yr.
Nominal engine power	12,000 shp

Installed cost of duct	\$138,000
Fuel rate improvement	7.2%
Additional maintenance	None
Remaining economic life	20 years

Equivalent uniform annual cost = 138,000 (A/P, .08, 20) = 138,000 (.1019) = \$14,062

Equivalent uniform annual benefit = (350 − 20 days) (24 hrs/day) (.3550 − .3311 lbs/shp–hr) ($^{10}/_{2,000}$ \$/lb) (12,000 shp) = (330) (24) (.0239) (.05) (12,000) = \$113,573

Net annual benefit = 113,573 − 14,062 = \$99,511

Present worth of annual gain = \$99,511 (P/A, .08, 20) = \$976,989

Example 13-10

A 250,000-dwt crude oil carrier is employed in a trade that is 4,000 nautical miles in the loaded direction. This is a 15-knot vessel which is operating at sea or in port 350 days per year. Maintenance procedures, cargo transfer rates, port clearance procedures, crew training, berthing priorities, and other operating factors determine the port time per voyage.

Table 13-3 illustrates the impact of 5 to 20 days' port time per voyage on the key performance measures for this tanker.

TABLE 13-3

Economic Impact of Port Time

Port Time, days / voyage	*Rate of Return, %*	*Voyages / yr.*	*Discount Freight Cost, \$ / ton*	*Cargo, million tons / yr.*
5	28.68	12.86	4.092	3.099
10	24.60	10.86	4.580	2.619
15	21.15	9.40	5.047	2.267
20	18.11	8.29	5.493	1.999

Example 13-11

A typical 245,000-dwt double-hull tanker operating at 13.2 knots on an 8,000- mile round-trip hire can be expected to move 2,428,000 tons of crude oil per year at a cost of \$4,853 per ton if the ship is maintained in service 355 days per year. Table 13.4 illustrates the deterioration of this performance as the days in operation are reduced from 355 to 330 days.

TABLE 13-4
Economic Impact of Annual Days in Operation

Days in Operation/yr.	*Rate of Return, %*	*Voyages/yr.*	*Discount Freight Cost, $/ton*	*Amount of Cargo, million tons/yr.*
355	25.42	10.07	4.853	2.428
350	25.13	9.93	4.903	2.394
345	24.85	9.79	4.954	2.359
340	24.56	9.65	5.008	2.325
335	24.26	9.50	5.062	2.291
330	23.97	9.36	5.118	2.257

Spare Parts and Material Management Principles

Spare parts carried aboard a vessel or stored ashore are an asset of the corporation. Despite the large investment in such items, they have no operating value until the moment of needing them to avoid disruption of a revenue-generating asset. Materials are a period expense which contribute nothing to vessel operations until consumed. In fact, inventories of spare parts and materials cause expenses such as storage cost, purchasing costs, interest on investment, handling cost, and revenue reductions.

Classification society rules include recommendations for onboard spare parts. However, it is the shipowner's responsibility to determine the specific spares to carry and/or to stock ashore. This determination should be based on an analysis of spare part economic variables, including prices, inventory cost, availability, revenue loss due to shortage, operating cost, distances, part deadweight revenue loss, vendor part stocking service, etc.

In an ideal world, the fleet owner would be able to purchase parts and materials in the correct quantity at the moment of need. In the real world, purchasing and stocking in advance assures that the items will be available when needed to avoid a disruption in vessel operation. Establishing an economic order quantity and the optimum reorder point, i.e., minimum stock level, assures that the stock costs involved are balanced with the out-of-stock penalty cost to minimize overall costs. One of the stocking costs involved in shipboard storage of spare parts is the reduction in revenue due to loss of cargo deadweight equal to the spares' deadweight.

Improvement of inventory and ordering of spare parts is normally included in the maintenance development program. However, separate projects might be undertaken which are aimed at reduction of inventory, and especially at reduction of ship's delays related to delivery of spares.

Major shipping or fishing companies with a fleet that includes series of similar design ships might create a shore storage of certain bulky and expensive spares, which eliminates the duplication in the ships' inventories and also allows them to store certain spare parts, especially those with a long lead time, which are normally ordered as necessary. Although increasing the assortment of spares, it normally decreases their combined cost due to the elimination of duplication. As a result, there is a savings in the cost of spares stored and also in the delays on delivery.

Computer software has been written to manage repair parts and materials so that the overall system operates at the minimum total cost. Wemmerlov, Prichard, and Giffin (see References) describe the numerous models of relationships needed for developing software for the optimization and management of a repair parts stocking system. The models are both deterministic and probabilistic. The former is popular for manufacturing systems where there is little random variation in demand. In this text, we will consider one probabilistic model which generally fits the needs of a maritime spare parts system.

Giffin and Prichard describe an inventory model developed by researchers for a fixed reorder point. In this model, the order quantity is the annual average demand divided by the cycle periods per year, the demand during a cycle is a Poisson distributed random integer, the spare part is used continuously over time, the vendor of the part has a fixed delivery time, a modern communication system exists so that a central manager knows immediately when a part is used, back orders are permitted so that all demands are eventually filled, and, finally, a reorder point is determined so that total costs are minimized. Figure 13-5 illustrates the operation of

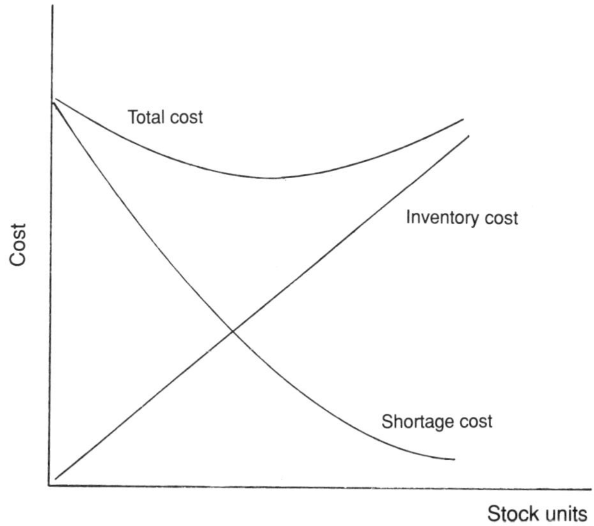

Figure 13-5. Inventory systems cost

such a system over time. Figure 13-6 shows the relations between the stocking costs (which increase with the units stocked), the shortage cost (which decreases with the units stocked), and the total costs (which is the sum of the former two costs).

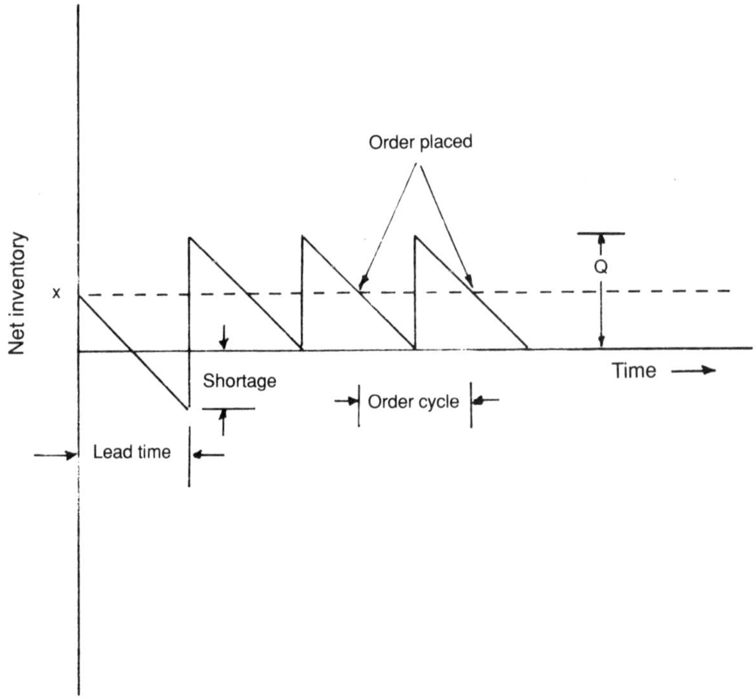

Figure 13-6. Inventory versus time for an inventory system

$$X = s + dl$$

where X = reorder point stock level, units
s = safety stock level, units
d = part demand/ yr.
l = lead time, years

The safety stock for a given reorder point is $s = X - dl$ and the cost of carrying this safety stock during one order cycle is:

$$\text{Carrying cost} = \frac{acsq}{d} = \frac{ac\,(X - dl)\,q}{d}$$

where a = stocking cost/year, fraction of purchase cost
c = cost of part, \$
q = purchase quantity

and

$$\text{Cost of a shortage} = b\,[1 - P(x)]$$

where b = cost of a stock shortage
P(x) = probability of x or less during lead time

Therefore, the total cost is

$$t(x) = \frac{ac\,(X - dl)\,q}{d} + b\,[1 - P(x)]$$

where x = demand during lead time
P(x) = cumulative probability of x during lead time

The minimum total cost reorder point occurs at X_*. The difference between the minimum cost reorder point and the cost for the next higher X approaches zero. Therefore,

$$t_{X_*} - t_{X_* - 1} \rightarrow 0$$

and substituting

$$t_{X_*} = t_{X_* - 1}$$

into the relationship for tX yields the following:

$$\frac{ac\,(X_* - dl)\,q}{d} - \frac{ac\,(X_* - dl - 1)}{d}\,a + b - bP(X_*) - b + bP(X_* - 1) = 0$$

$$\frac{acq}{d} - b\,[\,P(X_*) - P(X_* - 1)\,] = 0$$

since

$$P(X_*) = \sum_{0}^{X_*} p(x)$$

$$P(X_* - 1) = \sum_{0}^{X_* - 1} p(x)$$

$$P(X_*) - P(X_* - 1) = p(X_*)$$

Therefore:

$$\frac{acq}{d} - bp(X_*) = 0$$

$$p(X_*) = \frac{acq}{bd}$$

and

$$p(X_* + 1) = \frac{acq}{bd}$$

and the least cost reorder point is the integer which is:

$$p(X_*) > \frac{acq}{bd} \geq p(X_* + 1)$$

Since the demand x is Poisson distributed, the density function is:

$$p(X_*) = \frac{e^{-\lambda} \lambda^x}{X!}$$

$$p(X_*) = \sum_{x=0}^{n} \frac{e^{-\lambda} \lambda^x}{X!}$$

where λ is the distribution parameter which for the Poisson is equal to the mean and the variance. Table 13-5 contains a list of Poisson probabilities for $\lambda = 9$.

TABLE 13-5

Poisson Distribution for $\lambda = 9$

x	*p(x)*	*P(x)*	*1-P(x)*
0	.0001	.0001	.9999
1	.0011	.0012	.9988
2	.0050	.0062	.9938
3	.0150	.0212	.9788
4	.0338	.0550	.9450
5	.0607	.1157	.8843
6	.0911	.2068	.7932
7	.1171	.3239	.6761
8	.1338	.4577	.5423
9	.1297	.5874	.4216

11	.0970	.8030	.1970
12	.0728	.8758	.1242
13	.0503	.9261	.0738
14	.0324	.9585	.0415
15	.0195	.9780	.0220
16	.0109	.9889	.0111
17	.0058	.9947	.0053
18	.0029	.9976	.0024
19	.0013	.9989	.0011
20	.0007	.9996	.0004
21	.0002	.9998	.0002
22	.0001	.9999	.0001
23	.0001	1.0000	.0000

Example 13-12

A cruise liner shipowner operates three vessels from a Florida port on a 10-day cruise turnaround. The vessels all have three medium-speed diesel-generator sets and four diesel generator main propulsion units. All of these engines are the same model and utilize identical spare parts. The engines of the three vessels are eight cylinder and the fleet has 168 fuel injectors in service. The annual replacement rate of the fuel injectors averages 36. The ship operator stocks the spare fuel injectors. Spare part repair orders are placed each quarter. The cost of a fuel injector is $1,600 per unit and the stocking cost is 50 percent of the purchase price per year.

The measure of a shortage is the shortage of a single injector spare since that will cause an engine to be unavailable. The operators have found that the average cost of an engine temporarily out of service is $3,800. Determine the economic reorder point.

d = 36 injectors per year
l = ¼ year
c = $1,600
a = .5 per year
q = 36/4 = 9 units
b = $3,800

Table 13-6 contains the tabulation of the total cost of stocking the fuel injector.

TABLE 13-6

Total Cost for Various Reorder Points

Safety Stock(s)	*Holding Cost, $*	*Reorder Point, X*	*Shortage Cost*	*Total Cost, t(X)*
0	0	9	1,568	1,568
1	200	10	1,117	1,317
2	400	11	749	1,149
3	600	12	472	1,072*
4	800	13	281	1,081
5	1,000	14	158	1,158
6	1,200	15	84	1,284
7	1,400	16	42	1,442

*minimum cost reorder point

$$p(X_*) > \frac{acq}{bd} > p(X_* + 1)$$

$$p(X_*) > \frac{(.5)\ (1{,}600)\ (9)}{(3{,}800)\ (36)} \geq p(X_* + 1)$$

$$p(X_*) > .0526 \geq p(X_* + 1)$$

Therefore, to satisfy the relationship, use Table 13-5 to find

$$p(13) = .0503 \text{ and } p(14) = .0324$$

and the economic reorder point occurs when stock drops to 13, providing a safety stock of

$$s = X - dl = 13 - (36)\ (1/4) = 4$$

REFERENCES AND BIBLIOGRAPHY

Bowker, Albert H. and Gerald J. Lieberman. "Engineering Statistics," Englewood, N.J.: Prentice-Hall, Inc.; 1959.

Bush, A.D., and O'Leary, J.R. "Full-Scale Testing of a Wake Improvement Duct on a 31,000-dwt Parcel Tanker." Glen Cove, N.Y.: Webb Institute, 1990.

Comstock, John J., ed. Principles of Naval Architecture. SNAME, 1967.

"Fouling-free Fortunes," Marine Engineers Review, July 1991.

Giffin, Walter C. *Introduction to Operations Engineering.:* Richard D. Irwin, 1971.

Hunt, E.C., ed.-in-chief. *Modern Marine Engineers Manual,* Vol. II, 2d ed. Centreville, Md.: Cornell Maritime Press, 1991.

Hunt, E.C., ed. "Symposium Proceedings, Sustaining Design Thermal Performance of Ship Propulsion Machinery." Institute of Marine Engineering, February 1978.

Prichard, James W., and Robert H. Eagle. *Modern Inventory Management.* New York: John Wiley & Sons, 1965.

Rynn, Hugh S., and Rod Vulovic. "Re-Engining on the Run: New Generator Diesels for Sea-Land's Atlantic Class Vessels," SNAME Annual Meeting, October 1992.

Schneekluth, H. "Ship Design for Efficiency and Economy." London: Butterworth & Co. Ltd., 1987.

———"The Wake Duct, Old and New Aspects." Sixth Lips Propeller Symposium, May 1986.

Wemmerlov, Urban. *Inventory Management and Control: Handbook of Industrial Engineering.* New York: John Wiley & Sons, 1982.

QUESTIONS AND PROJECTS

1. Discuss the components of operational cost, their relative shares, and their impact on the selection of capital projects intended to improve performance and reduce the cost.
2. If the share of fuel cost in annual operational expenses is 50 percent, would it mean that the fuel conservation projects are the most profitable? Explain.
3. Select data from a published sea trial and prepare a curve of r/n^3 versus v/n with prediction intervals.
4. If the main engine fuel rate is increasing, how would you expect this to be reflected in the change of the I/n^3 versus v/n curve.
5. What change in the curve would you expect if the hull is deteriorating?
6. Explain how you would set up a system to monitor ship cost performance and identify the underlying problems.
7. What are the two principal types of economic results of capital investments in ship operation? Explain the impact of the improved performance of the ship on the operation results.
8. If the annual amount of cargo carried by the ship increases, how would you estimate the economic benefits?
9. Give examples of fuel conservation projects and explain the procedures and elements of their economic justification.
10. How would you approach a problem where the ship operators had sufficient reserve power to maintain speed as the hull resistance increased?
11. If the revenue per year was unknown, how would you approach the hull-cleaning problem using cost information? What cost information would be required?

12. Discuss a plan to collect data for the determination of the monthly speed deterioration between hull cleanings.
13. What design and operation conditions are necessary to consider a crew reduction? What type of initial expenses would normally be required?
14. Describe the components of maintenance and repair cost. What are the principal areas of possible reduction of maintenance and repair expenses?
15. Give examples of repairs which might be carried on board by riding repair teams; provide arguments to support your choice versus carrying the repairs at the shipyard.
16. Discuss Example 13-7 on the re-engining of ships while underway. What is your opinion regarding limitations of this approach from the economic standpoint?
17. What is the economic nature of the time factor in ship operation? How would you approach the problem of improvement of time utilization in operation? Give examples of projects.
18. Select a typical double-hull tanker and determine the impact on the internal rate of return for an initial fuel cost which may vary between \$170 and \$200 per metric ton. What is the difference between the equivalent annual cost of ship operations for fuel at \$170 compared to \$200 per metric ton?
19. Does the internal rate of return for a typical tanker vary linearly with the annual days in operation? Does it vary linearly with the port time per voyage? If not, why?
20. What are the factors considered when calculating the stocking or holding costs of shipboard repair parts?
21. What factors are considered when determining the cost of a shortage of a spare part? If a shortage of one spare part shuts down a machine, does the simultaneous shortage of two parts increase costs? If so, by how much?
22. There are many useful inventory models in the literature. What are the characteristics of the model described in this text?
23. What is the lead time? Is it the same as the cycle time?
24. How would you verify that the Poisson distribution is the best model for describing the demand for spare parts during an inventory cycle?
25. Why is the cost shortage equal to $b(1 - p(x))$. What does $1 - p(x)$ represent?

PROBLEMS

1. An engineering company was contracted by a shipowner to study performance-monitoring procedures onboard and to give proposals for their improvement. Total expected fuel savings if the proposals

were carried out was estimated to be $60,000 a year, and maintenance and repair costs would be reduced by $15,000. The total consultant's invoice was $45,000. Additional control and purification equipment with a total cost of $175,000 would be required, and cost of crew overtime for installation is estimated to be $35,000. The remaining service life of the ship is eight years. Assuming the rate of returns 15 percent, evaluate the proposal. Would you suggest the shipowner implement it?

2. Recalculate Example 13-2 for an annual saving of $25,000. Would you change the previous conclusion?
3. A former owner's cabin on board a midsize cruise ship might be retrofitted into a luxury suite while the ship is underway. Total cost of retrofitting is $95,000, and the revenue might grow by $40,000 a year. Additional passenger cost would be $15,000, and annual maintenance cost will grow by $2,000. The ship is expected to be operated for another 10 years. Assuming a 20 percent rate of return, evaluate the project. What might be maximum cost of construction?
4. If, in Example 13-4, the rate of return is 25 percent, will you still consider the project sound?
5. Analyze Example 13-7 and calculate the annual savings and equivalent uniform annual benefits for different remaining service lives (25, 20, 15, and 10), assuming a 12 percent rate of return, and for different rates of return (10, 12, 15 and 20), assuming 15 years' remaining service life. Plot the results.
6. Select a typical single hull tanker of about 100,000-dwt and determine the economic impact of varying the port time from 5 to 10 days per voyage. Assume a 9,000-nautical-mile round trip and 20-year economic life.
7. Select a typical double-hull tanker of about 100,000-dwt employed continuously on a 9,000-nautical-mile round trip and determine the impact on the equivalent annual operating cost for in-service periods of 345, 350, and 355 days per year during the 20-year economic life.
8. For the wake improvement duct in Example 13-9, determine the maximum amount the owner could pay for the installed duct at the break-even point.
9. For the wake improvement duct example, what is the rate of return?
10. For the wake improvement duct example, final the net annual gain for interest rates of 10 percent and 12 percent.
11. Redo the example problem for the Least Cost Reorder Point model, changing the cycle so that orders are placed each month. This change will increase a to .58.

Chapter 14

Ship Financing

Theodore A. Ulrich

Capital, viz., accumulated wealth employed with a view to profit.
—*Frederick Malthus,* Principles of Political Economy

Once a need for new, replacement, or used tonnage has been identified, a corporate body will need to become familiar with the current types of financing available unless it is able to purchase the asset with available cash. To be considered are traditional bank loans, leasing arrangements, and shipyard financing, as well as innovative mezzanine financing and shipping funds. The types of financing and the required documents and covenants are to a large extent standardized in the maritime industry, with changes and developments generally less revolutionary than in other industries, due in large measure to the history of the shipping industry. This chapter is based in part on Lenkowsky and Ulrich (see References).

BANK LOANS

It is not surprising to learn that the traditional approach to ship financing—borrowing from institutional or specialized ship lenders based on the value of the assets and/or the earning potential of one or more ships—is the rule rather than the exception. In part this is because the lender can take advantage of well-established and reliable legal systems for the creation, preservation, and enforcement of its rights and to manage the overall level of risk assumed. The degree to which a lender will become an investor varies, but few lenders consider shipping loans to be of insignificant risk, given their past history. As a result, for many shipowners this will still be the only viable source of funds.

Ship loans follow a fairly consistent pattern from one jurisdiction to the next. The lender is interested in the financial condition and track record of the borrower, the condition and age of the vessel, the business plan of the borrower for employment of the vessel, and the resulting quality and quantity of collateral securing its loan. The degree to which a lender will rely on the vessel's value may be eroded by the low market values of certain tonnage. However, higher market values do not necessarily satisfy a lender's concerns. If the vessel is not a new building, the trading history of the vessel and the circumstances of its sale to the borrower take on greater significance to the lender because, under maritime law in most jurisdictions, many claims will follow a vessel through successive owners, representing a financial drain on the borrower and a threat to the collateral value on which the lender predicated its decision to make the loan.

If the borrower will have obligations (such as debt service) payable in one currency and income (such as charter hire) payable in a different currency, the lender will insist on including currency swap or purchase arrangements so as to minimize or eliminate the exchange rate risk over the life of the loan.

Documentation

The documentation for the conventional ship loan generally includes at least the following documents:

1. *Loan agreement*. This document will set forth the basic terms of the loan, representations by the borrower (such as legal capacity and due authorization for the loan, lack of illegality or interference by the loan with other agreements or the laws applicable to the borrower, financial condition of the borrower, and compliance with all requirements of the lender), covenants of the borrower (such as maintenance of insurance, maintenance of the vessel, compliance with laws, retention of earnings, and limitations on investments, mergers, and dispositions of assets), events of default under the loan and the consequences thereof, and remedies that may be available to the lender upon such an event of default.
2. *Promissory note*. The requirement for, or significance of, a separate instrument evidencing the indebtedness of the borrower to the lender varies among jurisdictions. A negotiable promissory note may afford the lender a particular advantage in managing its portfolio of ship loans. Certain loans may be syndicated and other lenders will thus acquire a portion of the loan. If the borrower desires to deal with a specific lender, that lender should be retained as the "lead bank" in the syndication and the documents should so specify.
3. *Ship mortgage*. A valid and duly recorded ship mortgage, generally under the law of the vessel's flag after acquisition, is usually the principal security for the loan. The ship mortgage will represent an *in rem* claim on the vessel that will support the lender's taking

possession of the vessel, either to operate the vessel or to sell it (generally through a foreclosure sale), in the event that the loan is in default. The status of the ship mortgage as a perfected lien entitled to priority over other claims is a matter of the law of the jurisdiction of the flag of the vessel and the recognition offered to that law by the law of the jurisdiction in which the foreclosure proceedings are pursued. While the ship mortgage is important to the lender, the lender cannot rely on it exclusively because the market value of the vessel will fluctuate, frequently in cycles that similarly affect the business of the borrower. Furthermore, the priority accorded the ship mortgage against the claims of other creditors of the vessel may vary from time to time by the mere fortuity of the jurisdiction in which the vessel is located at the time of foreclosure as well as the secret nature of maritime liens which need not be recorded to be entitled to priority. Thus, lenders invariably require additional security as noted below.

4. *Assignments of charter and charter hire, insurances, freights, etc.* The borrower will generally be required to assign, as additional collateral, all ordinary and extraordinary cash flows relating to the vessel. These would include periodic charter hire payments, event of loss payments under hull insurance, off-hire insurance reimbursements, and government requisition payments. Typically, these cash flows will also be made payable directly to the lender or may be required to be routed through an account under the control of the lender. In order to assure itself of repayment in the ordinary course, the lender will place a high premium on known employment for the vessel with creditworthy charterers. Commitments of charterers, through "hell-or-high-water" or similar undertakings, that assure the owner (and thus the lender as assignee of the owner's rights) of payments sufficient to satisfy debt service and other undertakings in favor of the lender are most desirable. More commonly, lenders must evaluate the risk of off hire, the potential for early termination of any long-term charters, the inability or unwillingness of insurance to respond to claims, the imposition of new legal requirements on the vessel during its life which may require significant expenditures, and the unexpected casualty or breakdown that impacts upon the owner's profit from the vessel.
5. *Guaranties, third-party supports, and additional collateral.* The lender's overriding interest in any ship financing transaction is ensuring the timely repayment of its loan. If the financial condition of the shipowner itself or the cash flows reasonably expected from the employment of the vessel do not provide assurance of repayment, including the lender prudently discounting any certainty of repayment from an exercise of its remedies under a ship mortgage, the

lender will require a creditworthy person (for example, one or more of the shareholders or principals of the shipowner, or another financial institution) to guarantee the obligations of the shipowner to the lender through a guaranty or similar undertaking. Lenders generally will expect a "guaranty of payment" or its equivalent, which assures the lender that the guarantor will make each payment that the shipowner does not, rather than a "guaranty of collection" which assures only that, if the lender cannot recover payment after exhaustion of all remedies against the borrower, the guarantor will respond. Alternatively, another party associated with the vessel (for example, a long-term charterer) may provide such direct assurance to the lender that some portion or all of the loan will be paid. Such direct assurances by third parties may not be possible in certain forms of ship financing; for example, in U.S. tax-advantaged leveraged-lease financing, the charterer of the vessel cannot undertake to pay the shipowner's debt obligations without destroying the essential characteristic of the lease financing. The leveraged-lease structure addresses this issue by requiring the charterer to pay charter hire that is always in an amount sufficient to pay the shipowner's debt service; however, once paid, the charterer need not pay it again even though the lender may not receive the payment due to claims of third parties. A lender not presented with a long-term charter for the vessel being acquired may insist upon mortgages on additional vessels or assignments of unrelated cash flows, such as charter hire on sister ships, to support repayment of the loan. A long-term lender may also require that the borrower obtain some level of residual value insurance, when available, that will assure the lender of a certain minimum amount in the event that the vessel must be sold by the lender upon a default by the borrower. Letters of credit from other banks may also be utilized to provide standby coverage for defaults in payment by the borrower.

LEASE FINANCING

As an alternative to a traditional loan financing, shipping companies may find that lease financing is attractive. In a leasing structure, a separate company owns the vessel and charters the vessel to the shipping company for an appropriate period which is often, with respect to specialized vessels, a substantial portion of the economic useful life of the vessel. The advantage to the shipping company lies principally in the ability of the lessor (which often has substantial taxable income in its affiliated group of companies from unrelated business operations) to take better advantage of the available tax deductions and credits available in one or more of the taxing jurisdictions and associated with the ownership of the vessel, such as deductions for depreciation, interest expense, and capital investment incentive programs (for example, an investment tax credit). It then theoretically passes along some of those savings to the shipping company through reduced charter hire. Frequently, the shipping company is unable to take advantage of the opportunity to reduce its own tax liability due to poor current or anticipated earnings, recent or anticipated losses, recent capital expenditures for other items (generating their own tax advantages), incompatible tax attributes of its other operations, or non-tax driven limitations such as credit agreement restrictions, legal requirements for ownership of vessels, or other factors. However, the lessor is also interested in the potential for profit from the rechartering or resale of the vessel after expiration of the initial lease, since the residual value in the vessel at that time, absent a sharing or other agreement, belongs to the lessor as owner.

In the typical long-term lease financing, the lessor will acquire the vessel desired by the lessee (either a new building or a used vessel) from funds derived either from its own sources of equity or from a third-party lender. The ratio of equity to debt is determined by a variety of factors, including the interest rate required by the lender, the rate of return required by the lessor on its investment in the vessel (which is a function of cash payments, tax advantages, and anticipated residual value of the vessel), and the speed with which the lender desires to be repaid during the term of the lease. The lessee should evaluate the economics of a leasing structure in terms of the effective interest rate that will be paid by the lessee over time through payments of charter hire for the extension of credit to the lessee (represented by the investment of the lessor in the vessel) and the extent to which, on a net present value basis, the aggregate charter hire to be paid during the term of the lease is appreciably less than would be the cost to the lessee of owning the vessel directly. How much less will often depend on the willingness of the lessor to accept a portion of its return by way of the residual value of the vessel. If it is unwilling to do so, the lessor will expect higher returns (generally meaning higher cash payments by the lessee in the form of charter hire) during the term of the lease, and maybe greater payments earlier in the term rather than later. When market values of vessels are rising, lessors are normally more interested in the potential for profit upon realization of the residual value of the vessel at the end of the lease. Thus this type of financing is more popular during those periods. Greater savings in the effective lease cost to the lessee may also be obtained through careful selection of more than one taxing jurisdiction within which tax advantages of ownership may be obtained. While a discussion of such "multiple dip" and cross-border transactions is beyond the scope of this chapter, such arrangements are not unknown in ship financing.

Lease financing adds additional dimensions to the concerns of a lender providing the financing (the "leverage" in a leveraged lease). In addition to the traditional credit, asset value, and operational risks associated with a shipping loan, the lender now has another party to worry about: the lessor may go bankrupt (this is also a concern for the vessel's charterer if a bankruptcy court were to reject the lease and demand the return of the vessel), or other creditors of the lessor may assert claims against the vessel. On the other hand, the lender may find attractive the rather large equity investment in the vessel by the lessor, especially if the charterer should be unable to pay full charter hire for the entire term.

Additional documentation for lease financing includes the lease, which may be a bareboat charter (in which the charterer has full use of the vessel; is responsible for all operating, maintenance, and administrative matters regarding the vessel; and has full possession and control) or a time charter (in which the owner or its agent is responsible for operation, maintenance, and administrative matters, and the charterer obtains the use of the vessel), or some variant in between. The lease will set forth the duration, the periodic charter-hire payments to be paid, maintenance and insurance requirements, operational restrictions, lessor remedies upon default by the lessee, renewal and purchase options, and the responsibilities for various expenses, ordinary and extraordinary. If the lessee's own credit is not sufficient to support the lease, there may be a third-party guaranty, a sub-lease to a creditworthy user, entry in a charter pool, a long-term contract of affreightment with a substantial customer of the lessee, or some other means of assuring cash flow to the lessor and its lenders. If the lease is a leveraged lease there will also be lender documentation similar to that in a traditional ship loan, with the important difference being that the loan is to the shipowner, not the charterer, although it is the charterer, not the shipowner, whose performance (credit, business, and operational) is important to the lender. The lease, or a separate agreement, will also contain far-reaching indemnification provisions that aim to reduce the operational risk to the owner and lender which is inherent in vessel ownership and, in the owner's case, will also cover the loss or reduction of the tax advantages that make the leasing structure attractive.

MEZZANINE FINANCING

Known by several different names, "mezzanine financing" is a hybrid form of debt financing where a party is willing to take a secondary position behind the senior debt in a ship-financing arrangement while remaining a step ahead of the equity in the borrowing company (which consists of the cash and assets brought to the company by its principals). Mezzanine financing has been frequently utilized for domestic corporate acquisitions. Because of its unusual position—usually unsecured, with payments deferred and subordinated to prior payment of the senior debt—mezzanine financing often bears a high interest rate reflecting the risky nature of repayment.

The attractiveness of mezzanine financing to an investor is a combination of the rate of return and other features, such as conversion rights or other devices, that allow the mezzanine lender to share in the benefit of an increase in value of the borrower's assets or of the borrower itself. For example, mezzanine financing may start out as a subordinated debenture, entitling the holder to a fixed rate of interest until maturity, but may be converted to common stock for a fully participating interest in the net profits and capital appreciation of the company, should the debenture holder determine that this type of investment is attractive. Alternatively, mezzanine capital may be structured as preferred stock (prior in right only to the common stock), with a periodic distribution at a specified rate from available cash flow of the company and a proportionate liquidating or redemption payment from the sale of specified assets. From a borrower's perspective, mezzanine financing (1) provides capital that may simply be unavailable from a senior lender that prefers greater equity in the company, (2) reduces the dependence of the borrower on a single source of all-invasive debt financing, (3) provides a more flexible repayment structure for borrowed funds than would be available from a conventional lender, and (4) reduces, at least in theory, the borrower's need to meet the typically stricter financial coverage tests and covenants applicable to senior debt. Although a high debt-to-equity ratio has been a traditional characteristic of shipping companies and may deliver high returns on equity when markets are good, high leverage is a disadvantage when the markets reverse, leaving many shipping companies with high long-term debt payments requiring regular service. Lower leverage means a lower overall return on equity but a greater ability to weather a downturn. Mezzanine financing can thus effectively reduce leverage without permanently diluting the interest of the equity investors.

As noted, mezzanine financing is not without cost, which is the generally high interest rate required on this subordinated debt. If shipping markets generally recover enough to make credible a conversion feature (for example, at the option of the holder, the subordinated bond could be converted: for a certain number of shares of preferred stock with a higher fixed dividend; or for common stock, with a variable dividend, the potential for capital appreciation, and a vote), this may result in a lower interest rate from the mezzanine lender. Since mezzanine financing is similar to equity in terms of risk of repayment, particularly in the early stages of the senior debt financing, mezzanine lenders may seek to be involved in the management of the borrower, either through actual stock ownership or through additional covenants in the subordinated loan agreements.

SHIPYARD FINANCING

Each of the above types of financing has been utilized in the past with the lender being a shipyard or other associated entity. Such financing is especially prominent outside the United States and may comprise the entire financing package or merely a portion of it. Shipyard financing may act as a type of mezzanine financing and involve the deferral of a portion of the construction price. One of the impediments to such foreign shipyard financing on U.S. flag vessels had been the necessity of a Westhampton trustee and the problems attendant upon any enforcement of a potential violation of the U.S. citizenship requirements for U.S. flag vessels; however, this problem has to some extent been alleviated by the recent recodification of the Ship Mortgage Act. Moreover, shipyards have been in the past reluctant to extend credit or become lender for any substantial period of time. While this has not been a serious problem with foreign flag vessels, where short-term financing is common, it has not made such financing attractive to U.S. owners except as construction period or bridge financing or, in some cases, to absorb the excess requirements not met by senior debt.

Shipyard financing involves certain issues unique to its source, such as the shipyard's concern that other potential customers are not sold the new tonnage upon completion (thus, "prohibition of sales" clauses), that disputes are dealt with expeditiously, and that the shipyard retains control of the design for future production. It should also be realized that shipyards are, in the main, unaccustomed to acting as lenders and this factor may delay or complicate communications from time to time. On the other hand, support of shipyard financing by foreign governments has been significant in the past and may take the form of incentives other than attractive rates, e.g., tax or duty reductions, labor cost support, and other operational incentives.

SHIPPING FUNDS

A recent development which has attracted much interest as a mechanism for reviving shipping markets are the publicly offered equity funds. While each fund varies, they have as a common purpose the acquisition, operation, and disposition of ships in a managed program. Some of the funds are directed toward profitable operation of a stable fleet of vessels over time, while others (generally referred to as self-liquidating) have as their goal the timely disposition of all vessels owned by the fund as values peak, and a prompt winding up. An important characteristic of publicly offered funds that has attracted investors is the advantage of liquidity: funds whose shares are listed and traded on a recognized stock exchange come with an escape route. They have been most attractive to investors seeking to capitalize on rapidly increasing market values (such as secondhand vessels) with a relatively small investment. While some of the funds emphasize low debt-equity ratios (some have been designed for nominal leverage), the returns to investors have generally been impressive, largely as a function of improving charter rates rather than, as yet, significant asset dispositions.

The general concept of the funds is strikingly simple: acquire vessels at bargain prices, operate them efficiently and conservatively, and sell when the market for the vessel has improved. Amounts raised from investors are applied to the purchase of the vessels and initial working capital requirements. Fund fleets are generally composed of medium-sized bulk carriers and product tankers, markets in which surplus shipping capacity is minimal and the purchase and sale of tonnage is regularized. Fund organizers typically will identify a segment of the market in which charter rates are increasing due to an undersupply of vessels at a time when new building activity is delayed. Because secondhand tonnage is less expensive than comparable new building and may be brought on line quickly, freight rates that may not have improved enough to support new building activity may nonetheless be profitable for the fund fleet. More recent fund efforts have attempted to extend the approach to crude oil carriers and other vessels. But due to volatility in the shipping and financial markets and, some commentators believe, investor saturation, recent offerings have been less successful than earlier ones.

The funds are generally organized as corporations in jurisdictions with tax laws beneficial to shipping companies. The ship-owning company will have limited staff and overhead, instead employing a management company to provide operational, technical, and financial services support for the activities of its vessels. The vessels may be chartered through an affiliated chartering pool or may be chartered directly in the market. Periodic cash distributions to shareholders of the fund will be made from net chartering revenues after operating and maintenance expenses, debt service, management fees, and provision for appropriate reserves. Special distributions will be made as vessels are sold. If vessels cannot be sold profitably due to poor market conditions, they may continue in operation until being scrapped.

Shipping funds offered to the public in the United States generally must comply with federal and state securities laws' disclosure and reporting requirements, including a registration statement concerning the shares of the fund being offered, under the Securities Act of 1933. The registration statement will describe the fund's formation, objectives, management, prior operating experience, and financial status and projections, and will disclose relevant information regarding risks of the investment, conflicts of interest that may be presented by competing opportunities of the fund's organizers, and material tax considerations that apply to the fund and investors in its shares. Periodically (generally quarterly, annually, and whenever a material event occurs), the fund will be required to report to its shareholders the results of operations and other events. An obvious boon to investors, compliance with these requirements can be a formidable task and is a clear break with the traditional reluctance of the shipping industry to invite public scrutiny. Not only is such compliance costly, but it also is the possible source of potential litigation concerning inadequate or misleading disclosure.

USED TONNAGE

Although each of the above techniques may be used in the context of acquiring a "used" vessel—indeed, the shipping funds' focus is on such tonnage—certain of the techniques may be of limited value in that they necessarily take into consideration tax benefits which are only available or are maximized upon the initial delivery of the vessel. Lease financing is one such example.

The potential purchaser and lender of used tonnage must consider in earnest a number of special factors beyond the absence or limited nature of tax advantages. Operating, maintenance, regulatory, and insurance considerations are critical in determining the asset value of a used ship. The lender will assess the possible uses of the vessel, taking into consideration potential or pending regulatory or statutory changes (for example, requirements for double bottoms). It will also take into account the potential exposure of the vessel to preexisting maritime liens and claims. These claims are generally secretive in nature and thus difficult to ascertain. As protection against such claims, which would have priority over a "new" mortgage, the lender will seek covenants and assurances as well as additional security (such as deposits) for coverage. The lender will also seek additional disclosures of claims or potential claims. There will, of course, necessarily be an impact upon the term of the debt or lease financing unless the lender or investor is convinced of the vessel's useful life and employment opportunities. The current utilization of the vessel, long-term charters currently arranged and fixed, renewal options, and other commitments will take on added significance. Market conditions regarding overtonnage and rate structure will also be factors.

REFERENCES AND BIBLIOGRAPHY

Curtis, Simon. *The Law of Shipbuilding Contracts.* London: Lloyds of London Press, 1991. Explanation of the various clauses usually found in shipbuilding contracts, including the forms thereof and case law interpretation.

Gilmore, Grant, and Charles L. Black, Jr. *The Law of Admiralty*, 2d. ed. New York: The Foundation Press, 1975. General treatise covering all aspects of maritime law, including mortgages, maritime liens, charter parties, and governmental policies.

Goldrein, Iain S., ed. *Ship Sale and Purchase*, 2d. ed. London: Lloyds of London Press, 1993. Coverage of secondhand vessel sales and purchases, including the commonly utilized forms. It also has a general explanation of the principles of financing merchant ships as well as shipbuilding contracts.

Lenkowsky, Steven, and Theodore A. Ulrich. "Ship Financing: Recent Developments." SNAME 1990 Ship Operations, Management and Economics Symposium, April 26, 1990.

Poole, Gordon L., Barbara B. Powell, and Donald T. Gray. "Financing of United States-Flag Vessels." *Tulane Law Review,* Vol. 56 (1982), pp. 1171-1284. General discussion of U.S. flag financing at that time.

Schoenbaum, Thomas J. *Admiralty and Maritime Law.* St. Paul, Minn.: West Publishing Co., 1987. General treatise on maritime law, including computer finding aids for the Westlaw system.

QUESTIONS AND PROJECTS

1. What is the traditional form of ship financing?
2. What are the concerns of a lender in a ship loan transaction?
3. What are the concerns of the borrower in a ship loan agreement?
4. Is it possible to quantify the lender's risk in a ship loan transaction? How would you approach this question?
5. To what degree should the lender and the borrower rely on a vessel's market value for loan collateral? How would you determine a vessel's market value? Why does the market value change?
6. How do claims against a vessel change the market value?
7. How do lenders deal with ship revenues in a different currency than the original loan?
8. What conditions make it impossible for a lender to depend solely on a mortgage agreement to protect its interests? What additional security may lenders require?
9. What is the role of third-party guarantees for ship loan repayment? What is the difference between "guarantee of payment" and "guarantee of collection"?
10. What are the principal documents in a ship-financing transaction? What is the function of each document?
11. What are the advantages of ship leasing to the lessor and the lessee?
12. What is mezzanine financing? What is the incentive for the investor with this type of financing?
13. What is the difference between an investor who takes an equity position and one who provides mezzanine financing?

14. Why would a shipowner resort to mezzanine financing for a new vessel?
15. Under what circumstances is shipyard financing most commonly employed?
16. What is the concept of a shipping fund?
17. If you were to compare various financing arrangements for the purchase of a new vessel on the basis of the net present worth, list the cash flow elements you would include each year during the economic life.
18. Outline a computer program which you might prepare to calculate net present worth, equivalent annual cash flow, and internal rate of return for a shipowner interested in evaluating alternatives for ship financing.
19. Machinery manufacturers sometimes participate in ship financing. Why do they do so?
20. Is it possible to issue bonds to raise money for ship financing? Is it possible to issue stock for this purpose? What is the difference?

Chapter 15

Multicriteria Decisions Using Paired Comparisons

Reason is just as cunning as she is powerful. Her cunning consists principally in her mediating activity, which, by causing objects to act and react on each other in accordance with their own nature, in this way, without any direct interference the process, carries out Reason's intentions.

—*Hegel,* Die Logik

Throughout this text, engineering and business decisions have been made by selecting the alternative with the most favorable economic measure-of-merit. This approach implies that it is feasible to incorporate all information pertinent to the decision within the selected measure-of-merit; this is usually the case. For example, criteria such as reliability and maintainability are included in the engineering decision model by using probability distributions to determine the expected annual costs of breakdowns, outages, and repairs. The annual fuel cost of a vessel is a function of sea state, wind, currents, days out of dry dock, and other variables. All of these can be incorporated into the measure-of-merit with the assistance of probability distributions. Large computer models for the analysis of shipping ventures, such as the program VENTURE, frequently include information that suggests that there are decision criteria other than the final measure-of-merit. If it is not feasible or convenient to incorporate multicriteria into the single measure-of-merit, a different decision process is required. One such method is analysis by pairwise comparisons of the available alternatives employing valid criteria.

Trurstone (see the References) introduced paired comparisons in 1927 as a method for assessing the strengths of several stimuli in psychophysical analysis when no meaningful absolute measurements were practical. Paired comparisons became the basis of a flexible and consistent multicriteria decision method developed by Saaty and others. The decision method is called the analytic hierarchy process. Computer software has been written to implement this method (see "Expert Choice" in the References), but any software capable of matrix manipulation is satisfactory.

REVIEW OF MATRIX OPERATIONS

An appreciation of elementary matrix algebra is necessary to use the analytic hierarchy process for multicriteria decision making.

A matrix is a rectangular array of numbers, symbols, or functions. The rows are usually identified by the letter *i* and the columns by the letter *j*. An element of a matrix may be defined by subscripts a_i, a_j, or a_{ij}. A general matrix may be represented as follows:

$$\begin{bmatrix} a_{11} & a_{12} & \cdots & a_{1m} \\ a_{21} & a_{22} & \cdots & a_{2m} \\ \cdot & \cdot & \cdots & \cdot \\ \cdot & \cdot & \cdots & \cdot \\ a_{n1} & a_{n2} & \cdots & a_{nm} \end{bmatrix}$$

An identity matrix is a matrix with all elements equal to zero, except the diagonal elements, which are equal to one. The letter *I* is used to represent the identity matrix.

$$I = \begin{bmatrix} 1 & 0 & 0 \\ 0 & 1 & 0 \\ 0 & 0 & 1 \end{bmatrix}$$

A vector is a single row or column matrix.

$$V = \begin{bmatrix} a_1 \\ a_2 \\ a_3 \end{bmatrix}^T = \begin{bmatrix} a_1 & a_2 & a_3 \end{bmatrix}$$

The letter *T* indicates the transpose of a matrix. In a transposed matrix, the columns become row elements and the row elements become columns.

Matrices may be added, subtracted, multiplied, and raised to powers. For example,

$$\begin{bmatrix} a & b \\ c & d \end{bmatrix} + \begin{bmatrix} e & f \\ g & h \end{bmatrix} = \begin{bmatrix} a+e & b+f \\ c+g & d+h \end{bmatrix}$$

$$\begin{bmatrix} a & b \\ c & d \end{bmatrix} \bullet \begin{bmatrix} e & f \\ g & h \end{bmatrix} = \begin{bmatrix} ae + bg & af + bh \\ ce + dg & cf + dh \end{bmatrix}$$

In multiplication of a matrix, say a y by z matrix times a p by q matrix, the operation is possible only for the case where $z = p$. In a later section, it will be necessary to multiply a $[7 \times 14]$ matrix times a $[14 \times 1]$ matrix to obtain a $[7 \times 1]$ vector.

Any n-rowed square matrix has at least one and at most n distinct eigenvalues. The eigenvalues of a square matrix $[A]$ are the roots of the corresponding characteristic equation. The characteristic equation is:

$$D(\lambda) = \det(A - \lambda I) = 0$$

where λ = eigenvalues
det = determinant
A = square matrix
I = identity matrix

For example:

$$A = \begin{bmatrix} 5 & 4 \\ 1 & 2 \end{bmatrix}$$

$$I = \begin{bmatrix} 1 & 0 \\ 0 & 1 \end{bmatrix}$$

$$\lambda I = \begin{bmatrix} \lambda & 0 \\ 0 & \lambda \end{bmatrix}$$

$$D(\lambda) = \begin{bmatrix} 5 - \lambda & 4 \\ 1 & 2 - \lambda \end{bmatrix} = 0$$

$$D(\lambda) = \lambda^2 - 7\lambda + 6 = 0$$

$$D(\lambda) = \lambda - 6 \ (\lambda - 1) = 0$$

and the roots or eigenvalues are

$$\lambda_1 = 6$$
$$\lambda_2 = 1$$

λ_{max} or λ_1 is the principal eigenvalue. There are frequently imaginary parts to these roots but they are of no interest to our purpose here.

To calculate the eigenvector, the principal eigenvalue is substituted in the following equation:

$$A X = \lambda X$$

For $\lambda_{max} = 6$:

$$-x_1 + 4x_2 = 0$$

$$x_1 + 4x_2 = 0$$

$$x = \begin{bmatrix} 4 \\ 1 \end{bmatrix}$$

where *X* is the eigenvector for the principal eigenvalue of 6. A slightly more complicated example is:

$$A = \begin{bmatrix} 2 & 2 & 1 \\ 1 & 3 & 1 \\ 1 & 2 & 2 \end{bmatrix}$$

By changing all the signs and subtracting the product λ I:

$$\begin{bmatrix} \lambda-2 & -2 & -1 \\ -1 & \lambda-3 & -1 \\ -1 & -2 & \lambda-2 \end{bmatrix} = 0$$

Therefore,

$$\lambda^3 - 7\lambda^2 + 11\lambda - 5 = 0$$

and

$$\lambda_1 = 5$$

$$\lambda_2 = 1$$

$$\lambda_3 = 1$$

In practical problems, it is necessary to find eigenvalues of much larger matrices. A computer solution of the eigenvalues and the eigenvectors is essential. There are several software packages available which will provide the solution. MATLAB (see references) is one such package. A program called EIGEN is included in this text for finding eigenvalues and eigenvectors for a matrix up to n = 20. See Appendix A.

THE METHOD

The Analytic Hierarchy Process, as described by Saaty, consists of three principles: decomposition, comparative judgment, and priority synthesis.

Decomposition is the description of the problem in a hierarchical form. The elements of each level are independent of succeeding levels. The hierarchical structure starts at the top with a statement of a decision goal. The next lower level contains the criteria by which the alternatives are measured by the degree to which they satisfy the goal. The next lower level contains the actual alternatives from which the one that best satisfies the goal is selected.

A simple example and incomplete example would be the selection of a warping winch for addition to a ship in service.

Level 1	Goal	Select the most satisfactory winch
Level 2	Selection criteria	Reliability
		Weatherproofing
		Maintainability
		Cost
		Operation
Level 3	Available designs	Design A
		Design B
		Design C

Comparative judgment is the construction of pairwise comparisons to establish the relative importance among the elements on a given level with respect to the shared criterion on the next level up. Absolute measurement on a scale of 1 to 9 is used to score the paired comparisons. After experimenting with a large number of scales, Saaty selected the 1-to-9 scale because of the high consistency it provided. Application of the scale requires clear definitions of the meaning of each scale number and unambiguous definitions of the goal and the criteria. The pairwise comparisons are scored and entered into a square matrix. Assume for the above example that definitions of the scale values and the criteria have been employed to prepare the following matrix:

		1	2	3	4	5
Reliability	1	1	3	2	4	3
Weatherproofing	2	⅓	1	½	2	5
Maintainability	3	½	2	1	2	⅓
Cost	4	¼	½	½	1	¼
Operation	5	⅓	⅕	3	4	1

In the above matrix, the importance of the five selection criteria has been compared pairwise and scored using the 1 to 9 scale to rank the importance of the first criterion to the second criterion. Since there are five criteria, the pairwise comparisons yield the 25 scores arranged in the matrix. For example, in comparing reliability to cost, the score of 4 has been selected as the analyst's measure of the importance of reliability relative to cost. If these two criteria were equal in importance, the score would be 1. When weatherproofing is compared to operation, the analyst assigned a score of 5.

If criterion g is scored a value s relative to criterion h, then the score of h relative to g must be 1/s.

The eigenvalues for the above matrix are next calculated and the eigenvector corresponding to the principal eigenvalue is calculated and normalized to form the criteria priority vector. The principal eigenvalue for the above matrix of criteria comparisons is 6.1455 and the normalized eigenvector is:

$$\begin{bmatrix} .3562 \\ .2372 \\ .1636 \\ .0648 \\ .1782 \end{bmatrix}$$

At the next lower level, pairwise comparisons are made of the three available designs for each of the five criteria. For example, assume the three designs are compared for the criterion reliability and scored with the following result:

		1	2	3
Design A	1	1	6	8
Design B	2	⅙	1	9
Design C	3	⅛	⅑	1

The principal normalized eigenvector for the above matrix is:

$$\begin{bmatrix} .8235 \\ .0991 \\ .0774 \end{bmatrix}$$

The third-level design pairwise comparisons and scoring are repeated for each of the criteria. The normalized eigenvectors may then be combined into a matrix of local priorities which are then weighted by multiplying the matrix by the priority vector from the second level. Assuming that this process has been carried out, the results are as follows:

[Local priorities matrix] • [Priority vector] = [Result vector]

$$\begin{bmatrix} .8235 & .0718 & .7446 & .2330 & .6735 \\ .0991 & .6502 & .0654 & .0051 & .1015 \\ .0774 & .2780 & .1810 & .7129 & .2260 \end{bmatrix} \bullet \begin{bmatrix} .3562 \\ .2372 \\ .1636 \\ .0648 \\ .1728 \end{bmatrix} = \begin{bmatrix} .5673 \\ .2186 \\ .2096 \end{bmatrix} \begin{matrix} \text{A} \\ \text{B} \\ \text{C} \end{matrix}$$

The final vector shows that design *A* is the best decision considering the pairwise comparison of the five criteria and the three design alternates.

The assignment of scores on the 1-to-9 scale for the pairwise comparisons can be effectively accomplished by three or more knowledgeable persons working together to develop a consensus for each of the scores assigned.

AN ARCTIC PROPULSION SYSTEM: MULTICRITERIA DECISION MAKING

In the 1980s, the U.S. Department of Transportation supported a research program to define and resolve problems associated with the operation of large tankers in the high Arctic region. Such ships would be expected to cruise in six-foot-thick ice and have the capability to break through ice ridges up to 80 feet thick. One of the minor questions was to define the type of main propulsion system which would be most appropriate for Arctic operation. This question is well suited for an analytic hierarchy process approach to its solution.

Ship Characteristics and Operating Scenario

A tanker designed to operate regularly in six-foot-thick ice would have the following typical characteristics:

Length, BP, ft.	931
Beam, ft.	124
Depth, ft.	65
Draft, ft.	46
Open water speed, kts.	13.5
Icebreaking speed, kts.	5.0
Deadweight, tons	107,071
Displacement, tons	134,410
Block coefficient	.872
Shaft horsepower	160,134
Propellers, fixed	3
Propeller diameter, ft.	25.6
Propeller rpm	96.7

The typical operating scenario by which the machinery selection was judged is as follows:

Activity	*Days per Year*	%
Loading/discharging/repairs	29	8
High-power operation in ice	127	35
Low-power operation in open water	194	53
Ridge-breaking operation	14	04
		100

Propulsion Plant Selection Criteria

The traditional marine engineering selection criteria (see Harrington in the References), with some slight modification, is appropriate to the Arctic tanker propulsion system question. These criteria are as follows:

1. *Installed Cost.* The total installed cost of all machinery systems including auxiliaries, propulsion, cargo pumping, and hull machinery.
2. *Reliability.* Probability of temporary loss of part or all of the propulsion capability.
3. *Maintainability.* Frequency of preventive maintenance procedures, cost of maintenance, requirements for maintenance personnel, spare parts requirements, special tooling needs, and downtime for maintenance.
4. *Space and arrangements.* Space requirements, arrangement flexibility, need for ducts, access problems.
5. *Weight requirements.* Total machinery weight.
6. *Fuel quality requirements.* Cost, sources, specifications, and availability of required fuels.
7. *Fuel consumption.* All-purpose fuel rate including cargo heating, auxiliary power generation, cargo pumping.
8. *Fractional power requirements.* Ability of plant to operate at low power levels for long periods of time, ability to operate at low levels with acceptable fuel rates.
9. *Transient performance.* Ability to respond quickly to changes in power demand.
10. *Reversing capability.* Ability to easily and quickly reverse propeller, time limitations on astern operation, power available for astern operation.
11. *Operating personnel.* Number and required skills of operating personnel, automation features, needed maintenance skills.
12. *Rating limitations.* Suitability of machinery type for application to very high power ratings.
13. *Low-temperature operating problems.* Special operating and maintenance problems in the Arctic environment.

14. *Suitability for torsional shocks.* Resistance to failure, wear, and breakdown due to torsional shocks to the transmission system.

Comparison Scale. The Saaty relative importance scale used to score the pairwise comparisons has the following definitions for each of the intensities or scores:

Intensity	*Definition*	*Explanation*
1	Equal importance	Two criteria or alternatives contribute equally to goal.
2	Compromise score between 1 and 3	
3	Moderate importance of one over another	Experience and judgment strongly favor first criterion over second.
4	Compromise score between 3 and 5	
5	Essential or strong	Experience and judgment slightly favor first criterion over second.
6	Compromise score between 5 and 7	
7	Very strong importance	Criterion or alternative is strongly favored and its dominance is demonstrated in practice.
8	Compromise score between 7 and 9	
9	Extreme importance	Evidence favoring first activity over second is of highest possible order.

Alternative Power Plants

There are seven propulsion plants included in this study. They are identified by letters A through G as follows:

A. *Steam turbine gear.* Three high-temperature, high-pressure, reheat single boiler propulsion plants; optimum feed heaters, low-pressure steam generators for cargo heating, locked-train double reduction gears, main shaft power takeoff for feedwater pumping and power generation, auxiliary steam turbine generators, and backup diesel generator; flexible arrangements available; lowest initial cost system; very good all-purpose fuel rate; ability to burn inexpensive fuel and perhaps crude oil; high reliability; low maintenance; ice ingestion in cooling water problem.

B. *Steam turbine electric.* Similar to *A* but speed reduction and control is accomplished by an AC electric drive system; 30 percent higher cost over geared turbine; greater weight; increased maintenance; full power available for astern operation; slightly higher fuel consumption compared to geared unit; excellent low-power operation and transient response; main generators can be used for cargo pumping.

C. *Gas turbine gear.* Aeroderivative gas turbine–geared system; double reduction locked-train gear; auxiliary boilers and heat recovery steam generators for space heating, cargo heating, cargo pumping and electrical power generation; system requires extensive space for inlet and outlet ducts; high reliability; on board spare engines requirement for maintenance change-out; high-quality high-cost fuel; poorer fuel rate compared to steam plant, which can be reduced by regenerative heating; reversing capability a serious problem requiring reversing gear and clutch system; low-power operation difficult.

D. *Gas turbine electric.* Similar to alternate *C* except the reversing problem is resolved by electric AC propulsion; poorer fuel rate compared to *C*; higher weight and cost than alternate *C*.

E. *Medium-speed diesel geared.* Two or three engines drive a single reduction gear through clutches for each of the three propulsion plants; ability to connect auxiliary electric generators to medium-speed engines for cargo pumping, hull machinery, etc; limits on low-speed operations; slower transient response; higher maintenance cost, spare parts, and overhaul costs than turbines; better engine fuel rate than turbines; unavailability of high torque at low to zero speed; high machinery weight; requirement for specification fuel.

F. *Medium-speed diesel electric.* Similar to *E* with the added cost, weight, space, and complexity of electric generators and propulsion motors; availability of main generators for cargo pumping; full power astern; high torque at zero speed.

G. *Slow-speed diesel.* One engine for each propeller; highest weight, cost and length of all alternatives; lack of high torque at low speed; desirability of specification fuel; lowest engine fuel rate; high maintenance cost but better than medium speed; torsional shock must be isolated from crankshaft; necessity of high-pressure air system for reversing; heat recovery steam generators can be used for source of heating steam; requirement for separate turbine or diesel generator sets; special maintenance tooling; expensive and heavy onboard spare parts; need for large auxiliary power plant for cargo pumping and auxiliary power.

Hierarchical Decomposition

Figure 15-1 shows the decomposition of the Arctic propulsion plant selection problem into a hierarchy. The problem is the selection of a power plant

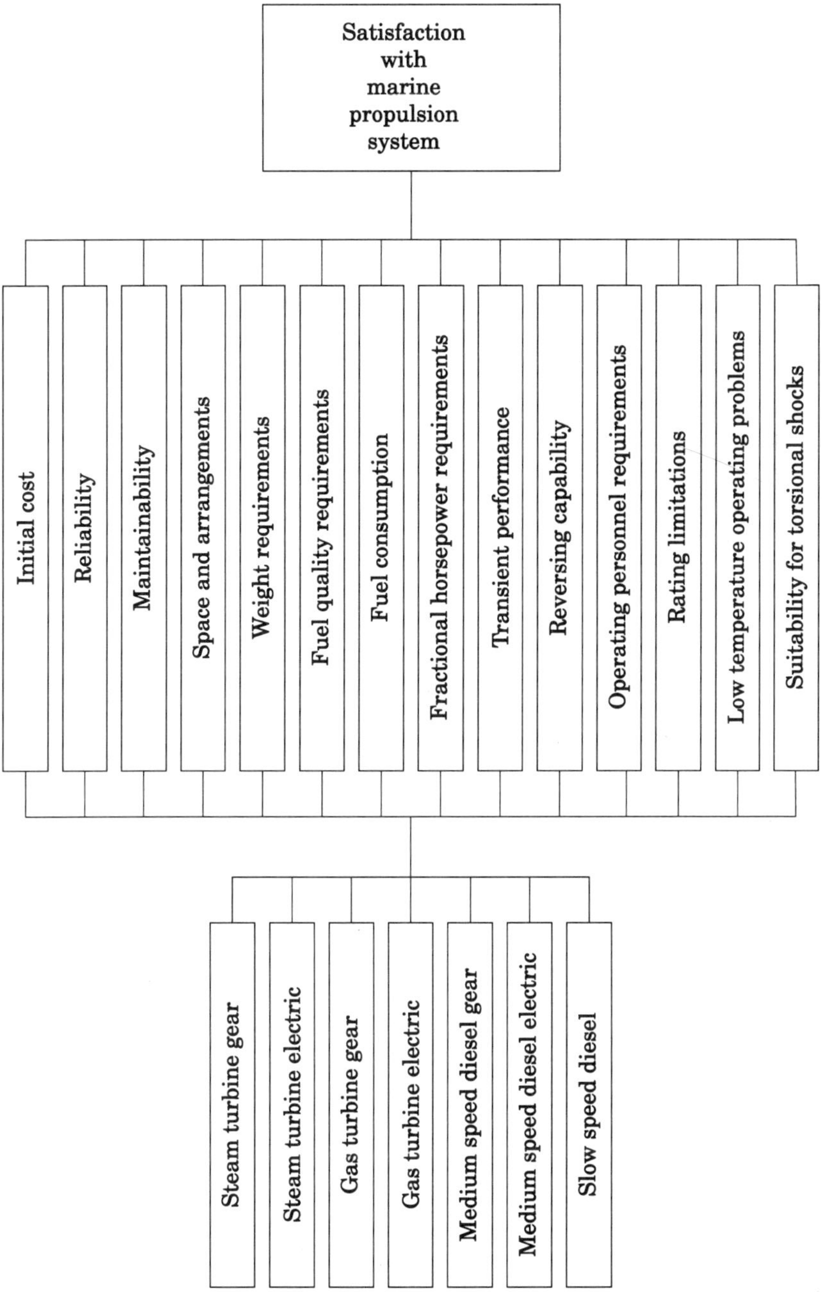

Figure 15-1. Decomposition of the Arctic propulsion system selection problem

from among the seven candidates. At the top level of the hierarchy is the goal, the satisfaction with the power plant system selected for an Arctic tanker. On the next level down are the 14 criteria that contribute to the goal. On the lowest level are the seven candidate power plants. It is obvious that additional candidates could have been included; however, they were eliminated because judgment suggested that they were not viable.

Comparative Judgment

The elements of the second level are arranged in a square matrix by assigning scores representing the relative importance of the pairwise comparisons among all of the elements. Note that the scores for a given comparison must be related as follows:

$$a_{kl} = \frac{1}{a_{lk}}$$

Table 15-1 is the matrix of the scored comparisons for the decision criteria. Using a software package, the principal eigenvalue was calculated as 17.067.

TABLE 15-1

Pairwise Comparisons and Scores in Matrix Format for Selection Criteria of Arctic Propulsion System Problem

	1	*2*	*3*	*4*	*5*	*6*	*7*	*8*	*9*	*10*	*11*	*12*	*13*	*14*
1	1	⅑	2	2	3	⅓	¼	¼	⅓	⅙	2	⅛	⅐	⅛
2	9	1	4	9	8	8	7	6	3	4	9	3	1	2
3	½	¼	1	7	8	9	4	2	1	1	9	2	1	3
4	½	⅑	⅐	1	1	2	⅓	½	⅛	⅐	1	⅐	⅑	⅑
5	⅓	⅛	⅛	1	1	2	⅑	⅓	¼	⅛	1	⅛	⅛	⅙
6	3	⅛	⅑	½	½	1	½	½	⅐	⅛	2	⅕	⅐	⅕
7	4	⅐	¼	3	9	2	1	⅙	⅐	⅐	3	⅙	⅕	⅕
8	4	⅙	½	2	3	2	6	1	2	1	1	2	3	4
9	3	⅓	1	8	4	7	7	½	1	1	7	1	1	1
10	6	¼	1	7	8	8	7	1	1	1	8	2	2	3
11	½	⅑	⅑	1	1	½	⅓	1	⅐	⅛	1	⅓	¼	⅕
12	8	⅓	½	7	8	5	6	½	1	½	3	1	½	½
13	7	1	1	9	8	7	5	⅓	1	½	4	2	1	3
14	8	½	⅓	9	6	5	5	¼	1	⅓	5	2	⅓	1

The principal eigenvector or criterion priority vector is then calculated and normalized to provide the following:

$$V = \begin{bmatrix} .0287 \\ .2048 \\ .1033 \\ .0151 \\ .0143 \\ .0196 \\ .0333 \\ .0999 \\ .0849 \\ .1165 \\ .0186 \\ .0753 \\ .1060 \\ .0797 \end{bmatrix}$$

Following the development of the criteria priority vector, the next step is to move to the lower level, to prepare pairwise comparison matrices and to calculate normalized local priority vectors for each of the alternate power plants under each criterion. The propulsion plants are identified by the letters *A* through *G* in Table 15-2. Table 15-3 shows a typical output of the MATLAB program for the alternate comparisons under the space and arrangement criterion of the Arctic Tanker problem.

TABLE 15-2
Alternate Comparison Matrices

1. Initial Cost λ_{max} = 8.3743

	A	*B*	*C*	*D*	*E*	*F*	*G*	*Local Vector*
A	1	4	5	3	5	6	8	.3826
B	1/4	1	2	5	5	6	8	.2182
C	1/5	1/2	1	5	6	7	8	.1936
D	1/3	1/5	1/5	1	2	1	9	.0731
E	1/5	1/5	1/6	1/2	1	3	9	.0659
F	1/6	1/6	1/7	1	1/3	1	9	.0502
G	1/8	1/8	1/8	1/9	1/9	1/9	1	.0164

2. Reliability λ_{max} = 7.2142

	A	*B*	*C*	*D*	*E*	*F*	*G*	*Local Vector*
A	1	2	1	3	4	5	6	.3028
B	1/2	1	1	2	3	4	5	.2001
C	1	1	1	2	1	2	3	.1773
D	1/3	1/2	1/2	1	2	3	2	.1175
E	1/4	1/3	1	1/2	1	2	1	.0866
F	1/5	1/4	1/2	1/3	1/2	1	1/2	.0502
G	1/6	1/5	1/3	1/2	1	1	1	.0559

3. Maintainability $\lambda_{max} = 7.5235$

	A	*B*	*C*	*D*	*E*	*F*	*G*	*Local Vector*
A	1	3	½	1	4	5	5	.2352
B	⅓	1	⅓	½	2	3	4	.1231
C	2	3	1	3	3	4	5	.3034
D	1	2	⅓	1	1	2	3	.1448
E	¼	½	½	1	1	2	½	.0768
F	⅕	⅓	¼	½	½	1	1	.0515
G	⅕	¼	⅕	⅓	2	1	1	.0602

4. Space and Arrangement $\lambda_{max} = 7.4609$

	A	*B*	*C*	*D*	*E*	*F*	*G*	*Local Vector*
A	1	3	5	6	1	2	4	.2881
B	⅓	1	2	4	½	2	3	.1506
C	⅕	½	1	3	¼	½	2	.0797
D	⅙	¼	⅓	1	⅓	½	3	.0598
E	1	2	4	3	1	4	4	.2692
F	½	½	2	2	¼	1	4	.1126
G	¼	⅓	½	⅓	⅕	¼	1	.0408

5. Weight Requirements $\lambda_{max} = 8.0354$

	A	*B*	*C*	*D*	*E*	*F*	*G*	*Local Vector*
A	1	4	⅕	¼	⅓	½	5	.0725
B	¼	1	⅙	⅓	⅙	⅕	6	.0445
C	5	6	1	3	4	6	8	.3825
D	4	3	⅓	1	2	4	7	.2030
E	3	6	¼	½	1	5	8	.1850
F	2	5	⅙	¼	⅕	1	7	.0930
G	⅕	⅙	⅛	⅐	⅛	⅐	1	.0192

6. Fuel Quality Requirement $\lambda_{max} = 8.2397$

	A	*B*	*C*	*D*	*E*	*F*	*G*	*Local Vector*
A	1	1	9	9	3	3	1	.2593
B	1	1	1	9	3	3	1	.2097
C	⅑	1	1	1	⅐	⅐	⅓	.0511
D	⅑	⅑	1	1	⅐	⅐	⅓	.0284
E	⅓	⅓	7	7	1	1	⅓	.1251
F	⅓	⅓	7	7	1	1	⅓	.1251
G	1	1	3	3	3	3	1	.2014

7. Fuel Consumption λ_{max} = *7.5200*

	A	*B*	*C*	*D*	*E*	*F*	*G*	*Local Vector*
A	1	2	3	4	1/5	1/4	1/8	.0707
B	1/2	1	1	2	1/6	1/5	1/9	.0389
C	1/3	1	1	3	1/4	1/5	1/7	.0454
D	1/4	1/2	1/3	1	1/6	1/7	1/8	.0259
E	5	6	4	6	1	4	1/2	.2726
F	4	5	5	7	1/4	1	1/3	.1657
G	8	9	7	8	2	3	1	.3808

8. Fractional Power Requirements λ_{max} = *7.2896*

	A	*B*	*C*	*D*	*E*	*F*	*G*	*Local Vector*
A	1	1/2	1	1/2	5	2	7	.1728
B	2	1	2	1	2	1	8	.2082
C	1	1/2	1	1/3	4	1	6	.1376
D	2	1	3	1	5	2	8	.2699
E	1/5	1/2	1/7	1/5	1	1/3	6	.0606
F	1/2	1	1/8	1/2	3	1	9	.1300
G	1/7	1/8	1/6	1/8	1/6	1/9	1	.0210

9. Transient Performance λ_{max} = *7.2722*

	A	*B*	*C*	*D*	*E*	*F*	*G*	*Local Vector*
A	1	1/2	1	1/2	4	2	9	.1623
B	2	1	2	1	5	4	9	.2794
C	1	1/2	1	1/2	3	1/2	8	.1263
D	2	1	2	1	4	2	8	.2345
E	1/4	1/5	1/3	1/4	1	1/3	3	.0481
F	1/2	1/4	2	1/2	3	1	7	.1289
G	1/9	1/9	1/8	1/8	1/3	1/7	1	.0201

10. Reversing Capability λ_{max} = *8.2152*

	A	*B*	*C*	*D*	*E*	*F*	*G*	*Local Vector*
A	1	1/3	9	1/2	1/2	3	5	.1451
B	3	1	8	1	4	7	9	.3488
C	1/9	1/8	1	1/5	1/7	1/8	1	.2046
D	2	1	5	1	4	1	8	.2046
E	2	1/4	7	1/4	1	1/5	5	.0997
F	1/3	1/7	8	1	5	1	8	.1588
G	1/5	1/9	1	1/8	1/5	1/8	1	.0215

11. Operating Personnel Requirements $\lambda_{max} = 8.5639$

	A	B	C	D	E	F	G	Local Vector
A	1	2	1	3	3	4	3	.2432
B	½	1	½	1	2	1	2	.1104
C	1	2	1	⅓	3	4	5	.2164
D	⅓	1	3	1	⅓	½	2	.1430
E	⅓	½	⅓	3	1	⅓	1	.0992
F	¼	1	¼	2	3	1	½	.1120
G	⅓	½	⅕	½	1	2	1	.0759

12. Rating Limitations $\lambda_{max} = 7.5532$

	A	B	C	D	E	F	G	Local Vector
A	1	4	2	4	3	4	5	.3240
B	¼	1	⅓	1	3	2	4	.1279
C	½	3	1	2	4	5	6	.2533
D	¼	1	½	1	⅓	2	3	.0950
E	⅓	⅓	¼	3	1	2	3	.1091
F	¼	½	⅕	½	½	1	2	.0347
G	⅕	¼	⅙	¼	⅓	½	1	.0347

13. Low-Temperature Operating Problems $\lambda_{max} = 7.2258$

	A	B	C	D	E	F	G	Local Vector
A	1	1	⅓	⅓	½	½	⅓	.0645
B	1	1	⅓	⅓	½	½	⅓	.0645
C	3	3	1	1	2	2	3	.2430
D	3	3	1	1	2	2	3	.2430
E	2	2	½	½	1	1	2	.1381
F	2	2	½	½	1	1	2	.1381
G	3	3	⅓	⅓	½	½	1	.1103

14. Suitability for Torsional Shocks $\lambda_{max} = 7.6165$

	A	B	C	D	E	F	G	Local Vector
A	1	3	1	3	2	3	4	.2590
B	⅓	1	⅓	1	½	⅓	2	.0722
C	1	3	1	3	2	½	4	.2046
D	⅓	1	⅓	1	2	1	3	.1194
E	½	2	½	½	1	2	4	.1398
F	⅓	3	2	1	½	1	5	.1659
G	¼	½	¼	⅓	¼	⅕	1	.0393

TABLE 15-3

Typical Input and Output for the MATLAB Program for Alternate Comparisons under the Space and Arrangement Criteria

Shown from top to bottom are the input, the eigenvalue value output D, and the eigenvector output V.

	1.0000	3.0000	5.0000	6.0000	1.0000	2.0000	4.0000
	.3333	1.0000	2.0000	4.0000	.5000	2.0000	3.0000
	.2000	.5000	1.0000	3.0000	.2500	.5000	2.0000
	.1666	.2500	.3333	1.0000	.3333	.5000	3.0000
	1.0000	2.0000	4.0000	3.0000	1.0000	4.0000	4.0000
	.5000	.5000	2.0000	2.0000	.2500	1.0000	4.0000
	.2500	.3333	.5000	.3333	.2000	.2500	1.0000

	7.4609 + .0000i	.0000 + .0000i	.0000 + .0000i	.0000 + .0000i	.0000 + .0000i	.0000 + .0000i	.0000 + .0000i
	.0000 + .0000i	−.0331+1.5793i	.0000 + .0000i	.0000 + .0000i	.0000 + .0000i	.0000 + .0000i	.0000 + .0000i
	.0000 + .0000i	.0000+ .0000i	−.0331 − 1.5793i	.0000 + .0000i	.0000 + .0000i	.0000 + .0000i	.0000 + .0000i
D =	.0000 + .0000i	.0000+ .0000i	.0000 + .0000i	−.0195 + .9508i	.0000 + .0000i	.0000 + .0000i	.0000 + .0000i
	.0000 + .0000i	.0000+ .0000i	.0000 + .0000i	.0000 + .0000i	−.0195 + .9508i	.0000 + .0000i	.0000 + .0000i
	.0000 + .0000i	.0000+ .0000i	.0000 + .0000i	.0000 + .0000i	.0000 + .0000i	.3582 − .0000i	.0000 + .0000i
	.0000 + .0000i	.0000+ .0000i	.0000 + .0000i	.0000 + .0000i	.0000 + .0000i	.0000 + .0000i	.0024 + .0000i

	.6243 + .1434i	.5261 + .8134i	.8897 − .3864i	−.5673 + .2936i	−.1591 − .6830i	1.8428 + 1.6516i	.1094 + .3831i
	.3264 + .0750i	.3600 + .0824i	−.0224 − .3691i	−.0565 − .3694i	−.2682 + .3105i	−.5003 − .4484i	.1908 + .6684i
	.1727 + .0397i	.3178 + .1720i	−.1178 − .3422i	−.1521 + .1549i	.0210 − .2375i	−.4366 − .3913i	−.0974 − .3412i
V =	.1277 + .0293i	−.1747 + .2658i	−.2912 + .1290i	−.0546 − .0691i	−.0967 + .0037i	.2173 + .1948i	.0028 + .0098i
	.5833 + .1340i	−.2490 − .5684i	.5207 + .3391i	.5204 − .8857i	−.3688 +1.0659i	−.0912 − .0818i	−.3849 − 1.3481i
	.2439 + .0560i	−.0992 − .0499i	.0973 + .0399i	.4442 + .2910i	.5596 + .1635i	.2773 + .2485i	.0381 + .1336i
	.0884 + .0203i	−.1411 − .0939i	.0697 + .1548i	−.0477 + .0534i	.0106 − .0778i	−.1467− .1314i	.0243 + .0852i

TABLE 15-4

Product of Local Priorities Matrix and Criteria Vector Equals Global Priorities Vector for Propulsion Machinery

$$
\begin{bmatrix}
.3826 & .3028 & .2353 & .2882 & .0725 & .2593 & .0707 & .1728 & .1625 & .1450 & .2431 & .3240 & .0645 & .2590 \\
.2182 & .2091 & .1232 & .1506 & .0445 & .2097 & .0389 & .2082 & .2794 & .3488 & .1104 & .1279 & .0645 & .0722 \\
.1936 & .1773 & .3035 & .0797 & .3825 & .0511 & .0454 & .1376 & .1265 & .0217 & .2164 & .2533 & .2424 & .2046 \\
.0731 & .1175 & .1449 & .0589 & .2031 & .0284 & .0259 & .2699 & .2345 & .2045 & .1430 & .0950 & .2424 & .1194 \\
.0659 & .0866 & .0769 & .2692 & .1851 & .1251 & .2726 & .0605 & .0481 & .0997 & .0992 & .1091 & .1381 & .1398 \\
.0502 & .0508 & .0515 & .1126 & .0931 & .1250 & .1657 & .1300 & .1289 & .1588 & .1120 & .0560 & .1381 & .1659 \\
.0164 & .0559 & .0602 & .0408 & .0192 & .2014 & .3808 & .0210 & .0210 & .0215 & .0759 & .0347 & .1100 & .0391
\end{bmatrix}
\bullet
\begin{bmatrix}
.0287 \\ .2048 \\ .1033 \\ .0151 \\ .0143 \\ .0196 \\ .0333 \\ .0999 \\ .0849 \\ .1165 \\ .0186 \\ .0753 \\ .1060 \\ .0797
\end{bmatrix}
=
\begin{bmatrix}
.2145 \\ .1796 \\ .1745 \\ .1621 \\ .1034 \\ .1048 \\ .0608
\end{bmatrix}
$$

Synthesis of Priorities

To establish the global priorities of the propulsion system alternatives, the local priorities vectors are assembled into a matrix. The criteria vector is multiplied by the local priorities matrix to yield the vector of global priorities for the propulsion plants as shown in Table 15-4.

Conclusion

The analytic hierarchy process results in a rational choice from among the alternatives. Engineers with similar knowledge and experience will arrive at the same conclusions.

For the Arctic propulsion system, the final priorities are:

Steam turbine gear	.2145
Steam turbine electric	.1796
Gas turbine gear	.1745
Gas turbine electric	.1621
Medium-speed diesel gear	.1034
Medium-speed diesel electric	.1048
Slow-speed diesel	.0608

REFERENCES

"Expert Choice" (software package). McLean, Va.: Decision Support Software, 1984.

Harrington, Roy, ed. *Marine Engineering*. SNAME, 1971.

Moler, Cleve. MATLAB Program. University of New Mexico, May 1981.

Saaty, Thomas L. *The Analytic Hierarchy Process*. New York: McGraw Hill, 1980.

——— "Axiomatic Foundation of the Analytical Hierarchy Process," *Management Science,* Vol. 32 (1986), No. 7: pp. 353-360.

———. *Mathematical Methods in Operations Research*. New York: Dover Publications, 1988.

Thurstone, L.L. "Psychophysical Analysis," *American Journal of Psychology,* Vol. 38 (1927): pp. 368-89.

QUESTIONS AND PROJECTS

1. What is an eigenvalue? How is it determined?
2. What is a principal eigenvalue?
3. What is an eigenvector? How is the eigenvector normalized?
4. What are the three principles of the analytic hierarchy process?
5. What criteria would you include in an analytic hierarchy process problem for the selection of the following shipboard machinery or systems?

a. Cargo refrigeration plant
b. Engine room fire-fighting system
c. Diesel engine fuel-processing system
d. Steering engine
e. Diesel electric generator
f. Main seawater circulating pump

6. Select one of the systems from question 5 and list the available alternatives. Draw an analytic hierarchy process diagram.
7. Select a shipboard system and prepare a complete analytic hierarchy process solution to the selection of the best alternate. If software is available, prepare a complete solution.
7. How would you employ an analytic hierarchy process to the determination of the best alternative where cost/benefit is the principal measure? Would two analytic hierarchy process solutions, one for cost and one for benefits, help?

PROBLEMS

1. Find the determinant of the following square matrix.

$$\begin{bmatrix} 2 & 3 & 5 \\ 1 & 0 & 1 \\ 2 & 1 & 2 \end{bmatrix}$$

2. Find the transpose of the following matrix.

$$\begin{matrix} .567 & .345 & .289 & .789 \\ .345 & .879 & .234 & .432 \\ .654 & .123 & .654 & .985 \\ .234 & .456 & .594 & .345 \end{matrix}$$

3. Find the product of the matrices below.

$$\begin{bmatrix} .12 & .45 & .53 & .78 \\ .45 & .67 & .89 & .12 \\ .23 & .32 & .29 & .87 \end{bmatrix} \begin{bmatrix} .43 \\ .67 \\ .34 \\ .56 \end{bmatrix}$$

4. Find the eigenvalue of the following vector. What is the principal eigenvalue?

$$\begin{bmatrix} 3 & 2 & 2 & -4 \\ 2 & 3 & 2 & -1 \\ 1 & 1 & 2 & -1 \\ 2 & 2 & 2 & -1 \end{bmatrix}$$

5. Find the eigenvector associated with the principal eigenvalue of the following matrix.

$$\begin{bmatrix} 2 & 1 & 1 \\ 1 & 2 & 1 \\ 0 & 0 & 1 \end{bmatrix}$$

6. Use available software to find the real part of the eigenvector associated with the principal eigenvalue of the following matrix. Normalize the eigenvector.

$$\begin{bmatrix} 1 & 1/9 & 2 & 2 & 2 & 3 \\ 9 & 1 & 4 & 9 & 8 & 8 \\ 1/2 & 1/4 & 1 & 7 & 8 & 9 \\ 1/2 & 1/9 & 1/7 & 1 & 1 & 2 \\ 1/3 & 1/8 & 1/8 & 1 & 1 & 2 \\ 3 & 1/8 & 1/9 & 1/2 & 1/2 & 1 \end{bmatrix}$$

Appendix A

Marine Engineering Economics and Cost Analysis Computer Programs

The following computer programs, written in Microsoft Fortran, compiled with a Microsoft Fortran Compiler Version 5.0, and suitable for running on PC, XT, AT, or compatible, are provided on the enclosed floppy disk.

1. Compound Interest Tables Program (TABLE)
2. Geometric Gradient Compound Interest Tables Program (GTABLE)
3. Elementary Decision Analysis Program (DECIANA)
4. Bayesian Decision Analysis Program (BAYES)
5. Tanker Ocean Shipping Venture Program (VENTURE)
6. Tanker Venture Variables Under Certainty Program (DESSEN)
7. Deterministic Required Freight Rate Program (RFRDET)
8. Required Freight Rate Simulation Program (RFRSIM)
9. Total Labor Estimate Program (LABEST)
10. Eigenvector for Multicriteria Decision Program (EIGEN)

These programs have been compiled to use a math coprocessor if available. If a coprocessor is not available, it may be necessary to use the DOS command, SET=NO87. The programs are in the executable file MARECON.EXE.

If you need assistance using these programs, write to Everett C. Hunt, c/o Cornell Maritime Press, Inc., P.O. Box 456, Centreville, Maryland 21617.

COMPOUND INTEREST TABLES PROGRAM

TABLE is a program to generate comprehensive listings of compound interest factors given any interest rate i and a range of periods, for example, n = 3 to n = 15. This program will be especially useful in real problem situations where interest rates are required to two decimal places. Such interest rates are rarely found in published tables. The program also has an option to print a table for a range of interest rates given a constant number of periods.

The compound interest factors included in the table are:

(F/P, i, n)	Compound amount factor
(P/F, i, n)	Present worth factor
(F/A, i, n)	Series compound amount factor
(A/F, i, n)	Sinking fund factor
(A/P, i, n)	Capital recovery factor
(P/A, i, n)	Series present-worth factor
(A/G, i, n)	Gradient uniform series factor
(P/G, i, n)	Gradient present-worth factor

GEOMETRIC GRADIENT INTEREST TABLES PROGRAM

GTABLE is a program which generates a compound interest factor table for the geometric gradient present-worth factor given an initial payment, an interest rate, and a geometric rate of increase. The table prints the present-worth factor for a range of geometric rate increases. The user must supply the minimum and maximum geometric rates of increase. The program prints a tabulation of: (P/A, r, i, n) for four values of *r*.

Since there are a limited number of print spaces available in the output format, the program may fail to produce results for very high values of *n* where the output numbers exceed the print space available. This should not be a problem for marine work since very long life of ships and ventures is unusual.

ELEMENTARY DECISION ANALYSIS PROGRAM

DECIANA is an elementary decision analysis program. The program accepts an *n* by *m* matrix of measures-of-merit such as present worth, internal rate of return, etc. There are *n* alternatives or acts and *m* future states or events for which the measures-of-merit have been calculated by hand or using a program such as RFRDET. If the user has determined the discrete probability distribution of the future events, this information may also be entered. The program provides three selections for output, all of which provide insights to the selection of the best alternative depending on the risk attitude of the decision maker. The outputs are as follows:

First output
Laplace criteria
Maximax criteria
Maximin criteria
Minimax regret criteria

Second output
Hurwicz criteria

Third output
Bayesian criteria *(requires probability distribution)*
Expected value
Opportunity loss
Cost of irrationality
Value of decision under certainty
Cost of uncertainty

Since the program prints out the date and time in the heading, the user should ensure that the computer date and time are updated. The user must establish a data file before running the program. (You may assign any name to the input data file, for example, ship1.dat; the program will ask the name of the file.)

The input data file must have the following form:

Line 1: Client name or blank
Line 2: Project i.d. or blank
Line 3: Identify measure of merit
Line 4: Six numbers separated by commas:
  1 for 1st output set or 0 to skip
  1 for 2nd output set or 0 to skip
  1 for 3rd output set or 0 to skip
  0 a zero is required in this position
  N = number of alternatives
  M = number of future states
Line 5: Enter payoff matrix starting with the first row, that is, i(1,1) and ending with i(n,m). Use as many lines as required. Final line probability vector, if required, in the same order as events in the payoff matrix.

All input data must be separated by commas. The following is a sample of an input file.

```
Mad River Shipping Company
fast ferry
pw of net cash flow dollars
1,1,1,0,5,12
–6793898.,–7815667.,–4130645.,–4897372.,–1588372.,–2320217.
1137393.,433208.,7076547.,6402180.,10517840.,9848627.
–2692997.,–3078392.,–373463.,–739751.,1962332.,1606188.
4583131.,4226988.,10126030.,9775254.,13566480.,13215700.
–6414921.,–7446790.,–3727673.,–4337562.,–1031311.,–1770393.
1680214.,969068.,7333338.,6652304.,10773780.,10093750.
–1627737.,–2442535.,260381.,–325292.,2093282.,1539390.
4157549.,3603657.,8419294.,7865403.,11171650.,10617760.
–12166790.,–13708050.,–9560396.,–11101650.,–7375895.,–8917148.
–4343764.,–5789471.,1386514.,361470.,4138871.,3121644.
.168,.042,.072,.018,.28,.07,.12,.03,.112,.028,.048,.012
```

Files DECI1.DAT and DECI2.DAT are sample input files.

BAYESIAN DECISION ANALYSIS PROGRAM

The Bayesian decision analysis program performs a simple but useful task for the decision maker confronted with selecting among two or more alternatives (or acts) for which payoffs have been calculated for a finite set of future events (or states of nature). The future events will usually be described by a discrete random variable which is the major factor in determining the degree of success of the selected alternative. For example, in a sea transportation setting, the factor might be annual cargo load factor, fuel cost increases, passenger load factor, trade growth factor or some combination of such factors.

An n by m matrix must be calculated in which the payoff (or measure--of-merit) for each alternative under each future state is provided. Associated with the matrix is an *a priori* discrete probability distribution which describes the decision maker's subjective view of the likelihood of each future event. This could be a uniform distribution in the unusual absence of a subjective view of the future.

The decision maker must also determine a discrete probability sample distribution for the random variable which is the major factor in the decision analysis. This sample distribution can be made by undertaking a survey or repeated surveys of the marketplace or other sources using established survey techniques. Given this information, the program will employ Bayes rule to calculate the *posterior* distribution which is the *a priori* distribution updated to include the information encoded in the sampling distribution.

The program then calculates the expected payoff, opportunity loss (or regret), and the cost of irrationality in units of the measure-of-merit for each alternative under the *a priori* distribution and the *posterior* distribution, leading to the selection of the best alternative for both the original view of the future and the updated view of the future based on the sample information. The cost of uncertainty under both *a priori* and *posterior* distributions is provided as an indication of the economic value of additional survey and study. If additional survey work is undertaken, the *posterior* distribution becomes the *a priori* distribution for a second run of the program.

If the user does not have a sample distribution based on survey or study but wishes to view the expected payoffs and the cost of uncertainty under an *a priori* distribution, the program DECIANA may be used for this purpose.

The user must establish an input data file and give it a name, for example, proj1.dat. The program will inquire the name of the input file. Since the program prints the time and date of the analysis, the user should update the PC time and date routines. The input file must have the following format:

Line 1: Client name
Line 2: Project identification
Line 3: Measure-of-merit
Line 4: Three numbers separated by commas
n = number of acts
m = number of events
0 = a zero
Line 5: First row of matrix payoffs starting at i(1,1). Use subsequent lines as needed to enter the entire matrix ending at i(n,m).
Line (n-1): Enter the *a priori* distribution in matrix event order.
Line *n:* Enter the sampling distribution in matrix event order.

The following is an example of an input file for five alternatives and 20 events.

Ajax Shipping Company
Terminal modification
Net gain millions of dollars
5,20,0
–15,–10,–7,–3.5,0,4,8,11,16,18,22,25,24,23,22,20,19,17,16,13
–10,–5,–1,4,8,12,16,20,25,24,23,22,21,20,19,18,17,16,15,12
–5,0,5,10,15,19,25,24,23,22,21,20,19,18,17,16,16,14,13,11
–20,–17,–13,–10,–7,–4,0,3,6,9,12,15,18,22,25,23,21,20,17,14
0,5,12,18,25,24,23,22,21,20,19,18,17,16,15,14,10,9,8,7
.01,.02,.02,.05,.06,.06,.08,.08,.13,.12,.09,.06,.05,.05,.05,
.03,.01,.01,.01,.01
.0034,.0062,.0108,.0182,.0291,.0443,.0638,.086,.108,.1251,
.1319,.1243,.1023,.0712,.0399,.0167,.0046,.0007,.0001,.00001

The following references may be useful in applying Bayesian methods:
Raiffa and Schlaifer, *Applied Statistical Decision Theory.*
Raiffa, *Decision Theory.*
Savage, *Foundations of Statistics.*

Files BAY1.DAT and BAY2.DAT are sample input files.

TANKER OCEAN SHIPPING VENTURE PROGRAM

The ocean shipping program VENTURE accepts data describing the characteristics of a tanker, cost information, operating data, financial information, and shipbuilding data. This information is used to find the ship characteristics and ship speed needed to provide the best internal rate of return (IRR). To accomplish this, the program calculates the ship characteristics, weights, costs, annual cash flows, etc., for a series of similar vessels at increasing operating speeds. The program prints out a tabulation of speed and internal rate of return from which the decision maker can select the best IRR. The program can then be instructed to print out the detailed characteristics, economics, and operating information for the best vessel in the particular tanker trade. The detailed output includes the following:

Ship design particulars
Steel weight and cost
Machinery weight and cost
Outfit weight and cost
Hydrodynamic data
Propulsion power
Ship cost estimate and labor-hours
Economic summary
Annual cash flow analysis for life
Ship operating data
Shipowner construction payment schedule

To run VENTURE, the user must prepare an input data file. The program will inquire for the input data file name which the user provides. The user may assign any name to the input file, for example tank2.dat. The input file has 51 lines of information listed below.

John Smith	Client name
Big Ship	Job identification
281.7	Length between perpen, meters
7.5	Length beam ratio
1.999	Beam depth ratio
0.995	Midship section coefficient, enter 0 for program-selected value
915.00	Cargo density, kilogram/cubic meter
1.0	Steel weight adjustment, double hull 1.08, double bottom 1.03
.57	Center of gravity above keel, decimal of depth
0.0004	Correlation allowance, enter 0 for program-selected value
1.020	Appendage factor, enter 0 for program-selected value
88.0	Propeller revolutions per minute
1.2	Service allowance factor, enter 0 for program selection
9000.0	Range, nautical miles in loaded direction
273.0	Specific fuel consumption, kilogram/mw-hr
1	Number of propellers
13.0	Design sea speed, first case to calculate, knots
9	Number of speed cases to calculate, maximum is nine
.250	Speed increment for calculation
.525	Ratio of cargo tons to displacement tons, enter 0 for program-selected value for a crude carrier
481.26	Steel cost, $/metric ton
35.0	Steel fabrication, labor-hours/metric ton
16.0	Labor rate including applied overhead, $/hour

.070	Shipbuilder profit on sales, decimal
15	Loan repayment period, years, integer
.1255	Interest rate on borrowed funds, decimal
.212	Owner equity, decimal
10	Ship depreciation period, years, integer
.315	Tax rate, decimal
.105	Salvage value, decimal of first cost
6.5	Port time per round trip, days
.049	Inflation rate, decimal
350.	Operating days per year
25	Number of crew members
.00599	Freight rate, $/metric tonne-mile
191.8	Cost of heavy bunker fuel, $/tonne
249.2	Cost of diesel fuel, $/tonne
6600000.	Shipyard fixed cost for ship, $
.031	Construction steel cost adjustment
.072	Steel cost adjustment
-.061	Machinery adjustment 1
0.021	Machinery adjustment 2
.120	Steel adjustment
.031	Steel cost adjustment 1
.0105	Steel cost adjustment 2
.1402	Productivity adjustment 1
.303	Productivity adjustment 2
.07504	Labor cost adjustment 1
0.001	Labor cost adjustment 2
1.0001	Steel index
1.0002	Outfit index
1.0002	Machinery index
.0201	Labor cost adjustment 1
.0301	Labor cost adjustment 2
20	Economic life up to 20 years, integer
1	Switch, 0 prints analysis with annual cash flows 1 prints IRR versus speed 2 prints analysis without annual cash flows

The speed-power analysis of VENTURE is based on Series 60 data. If the user or the program operates outside the nominal range of Series 60 data, a notice is supplied on the screen but no other action is taken. The user must decide if the output data have validity if basic parameters are outside Series 60. Nominal values are:

Length-beam ratio	5.0 to 8.6

Beam-depth ratio	2.3 to 3.6
Block coefficient	0.6 to 0.90
Midship section coefficient	.97 to .996
Prismatic coefficient	.595 to .856
Length-displacement ratio	4.5 to 7.6

If the user selects switch 0 for the input data while the input calls for a large number of speeds, the output file will have a massive amount of data. It is best to specify a single speed when switch 0 is used.

The program DESSEN (design and sensitivity of tanker variables under certainty) uses the same algorithm and input file format as VENTURE but allows the user to explore design and operating variables.

Teak1.dat, teak2.dat, teak3.dat, and teak4.dat are sample ASCII input files.

Applicability of this program depends on properly incorporating the cost structure, productivity, methods, etc., of a specific shipyard and ship operator. For student use, the program employment of typical relationships and mean values for key data provides a satisfactory learning vehicle. Some program modifications based on a specific shipyard and shipowner are necessary before this program can be successfully applied to a business situation.

TANKER VENTURE VARIABLES UNDER CERTAINTY PROGRAM

DESSEN is a program for the exploration of design variables or future operating variables in a decision under certainty environment for a single cargo tanker. The design variables which can be explored include the following:

Design sea speed
Power service allowance
Length between perpendiculars
Length/beam ratio
Beam/depth ratio

The operating variables which can be explored include:

Range, i.e., one-way voyage distance, n.m.
Annual inflation rate, decimal
Voyage port time, days/voyage
Freight rate, $/tonne-nautical mile
Heavy fuel cost, $/tonne
Operating days/year

For each variable, the VENTURE algorithm is employed to specify an optimum vessel for each design sea speed. Very large outputs and long run times will result from the selection of a large number of variables for study.

The required input is identical to the input file for VENTURE. DESSEN will ignore the unneeded variables included in the input file. However, these variables must be in the file. It is recommended that the user avoid using 0 for program-selected values when the input file is for DESSEN.

If the internal rate of return for the variables under study is less than zero or greater than 100 percent, then the table for design sea speed versus internal rate of return is not printed.

DETERMINISTIC REQUIRED FREIGHT RATE PROGRAM

RFRDET has a wide range of application to ocean shipping economic analysis of bulk and near bulk trades where the return trip is in ballast. Given the description of a ship, voyage data, and financial data, the program calculates a required freight rate or a present worth of the venture.

A useful application of RFRDET is the evaluation of a given vessel under a range of future operating and economic conditions. The measures-of-merit for a decision under risk or uncertainty can be generated with this program and used subsequently as the input for a decision analysis program such as BAYES or DECIANA. RFRDET does not modify the ship design. It only evaluates the economics of a given design for the external factors provided.

The input for RFRDET is read from an ASCII file prepared in advance which contains the following information. Files rfr1.dat, rfr2.dat, rfr3.dat and rfr4.dat are input file examples.

Variable Description	*Sample Input*
Client name	Ajax
Job	16,000 dwt
Cargo	Containers
Analyst	John Brown
One-Way voyage distance (n.m.)	6000
Number of fee-charging canals (0 to 3), integer	1
Canal codes (1 Panama, 2 Suez, 3 other), integer	1
Canal inflation factor (i.e., .03)	.025
Canal delay time (days/passage)	1
Canal charge, loaded, $/GRT	2.07
Canal charge, ballast, $/GRT	1.65
Ports per round trip, integer	4
Required freight rate (0 or RFR), $/long ton	0.
RFR annual escalation, % (i.e., 6.0 or 0.0)	3.0
Inflation for labor, decimal	.025
Inflation for material, decimal	.02

Inflation for fuel, decimal		.05
Corporate tax rate, decimal		.36
Cost of capital 1, decimal		.105
Cost of capital 2, decimal		.120
Cost of capital 3, decimal		.145
Contract cost of vessel, $		47000000
Licensed crew, integer		7
Unlicensed crew, integer		16
Average daily wages per licensed crew member, $		250.
Average daily wages per unlicensed crew member, $		150.
Crew subsistence, $/day/person		35.
Ship length, feet		530.
Ship breadth, feet		73.
Ship depth, feet		61.
SHP		16000.
Loaded displacement, long tons		23000.
Gross registered tons		20160.
Full load, DWT		16000.
Cargo, DWT		14000.
Bale, cubic feet		15000.
Ship availability (day/year)		345.
Cargo-handling rate, long tons/hour		300.
Full load design speed, knots		17.
Annual marine insurance cost (0 for calculated)		0.
Fuel consumption from design speed minus three knots to design speed plus three knots in one-knot steps		
Fuel consumption, full displacement, 14 to 20 knots, long tons/hour	14	2.02
	15	2.50
	16	3.03
	17	3.64
	18	4.25
	19	4.99
	20	5.82
Fuel consumption, ballast displacement, 14 to 20 knots, long tons/hour	14	1.47
	15	1.82
	16	2.20
	17	2.60
	18	3.20
	19	3.76
	20	4.39

Fuel price, propulsion, $/long tons	79.
Fuel price, auxiliary, $/long tons	192.
Auxiliary fuel consumption at sea, long tons/hour	0.
Main fuel consumption in port, long tons/hour	0.0
Auxiliary fuel consumption in port, long tons/day	1.90
Type one lube oil consumption at sea, gallons/day	20.
Type one lube oil consumption in port, gallons/day	5.
Type one lube oil cost, $/gallon	.80
Overhead expense, decimal of operation, $	.25
Backhaul revenue, annual, $	500000.
Other revenue, annual, $	100000.
Average port cost, $/day (0 calculated)	0.
Port entry exit cost, $/port (0 calculated)	0.
Other depreciable vessel cost	0.
Tax-deductible construction costs, decimal	.02
Other amortizable cost, $	75000.
Decimal of total cost, borrowed	.80
Interest rate, decimal	.10
Guarantee premium	0.
Length of loan, integer	10.
Decimal of total cost used as initial tax credit	0.
Depreciation years, economic life, integer	20
Salvage value, $	3000000.
Depreciation method 1, 2, 3, 4, integer	2

1 straight line
2 double declining balance
3 sum of the years digits
4 user-specified percent each year, j = 1 to n

Option 4 cost depreciated each year, decimal

The program output for each hull considered includes the following information:

- Project identification, date, and time
- Input description of vessel
- Input description of financial data
- For each speed and interest rate:
  Annual tons hauled
  Trips/year
  Required freight rate or net present value at three interest rates

Examples of input files are included on the disk. Example file names are rfr1.dat, rfr2.dat, rfr3.dat and rfr4.dat.

This program requires a wide carriage or condensed print pitch for successful printing of the output file.

REQUIRED FREIGHT RATE SIMULATION PROGRAM

RFRSIM is a compiled Fortran program which will calculate the required freight rate and the economic cost of transport when the inputs are expressed as normal or truncated-normal variants. The basic inputs are as follows:

Variable Number	*Variable Symbol*	*Variable*	*Typical Median Value*
0	Sd	Design speed, constant, knots	20.0
1	m	Round trip, n.m.	6000.0
2	s	Vessel speed made good, knots	20.0
3	t	Round-trip hours in port	65.28
4	d	Operating days per year	355.0
5	c	Unit capacity of vessel	1200.0
6	Of	Outbound load factor	.9
7	If	Inbound load factor	.8
8	cr	Annual crew cost, $	2042600.0
9	fr	Fuel rate, @ design speed, tons/day	36.3
10	l	Lube oil consumption, ton/day	0.25
11	fp	Inport fuel rate, tons/day	3.0
12	fc	Fuel cost, $/ton	100.0
13	lc	Lube oil cost, $/ton	750.0
14	mc	Annual maintenance & repair, $	1749000.0
15	Ic	Annual insurance cost, $	455300.0
16	Pc	Annual port, terminal, canal fee, $	1195667.0
17	Sc	Annual stevedoring costs, $	7823000.0
18	Ga	Annual general & administrative expense, $	666000.0
19	Ew	Owner equity, $	14000000.0
20	Re	Rate of return on equity, decimal	0.12
21	LP	Loan payments, interest & principal, $	9677434.0
22	Ccg	Cargo cost per unit, $	6000.0
23	Cai	Cargo annual interest rate	0.10

Output variables are as follows:

1	Vy	Voyages per year	23.3
2	Ca	Annual cargo throughput	400000.0
3	Fd	Fuel consumption, tons/day	33.0
4	Fac	Annual fuel cost, $	2380700.0
5	Lac	Annual lube oil cost, $	265400.0
6	Cao	Annual operating cost, $	16639400.0
7	Ctc	Annual total cost, $	26340000.0

8	Rfr	Required freight rate, $	601.7
9	ECT	Economic cost of transport, $	625.7

The basic relationships are:

Vy = 24d/(m/s + t)	Trips/year
Ca = Vy C (Of + If)	Units/year-ship
Fd = fr (s/Sd) Exp 3.	fuel, tons/day
Fac= fc ((d - t/24) fr + fp x t/24)	$/year
Lac= (d - t/24) l lc	$/year
Cao= Cr + Fac + Lac + Mc + Ic + Pc + Sc + Ga	$/year
Ctc = Cao + LP	$/year
Rfr = (Ctc + Ew * Re)/Ca	$/unit-year
ECT = RFR + (Ccg × 360./vy × Cai/360.)	$/unit-year

The output has four alternative formats:

(1) Summary statistics
(2) Summary statistics, prediction intervals
(3) Summary statistics, histogram
(4) Summary statistics, intervals, histogram

The user must prepare a data file with the following sample format:

Company ID (or blank)
Project ID (or blank)
Name of analyst (or blank)
20. (vessel design speed, a constant)
XLOW, XMED, XHIGH, PTAIL — This information is tabulated for each of the 23 input variables in the same order as listed above. For probability in the tails of the distribution, the following values are acceptable: (1) Any integer between 1 and 49 which represents the area in each of the tails, (2) .135 if the high and low estimates are at three standard deviations, and (3) 15.8655 if the high and low estimates are at one standard deviation.

Included on the program disk are two sample input files: SIM2.DAT and SIM4.DAT. The user will wish to review these files.

An option desirable in business calculation is the truncation of the probability distribution so that no value above or below the truncated value may enter the calculation. If the input format is revised as follows, truncated input variables are acceptable:

Company ID (or blank)
Project ID (or blank)
Name of analyst (or blank)

20.	(vessel design speed, a constant)
3	(Number of variables to be truncated)
1	(ID number of first truncated variable)
2	(ID number of second truncated variable)
Etc.	
XLOW XMED XHIGH PTAILS	For conventional variables
XLOW XMED XHIGH PTAILS Low Truncation Value High Truncation Value	For each of the variables identified as truncated

The maximum number of Monte Carlo iterations with this program is 600.

TOTAL LABOR ESTIMATE PROGRAM

The total labor estimate program accepts work task labor-hour estimates defined by five values:

Low estimate LL
Left tail cum probability of low estimate P(LL)
Mode estimate MODE
Right tail cum probability of high estimate P(HL)
High estimate HL

The three estimates are fitted to the distribution (Weibull or Normal) which provides the best fit. The mean and standard deviation of the fitted distribution are calculated and printed in the output file along with the input estimates. The procedure is followed for each of the labor task inputs provided by the user. The program calculates the left tail cumulative distribution of the total labor estimate. The Central Limit Theorem assures that the total labor estimate is normally distributed. The program employs a Monte Carlo routine to estimate the probability of negative labor hours when fitting a truncated normal distribution. Therefore, subsequent runs with the same data input will produce slightly different total labor estimates.

The program user must create an input file of the task labor estimating data. The first line of the file is an integer for the number of tasks. The subsequent lines in the data file contain a less-than-four-digit-task identification number followed by a space, then the above five values in the order listed, e.g., 25 25.,.05,40.,50.,10. To make the data file readable, the line numbers may be related to the task numbers in a proposal, or the file lines may be numbered successively as in the sample input file below.

```
6
1 10.,.02,12.,40.,.02
2 10.,.02,16.,40.,.02
3 10.,.02,20.,40.,.05
4 10.,.06,25.,40.,.10
5 10.,.10,30.,40.,.15
6 10.,.25,35.,40.,.20
```

The program aborts if an inconsistency or error is located in the input file. If there are a large number of tasks, it is inconvenient and time consuming to find the input error. An input file check program is available to run on query from the screen to assure that the data in the input file have no errors. Any mistake in the input format will cause a math or other error. If the sum of the two probability values is greater than .40, this may cause a difficulty. If the mode estimate is too close to either the LL or HL values, the program may fail to complete the run. No space after the line number will cause a failure.

EIGENVECTOR FOR MULTICRITERIA DECISION PROGRAM

This program, which calculates the eigenvalues and normalized eigenvector for a square matrix, was prepared to provide the maximum eigenvalue and the associated eigenvector needed in the analytical hierarchy process of multicriteria decision making described in Chapter 15.

To use the program, open a data file and type the square matrix proceeded by the order of the matrix. For example, the data file for the reliability matrix of Table 15-2 would be entered as follows:

```
7
1,2,1,3,4,5,6
.5,1,1,2,3,4,5
1,1,1,2,1,2,3
.333,.5,.5,1,2,3,2
.25,.333,1,.5,1,2,1
.2,.25,.2,.333,.5,1,.5
.167,.2,.333,.5,1,1,1
```

The program will ask for the name of the prepared input file and a name for the output file. The program output is a list of the n eigenvalues. The maximum eigenvalue and the normalized values of the eigenvector associated with the maximum value are selected for use in the analytical hierarchy process. A typical input file MATRIX.DAT is provided on the floppy disk. The program will handle a square matrix of up to n = 20.

The following is a sample of the program output with comments:

Eigenvalues, Real and Imaginary Parts

J	*Real*	*Imaginary*	
1	.721368000E+01	.000000000E+00	[maximum
2	–.195041600E–01	.138508900E+01	eigenvalue
3	–.195041600E–01	–.138508900E+01	is at j = 1]
4	–.407054600E–01	.318747200E+00	
5	–.407054600E–01	–.318747200E+00	
6	–.466311500E–01	.845234900E–01	
7	–.466311500E–01	–.845234900E–01	

Eigenvector; Real, Normalized, Imaginary Parts

	Real	*Normalized*	*Imaginary*
J = 1		[associated normalized eigenvector]	
	–.687365500E+00	.302815000E+00	.000000000E+00
	–.476146400E+00	.209763600E+00	.000000000E+00
	–.402569600E+00	.177349700E+00	.000000000E+00
	–.266616100E+00	.117456200E+00	.000000000E+00
	–.196497700E+00	.865659300E–01	.000000000E+00
	–.113974500E+00	.502108100E–01	.000000000E+00
	–.126749400E+00	.558387200E–00	.000000000E+00
J = 2			
	–.691908700E+00	.701411700E+00	–.931792900E+00
	.586594000E–02	–.594650500E–02	–.630306300E+00
	–.107169200E+01	.108641100E+01	.320289700E+00
	.296439000E+00	–.300510400E+00	–.386974700E+00
	.317500700E+00	–.321861400E+00	.408678100E+00
	.474028800E–01	–.480539300E–01	.225255100E+00
	.109940700E+00	–.111450700E+00	–.111354800E+00
J = 3			
	–.691908700E+00	.701411700E+00	.931792900E+00
	.586594000E–02	–.594650500E–02	.630306300E+00
	–.107169200E+01	.108641100E+01	–.320289700E+00
	.296439000E+00	–.300510400E+00	.386974700E+00
	.317500700E+00	–.321861400E+00	–.408678100E+00
	.474028800E–01	–.480539300E–01	–.225255100E+00
	.109940700E+00	–.111450700E+00	.111354800E+00

Eigenvector; Real, Normalized, Imaginary Parts

	Real	*Normalized*	*Imaginary*
J = 4			
	–.570124300E+00	.968201000E+00	.855348900E+00
	–.672988200E+00	.114288700E+01	–.550596800E+00
	.148542000E+00	–.252258100E+00	–.325761500E+00
	.595699400E+00	–.101163400E+01	.109970300E–01
	–.139735600E+00	.237302900E+00	.230061000E–01
	.857706400E–02	–.145658100E–01	.156814500E+00
	.411805600E–01	–.699339900E–01	–.923375000E–01
J = 5			
	–.570124300E+00	.968201000E+00	–.855348900E+00
	–.672988200E+00	.114288700E+01	.550596800E+00
	.148542000E+00	–.252258100E+00	.325761500E+00
	.595699400E+00	–.101163400E+01	–.109970300E–01
	–.139735600E+00	.237302900E+00	–.230061000E–01
	.857706400E–02	–.145658100E–01	–.156814500E+00
	.411805600E–01	–.699339900E–01	.923375000E–01
J = 6			
	–.277301700E+01	–.109063400E+03	–.494410700E+01
	.122789600E+01	.482934500E+02	–.123356100E+00
	.817899800E+00	.321681900E+02	.622222700E+00
	.142395000E+01	.560043000E+02	.368335000E+01
	.378028000E+00	.148679300E+02	–.393717800E+00
	–.558555400E+00	–.219681100E+02	.648335800E–01
	–.490776200E+00	–.193023400E+02	–.872431300E+00
J = 7			
	–.277301700E+01	–.109063400E+03	.494410700E+01
	.122789600E+01	.482934500E+02	.123356100E+00
	.817899800E+00	.321681900E+02	–.622222700E+00
	.142395000E+01	.560043000E+02	–.368335000E+01
	.378028000E+00	.148679300E+02	.393717800E+00
	–.558555400E+00	–.219681100E+02	–.648335800E–01
	–.490776200E+00	–.193023400E+02	.872431300E+00

Appendix B

Standard Normal Distribution and Compound Interest Tables

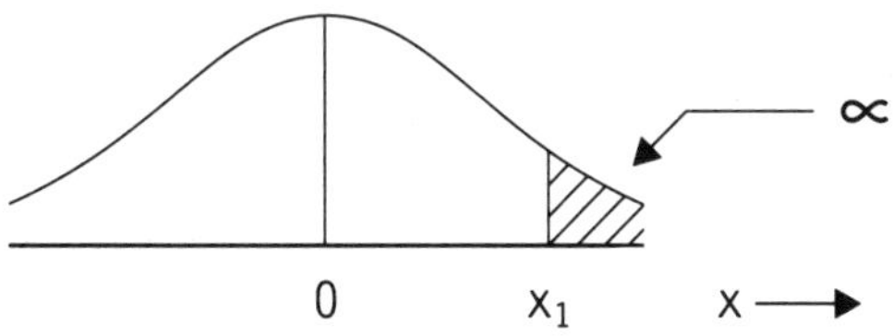

x	*.00*	*.01*	*.02*	*.03*	*.04*	*.05*	*.06*	*.07*	*.08*	*.09*
0.0	.5000	.4960	.4920	.4880	.4840	.4801	.4761	.4721	.4681	.4641
0.1	.4602	.4562	.4522	.4483	.4443	.4404	.4364	.4325	.4286	.4247
0.2	.4207	.4168	.4129	.4090	.4052	.4013	.3974	.3936	.3897	.3859
0.3	.3821	.3783	.3745	.3707	.3669	.3632	.3594	.3557	.3520	.3483
0.4	.3446	.3409	.3372	.3336	.3300	.3264	.3228	.3192	.3156	.3121
0.5	.3085	.3050	.3015	.2981	.2946	.2912	.2877	.2843	.2810	.2776
0.6	.2743	.2709	.2676	.2643	.2611	.2578	.2546	.2514	.2483	.2451
0.7	.2420	.2389	.2358	.2327	.2296	.2266	.2236	.2206	.2177	.2148
0.8	.2119	.2090	.2061	.2033	.2005	.1977	.1949	.1922	.1894	.1867
0.9	.1841	.1814	.1788	.1762	.1736	.1711	.1685	.1660	.1635	.1611
1.0	.1587	.1562	.1539	.1515	.1492	.1469	.1446	.1423	.1401	.1379
1.1	.1357	.1335	.1314	.1292	.1271	.1251	.1230	.1210	.1190	.1170
1.2	.1151	.1131	.1112	.1093	.1075	.1056	.1038	.1020	.1003	.0985
1.3	.0968	.0951	.0934	.0918	.0901	.0885	.0869	.0853	.0838	.0823
1.4	.0808	.0793	.0778	.0764	.0749	.0735	.0721	.0708	.0694	.0681
1.5	.0668	.0655	.0643	.0630	.0618	.0606	.0594	.0582	.0571	.0559
1.6	.0548	.0537	.0526	.0516	.0505	.0495	.0485	.0475	.0465	.0455
1.7	.0446	.0436	.0427	.0418	.0409	.0401	.0392	.0384	.0375	.0367
1.8	.0359	.0351	.0344	.0336	.0329	.0322	.0314	.0307	.0301	.0294

1.9	.0287	.0281	.0274	.0268	.0262	.0256	.0250	.0244	.0239	.0233
2.0	.0288	.0222	.0217	.0212	.0207	.0202	.0197	.0192	.0188	.0183
2.1	.0179	.0174	.0170	.0166	.0162	.0158	.0154	.0150	.0146	.0143
2.2	.0139	.0136	.0132	.0129	.0125	.0122	.0119	.0116	.0113	.0110
2.3	.0107	.0104	.0102	.00990	.00964	.00939	.00914	.00889	.00866	.00842
2.4	.00820	.00798	.00776	.00755	.00734	.00714	.00695	.00676	.00657	.00639
2.5	.00621	.00604	.00587	.00570	.00554	.00539	.00523	.00508	.00494	.00480
2.6	.00466	.00453	.00440	.00427	.00415	.00402	.00391	.00379	.00368	.00357
2.7	.00347	.00336	.00326	.00317	.00307	.00298	.00289	.00280	.00272	.00264
2.8	.00256	.00248	.00240	.00233	.00226	.00219	.00212	.00205	.00199	.00193
2.9	.00187	.00181	.00175	.00169	.00164	.00159	.00154	.00149	.00144	.00139

Right Tail Area of Standardized Normal Distribution

$$p\,(x \ge x_1) = \int_{x_1}^{\infty} \frac{1}{\sqrt{2\,\pi}}\; e^{-x^2/2}\, dx = \alpha$$

COMPOUND INTEREST FACTOR TABLES

INTEREST RATE 3.000 PERCENT

N	F/P	P/F	F/A	A/F	A/P	P/A	A/G	P/G	N
1	1.030	.97087	1.000	1.00000	1.03000	.9709	.0000	.000	1
2	1.061	.94260	2.030	.49261	.52261	1.9135	.4926	.943	2
3	1.093	.91514	3.091	.32353	.35353	2.8286	.9803	2.773	3
4	1.126	.88849	4.184	.23903	.26903	3.7171	1.4631	5.438	4
5	1.159	.86261	5.309	.18835	.21835	4.5797	1.9409	8.889	5
6	1.194	.83748	6.468	.15460	.18460	5.4172	2.4138	13.076	6
7	1.230	.81309	7.662	.13051	.16051	6.2303	2.8819	17.955	7
8	1.267	.78941	8.892	.11246	.14246	7.0197	3.3450	23.481	8
9	1.305	.76642	10.159	.09843	.12843	7.7861	3.8032	29.612	9
10	1.344	.74409	11.464	.08723	.11723	8.5302	4.2565	36.309	10
11	1.384	.72242	12.808	.07808	.10808	9.2526	4.7049	43.533	11
12	1.426	.70138	14.192	.07046	.10046	9.9540	5.1485	51.248	12
13	1.469	.68095	15.618	.06403	.09403	10.6350	5.5872	59.420	13
14	1.513	.66112	17.086	.05853	.08853	11.2961	6.0210	68.014	14
15	1.558	.64186	18.599	.05377	.08377	11.9379	6.4500	77.000	15
16	1.605	.62317	20.157	.04961	.07961	12.5611	6.8742	86.348	16
17	1.653	.60502	21.762	.04595	.07595	13.1661	7.2936	96.028	17
18	1.702	.58739	23.414	.04271	.07271	13.7535	7.7081	106.014	18
19	1.754	.57029	25.117	.03981	.06981	14.3238	8.1179	116.279	19
20	1.806	.55368	26.870	.03722	.06722	14.8775	8.5229	126.799	20
21	1.860	.53755	28.676	.03487	.06487	15.4150	8.9231	137.550	21
22	1.916	.52189	30.537	.03275	.06275	15.9369	9.3186	148.509	22
23	1.974	.50669	32.453	.03081	.06081	16.4436	9.7093	159.657	23
24	2.033	.49193	34.426	.02905	.05905	16.9355	10.0954	170.971	24
25	2.094	.47761	36.459	.02743	.05743	17.4131	10.4768	182.434	25

COMPOUND INTEREST FACTOR TABLES

INTEREST RATE 4.000 PERCENT

N	F/P	P/F	F/A	A/F	A/P	P/A	A/G	P/G	N
1	1.040	.96154	1.000	1.00000	1.04000	.9615	.0000	.000	1
2	1.082	.92456	2.040	.49020	.53020	1.8861	.4902	.925	2
3	1.125	.88900	3.122	.32035	.36035	2.7751	.9739	2.703	3
4	1.170	.85480	4.246	.23549	.27549	3.6299	1.4510	5.267	4
5	1.217	.82193	5.416	.18463	.22463	4.4518	1.9216	8.555	5
6	1.265	.79031	6.633	.15076	.19076	5.2421	2.3857	12.506	6
7	1.316	.75992	7.898	.12661	.16661	6.0021	2.8433	17.066	7
8	1.369	.73069	9.214	.10853	.14853	6.7327	3.2944	22.181	8
9	1.423	.70259	10.583	.09449	.13449	7.4353	3.7391	27.801	9
10	1.480	.67556	12.006	.08329	.12329	8.1109	4.1773	33.881	10
11	1.539	.64958	13.486	.07415	.11415	8.7605	4.6090	40.377	11
12	1.601	.62460	15.026	.06655	.10655	9.3851	5.0343	47.248	12
13	1.665	.60057	16.627	.06014	.10014	9.9856	5.4533	54.455	13
14	1.732	.57748	18.292	.05467	.09467	10.5631	5.8659	61.962	14
15	1.801	.55526	20.024	.04994	.08994	11.1184	6.2721	69.736	15
16	1.873	.53391	21.825	.04582	.08582	11.6523	6.6720	77.744	16
17	1.948	.51337	23.698	.04220	.08220	12.1657	7.0656	85.958	17
18	2.026	.49363	25.645	.03899	.07899	12.6593	7.4530	94.350	18
19	2.107	.47464	27.671	.03614	.07614	13.1339	7.8342	102.893	19
20	2.191	.45639	29.778	.03358	.07358	13.5903	8.2091	111.565	20
21	2.279	.43883	31.969	.03128	.07128	14.0292	8.5779	120.341	21
22	2.370	.42196	34.248	.02920	.06920	14.4511	8.9407	129.202	22
23	2.465	.40573	36.618	.02731	.06731	14.8568	9.2973	138.128	23
24	2.563	.39012	39.083	.02559	.06559	15.2470	9.6479	147.101	24
25	2.666	.37512	41.646	.02401	.06401	15.6221	9.9925	156.104	25

COMPOUND INTEREST FACTOR TABLES

INTEREST RATE 5.000 PERCENT

N	F/P	P/F	F/A	A/F	A/P	P/A	A/G	P/G	N
1	1.050	.95238	1.000	1.00000	1.05000	.9524	.0000	.000	1
2	1.102	.90703	2.050	.48781	.53781	1.8594	.4878	.907	2
3	1.158	.86384	3.152	.31721	.36721	2.7232	.9675	2.635	3
4	1.216	.82270	4.310	.23201	.28201	3.5459	1.4390	5.103	4
5	1.276	.78353	5.526	.18097	.23097	4.3295	1.9025	8.237	5
6	1.340	.74622	6.802	.14702	.19702	5.0757	2.3579	11.968	6
7	1.407	.71068	8.142	.12282	.17282	5.7864	2.8052	16.232	7
8	1.477	.67684	9.549	.10472	.15472	6.4632	3.2445	20.970	8
9	1.551	.64461	11.027	.09069	.14069	7.1078	3.6758	26.127	9
10	1.629	.61391	12.578	.07950	.12950	7.7217	4.0991	31.652	10
11	1.710	.58468	14.207	.07039	.12039	8.3064	4.5144	37.499	11
12	1.796	.55684	15.917	.06283	.11283	8.8633	4.9219	43.624	12
13	1.886	.53032	17.713	.05646	.10646	9.3936	5.3215	49.988	13
14	1.980	.50507	19.599	.05102	.10102	9.8986	5.7133	56.554	14
15	2.079	.48102	21.579	.04634	.09634	10.3797	6.0973	63.288	15
16	2.183	.45811	23.657	.04227	.09227	10.8378	6.4736	70.160	16
17	2.292	.43630	25.840	.03870	.08870	11.2741	6.8423	77.140	17
18	2.407	.41552	28.132	.03555	.08555	11.6896	7.2034	84.204	18
19	2.527	.39573	30.539	.03275	.08275	12.0853	7.5569	91.327	19
20	2.653	.37689	33.066	.03024	.08024	12.4622	7.9030	98.488	20
21	2.786	.35894	35.719	.02800	.07800	12.8212	8.2416	105.667	21
22	2.925	.34185	38.505	.02597	.07597	13.1630	8.5730	112.846	22
23	3.072	.32557	41.430	.02414	.07414	13.4886	8.8971	120.009	23
24	3.225	.31007	44.502	.02247	.07247	13.7986	9.2140	127.140	24
25	3.386	.29530	47.727	.02095	.07095	14.0939	9.5238	134.227	25

COMPOUND INTEREST FACTOR TABLES

INTEREST RATE 6.000 PERCENT

N	F/P	P/F	F/A	A/F	A/P	P/A	A/G	P/G	N
1	1.060	.94340	1.000	1.00000	1.06000	.9434	.0000	.000	1
2	1.124	.89000	2.060	.48544	.54544	1.8334	.4854	.890	2
3	1.191	.83962	3.184	.31411	.37411	2.6730	.9612	2.569	3
4	1.262	.79209	4.375	.22859	.28859	3.4651	1.4272	4.946	4
5	1.338	.74726	5.637	.17740	.23740	4.2124	1.8836	7.935	5
6	1.419	.70496	6.975	.14336	.20336	4.9173	2.3304	11.459	6
7	1.504	.66506	8.394	.11914	.17914	5.5824	2.7676	15.450	7
8	1.594	.62741	9.897	.10104	.16104	6.2098	3.1952	19.842	8
9	1.689	.59190	11.491	.08702	.14702	6.8017	3.6133	24.577	9
10	1.791	.55839	13.181	.07587	.13587	7.3601	4.0220	29.602	10
11	1.898	.52679	14.972	.06679	.12679	7.8869	4.4213	34.870	11
12	2.012	.49697	16.870	.05928	.11928	8.3838	4.8113	40.337	12
13	2.133	.46884	18.882	.05296	.11296	8.8527	5.1920	45.963	13
14	2.261	.44230	21.015	.04758	.10758	9.2950	5.5635	51.713	14
15	2.397	.41727	23.276	.04296	.10296	9.7122	5.9260	57.555	15
16	2.540	.39365	25.673	.03895	.09895	10.1059	6.2794	63.459	16
17	2.693	.37136	28.213	.03544	.09544	10.4773	6.6240	69.401	17
18	2.854	.35034	30.906	.03236	.09236	10.8276	6.9597	75.357	18
19	3.026	.33051	33.760	.02962	.08962	11.1581	7.2867	81.306	19
20	3.207	.31180	36.786	.02718	.08718	11.4699	7.6051	87.230	20
21	3.400	.29416	39.993	.02500	.08500	11.7641	7.9151	93.114	21
22	3.604	.27751	43.392	.02305	.08305	12.0416	8.2166	98.941	22
23	3.820	.26180	46.996	.02128	.08128	12.3034	8.5099	104.701	23
24	4.049	.24698	50.816	.01968	.07968	12.5504	8.7951	110.381	24
25	4.292	.23300	54.865	.01823	.07823	12.7834	9.0722	115.973	25

COMPOUND INTEREST FACTOR TABLES

INTEREST RATE 7.000 PERCENT

N	F/P	P/F	F/A	A/F	A/P	P/A	A/G	P/G	N
1	1.070	.93458	1.000	1.00000	1.07000	.9346	.0000	.000	1
2	1.145	.87344	2.070	.48309	.55309	1.8080	.4831	.873	2
3	1.225	.81630	3.215	.31105	.38105	2.6243	.9549	2.506	3
4	1.311	.76290	4.440	.22523	.29523	3.3872	1.4155	4.795	4
5	1.403	.71299	5.751	.17389	.24389	4.1002	1.8650	7.647	5
6	1.501	.66634	7.153	.13980	.20980	4.7665	2.3032	10.978	6
7	1.606	.62275	8.654	.11555	.18555	5.3893	2.7304	14.715	7
8	1.718	.58201	10.260	.09747	.16747	5.9713	3.1465	18.789	8
9	1.838	.54393	11.978	.08349	.15349	6.5152	3.5517	23.140	9
10	1.967	.50835	13.816	.07238	.14238	7.0236	3.9461	27.716	10
11	2.105	.47509	15.784	.06336	.13336	7.4987	4.3296	32.466	11
12	2.252	.44401	17.888	.05590	.12590	7.9427	4.7025	37.351	12
13	2.410	.41496	20.141	.04965	.11965	8.3577	5.0648	42.330	13
14	2.579	.38782	22.550	.04434	.11434	8.7455	5.4167	47.372	14
15	2.759	.36245	25.129	.03979	.10979	9.1079	5.7583	52.446	15
16	2.952	.33873	27.888	.03586	.10586	9.4466	6.0897	57.527	16
17	3.159	.31657	30.840	.03243	.10243	9.7632	6.4110	62.592	17
18	3.380	.29586	33.999	.02941	.09941	10.0591	6.7225	67.622	18
19	3.617	.27651	37.379	.02675	.09675	10.3356	7.0242	72.599	19
20	3.870	.25842	40.995	.02439	.09439	10.5940	7.3163	77.509	20
21	4.141	.24151	44.865	.02229	.09229	10.8355	7.5990	82.339	21
22	4.430	.22571	49.006	.02041	.09041	11.0612	7.8725	87.079	22
23	4.741	.21095	53.436	.01871	.08871	11.2722	8.1369	91.720	23
24	5.072	.19715	58.177	.01719	.08719	11.4693	8.3923	96.255	24
25	5.427	.18425	63.249	.01581	.08581	11.6536	8.6391	100.676	25

COMPOUND INTEREST FACTOR TABLES

INTEREST RATE 8.000 PERCENT

N	F/P	P/F	F/A	A/F	A/P	P/A	A/G	P/G	N
1	1.080	.92593	1.000	1.00000	1.08000	.9259	.0000	.000	1
2	1.166	.85734	2.080	.48077	.56077	1.7833	.4808	.857	2
3	1.260	.79383	3.246	.30803	.38803	2.5771	.9487	2.445	3
4	1.360	.73503	4.506	.22192	.30192	3.3121	1.4040	4.650	4
5	1.469	.68058	5.867	.17046	.25046	3.9927	1.8465	7.372	5
6	1.587	.63017	7.336	.13632	.21632	4.6229	2.2763	10.523	6
7	1.714	.58349	8.923	.11207	.19207	5.2064	2.6937	14.024	7
8	1.851	.54027	10.637	.09401	.17401	5.7466	3.0985	17.806	8
9	1.999	.50025	12.488	.08008	.16008	6.2469	3.4910	21.808	9
10	2.159	.46319	14.487	.06903	.14903	6.7101	3.8713	25.977	10
11	2.332	.42888	16.645	.06008	.14008	7.1390	4.2395	30.266	11
12	2.518	.39711	18.977	.05270	.13270	7.5361	4.5957	34.634	12
13	2.720	.36770	21.495	.04652	.12652	7.9038	4.9402	39.046	13
14	2.937	.34046	24.215	.04130	.12130	8.2442	5.2731	43.472	14
15	3.172	.31524	27.152	.03683	.11683	8.5595	5.5945	47.886	15
16	3.426	.29189	30.324	.03298	.11298	8.8514	5.9046	52.264	16
17	3.700	.27027	33.750	.02963	.10963	9.1216	6.2037	56.588	17
18	3.996	.25025	37.450	.02670	.10670	9.3719	6.4920	60.843	18
19	4.316	.23171	41.446	.02413	.10413	9.6036	6.7697	65.013	19
20	4.661	.21455	45.762	.02185	.10185	9.8181	7.0369	69.090	20
21	5.034	.19866	50.423	.01983	.09983	10.0168	7.2940	73.063	21
22	5.437	.18394	55.457	.01803	.09803	10.2007	7.5412	76.926	22
23	5.871	.17032	60.893	.01642	.09642	10.3711	7.7786	80.673	23
24	6.341	.15770	66.765	.01498	.09498	10.5288	8.0066	84.300	24
25	6.848	.14602	73.106	.01368	.09368	10.6748	8.2254	87.804	25

COMPOUND INTEREST FACTOR TABLES

INTEREST RATE 9.000 PERCENT

N	F/P	P/F	F/A	A/F	A/P	P/A	A/G	P/G	N
1	1.090	.91743	1.000	1.00000	1.09000	.9174	.0000	.000	1
2	1.188	.84168	2.090	.47847	.56847	1.7591	.4785	.842	2
3	1.295	.77218	3.278	.30505	.39505	2.5313	.9426	2.386	3
4	1.412	.70843	4.573	.21867	.30867	3.2397	1.3925	4.511	4
5	1.539	.64993	5.985	.16709	.25709	3.8897	1.8282	7.111	5
6	1.677	.59627	7.523	.13292	.22292	4.4859	2.2498	10.092	6
7	1.828	.54703	9.200	.10869	.19869	5.0330	2.6574	13.375	7
8	1.993	.50187	11.028	.09067	.18067	5.5348	3.0512	16.888	8
9	2.172	.46043	13.021	.07680	.16680	5.9952	3.4312	20.571	9
10	2.367	.42241	15.193	.06582	.15582	6.4177	3.7978	24.373	10
11	2.580	.38753	17.560	.05695	.14695	6.8052	4.1510	28.248	11
12	2.813	.35553	20.141	.04965	.13965	7.1607	4.4910	32.159	12
13	3.066	.32618	22.953	.04357	.13357	7.4869	4.8182	36.073	13
14	3.342	.29925	26.019	.03843	.12843	7.7862	5.1326	39.963	14
15	3.642	.27454	29.361	.03406	.12406	8.0607	5.4346	43.807	15
16	3.970	.25187	33.003	.03030	.12030	8.3126	5.7245	47.585	16
17	4.328	.23107	36.974	.02705	.11705	8.5436	6.0024	51.282	17
18	4.717	.21199	41.301	.02421	.11421	8.7556	6.2687	54.886	18
19	5.142	.19449	46.018	.02173	.11173	8.9501	6.5236	58.387	19
20	5.604	.17843	51.160	.01955	.10955	9.1285	6.7674	61.777	20
21	6.109	.16370	56.765	.01762	.10762	9.2922	7.0006	65.051	21
22	6.659	.15018	62.873	.01590	.10590	9.4424	7.2232	68.205	22
23	7.258	.13778	69.532	.01438	.10438	9.5802	7.4357	71.236	23
24	7.911	.12640	76.790	.01302	.10302	9.7066	7.6384	74.143	24
25	8.623	.11597	84.701	.01181	.10181	9.8226	7.8316	76.926	25

COMPOUND INTEREST FACTOR TABLES

INTEREST RATE 10.000 PERCENT

N	F/P	P/F	F/A	A/F	A/P	P/A	A/G	P/G	N
1	1.100	.90909	1.000	1.00000	1.10000	.9091	.0000	.000	1
2	1.210	.82645	2.100	.47619	.57619	1.7355	.4762	.826	2
3	1.331	.75131	3.310	.30211	.40211	2.4869	.9366	2.329	3
4	1.464	.68301	4.641	.21547	.31547	3.1699	1.3812	4.378	4
5	1.611	.62092	6.105	.16380	.26380	3.7908	1.8101	6.862	5
6	1.772	.56447	7.716	.12961	.22961	4.3553	2.2236	9.684	6
7	1.949	.51316	9.487	.10541	.20541	4.8684	2.6216	12.763	7
8	2.144	.46651	11.436	.08744	.18744	5.3349	3.0045	16.029	8
9	2.358	.42410	13.579	.07364	.17364	5.7590	3.3724	19.421	9
10	2.594	.38554	15.937	.06275	.16275	6.1446	3.7255	22.891	10
11	2.853	.35049	18.531	.05396	.15396	6.4951	4.0641	26.396	11
12	3.138	.31863	21.384	.04676	.14676	6.8137	4.3884	29.901	12
13	3.452	.28966	24.523	.04078	.14078	7.1034	4.6988	33.377	13
14	3.797	.26333	27.975	.03575	.13575	7.3667	4.9955	36.800	14
15	4.177	.23939	31.772	.03147	.13147	7.6061	5.2789	40.152	15
16	4.595	.21763	35.950	.02782	.12782	7.8237	5.5493	43.416	16
17	5.054	.19784	40.545	.02466	.12466	8.0216	5.8071	46.582	17
18	5.560	.17986	45.599	.02193	.12193	8.2014	6.0526	49.640	18
19	6.116	.16351	51.159	.01955	.11955	8.3649	6.2861	52.583	19
20	6.727	.14864	57.275	.01746	.11746	8.5136	6.5081	55.407	20
21	7.400	.13513	64.002	.01562	.11562	8.6487	6.7189	58.110	21
22	8.140	.12285	71.403	.01401	.11401	8.7715	6.9189	60.689	22
23	8.954	.11168	79.543	.01257	.11257	8.8832	7.1085	63.146	23
24	9.850	.10153	88.497	.01130	.11130	8.9847	7.2881	65.481	24
25	10.835	.09230	98.347	.01017	.11017	9.0770	7.4580	67.696	25

COMPOUND INTEREST FACTOR TABLES

INTEREST RATE 11.000 PERCENT

N	F/P	P/F	F/A	A/F	A/P	P/A	A/G	P/G	N
1	1.110	.90090	1.000	1.00000	1.11000	.9009	.0000	.000	1
2	1.232	.81162	2.110	.47393	.58393	1.7125	.4739	.812	2
3	1.368	.73119	3.342	.29921	.40921	2.4437	.9306	2.274	3
4	1.518	.65873	4.710	.21233	.32233	3.1024	1.3700	4.250	4
5	1.685	.59345	6.228	.16057	.27057	3.6959	1.7923	6.624	5
6	1.870	.53464	7.913	.12638	.23638	4.2305	2.1976	9.297	6
7	2.076	.48166	9.783	.10222	.21222	4.7122	2.5863	12.187	7
8	2.305	.43393	11.859	.08432	.19432	5.1461	2.9585	15.225	8
9	2.558	.39092	14.164	.07060	.18060	5.5370	3.3144	18.352	9
10	2.839	.35218	16.722	.05980	.16980	5.8892	3.6544	21.522	10
11	3.152	.31728	19.561	.05112	.16112	6.2065	3.9788	24.695	11
12	3.498	.28584	22.713	.04403	.15403	6.4924	4.2879	27.839	12
13	3.883	.25751	26.212	.03815	.14815	6.7499	4.5822	30.929	13
14	4.310	.23199	30.095	.03323	.14323	6.9819	4.8619	33.945	14
15	4.785	.20900	34.405	.02907	.13907	7.1909	5.1275	36.871	15
16	5.311	.18829	39.190	.02552	.13552	7.3792	5.3794	39.695	16
17	5.895	.16963	44.501	.02247	.13247	7.5488	5.6180	42.409	17
18	6.544	.15282	50.396	.01984	.12984	7.7016	5.8439	45.007	18
19	7.263	.13768	56.939	.01756	.12756	7.8393	6.0574	47.486	19
20	8.062	.12403	64.203	.01558	.12558	7.9633	6.2590	49.842	20
21	8.949	.11174	72.265	.01384	.12384	8.0751	6.4491	52.077	21
22	9.934	.10067	81.214	.01231	.12231	8.1757	6.6283	54.191	22
23	11.026	.09069	91.148	.01097	.12097	8.2664	6.7969	56.186	23
24	12.239	.08170	102.174	.00979	.11979	8.3481	6.9555	58.066	24
25	13.585	.07361	114.413	.00874	.11874	8.4217	7.1045	59.832	25

COMPOUND INTEREST FACTOR TABLES

INTEREST RATE 12.000 PERCENT

N	F/P	P/F	F/A	A/F	A/P	P/A	A/G	P/G	N
1	1.120	.89286	1.000	1.00000	1.12000	.8929	.0000	.000	1
2	1.254	.79719	2.120	.47170	.59170	1.6901	.4717	.797	2
3	1.405	.71178	3.374	.29635	.41635	2.4018	.9246	2.221	3
4	1.574	.63552	4.779	.20923	.32923	3.0373	1.3589	4.127	4
5	1.762	.56743	6.353	.15741	.27741	3.6048	1.7746	6.397	5
6	1.974	.50663	8.115	.12323	.24323	4.1114	2.1720	8.930	6
7	2.211	.45235	10.089	.09912	.21912	4.5638	2.5515	11.644	7
8	2.476	.40388	12.300	.08130	.20130	4.9676	2.9131	14.471	8
9	2.773	.36061	14.776	.06768	.18768	5.3282	3.2574	17.356	9
10	3.106	.32197	17.549	.05698	.17698	5.6502	3.5847	20.254	10
11	3.479	.28748	20.655	.04842	.16842	5.9377	3.8953	23.129	11
12	3.896	.25668	24.133	.04144	.16144	6.1944	4.1897	25.952	12
13	4.363	.22917	28.029	.03568	.15568	6.4235	4.4683	28.702	13
14	4.887	.20462	32.393	.03087	.15087	6.6282	4.7317	31.362	14
15	5.474	.18270	37.280	.02682	.14682	6.8109	4.9803	33.920	15
16	6.130	.16312	42.753	.02339	.14339	6.9740	5.2147	36.367	16
17	6.866	.14564	48.884	.02046	.14046	7.1196	5.4353	38.697	17
18	7.690	.13004	55.750	.01794	.13794	7.2497	5.6427	40.908	18
19	8.613	.11611	63.440	.01576	.13576	7.3658	5.8375	42.998	19
20	9.646	.10367	72.052	.01388	.13388	7.4694	6.0202	44.968	20
21	10.804	.09256	81.699	.01224	.13224	7.5620	6.1913	46.819	21
22	12.100	.08264	92.503	.01081	.13081	7.6446	6.3514	48.554	22
23	13.552	.07379	104.603	.00956	.12956	7.7184	6.5010	50.178	23
24	15.179	.06588	118.155	.00846	.12846	7.7843	6.6406	51.693	24
25	17.000	.05882	133.334	.00750	.12750	7.8431	6.7708	53.105	25

COMPOUND INTEREST FACTOR TABLES

INTEREST RATE 13.000 PERCENT

N	F/P	P/F	F/A	A/F	A/P	P/A	A/G	P/G	N
1	1.130	.88496	1.000	1.00000	1.13000	.8850	.0000	.000	1
2	1.277	.78315	2.130	.46948	.59948	1.6681	.4695	.783	2
3	1.443	.69305	3.407	.29352	.42352	2.3612	.9187	2.169	3
4	1.630	.61332	4.850	.20619	.33619	2.9745	1.3479	4.009	4
5	1.842	.54276	6.480	.15431	.28431	3.5172	1.7571	6.180	5
6	2.082	.48032	8.323	.12015	.25015	3.9975	2.1468	8.582	6
7	2.353	.42506	10.405	.09611	.22611	4.4226	2.5171	11.132	7
8	2.658	.37616	12.757	.07839	.20839	4.7988	2.8685	13.765	8
9	3.004	.33288	15.416	.06487	.19487	5.1317	3.2014	16.428	9
10	3.395	.29459	18.420	.05429	.18429	5.4262	3.5162	19.080	10
11	3.836	.26070	21.814	.04584	.17584	5.6869	3.8134	21.687	11
12	4.335	.23071	25.650	.03899	.16899	5.9176	4.0936	24.224	12
13	4.898	.20416	29.985	.03335	.16335	6.1218	4.3573	26.674	13
14	5.535	.18068	34.883	.02867	.15867	6.3025	4.6050	29.023	14
15	6.254	.15989	40.417	.02474	.15474	6.4624	4.8375	31.262	15
16	7.067	.14150	46.672	.02143	.15143	6.6039	5.0552	33.384	16
17	7.986	.12522	53.739	.01861	.14861	6.7291	5.2589	35.388	17
18	9.024	.11081	61.725	.01620	.14620	6.8399	5.4491	37.271	18
19	10.197	.09806	70.749	.01413	.14413	6.9380	5.6265	39.037	19
20	11.523	.08678	80.947	.01235	.14235	7.0248	5.7917	40.685	20
21	13.021	.07680	92.470	.01081	.14081	7.1016	5.9454	42.221	21
22	14.714	.06796	105.491	.00948	.13948	7.1695	6.0881	43.649	22
23	16.627	.06014	120.205	.00832	.13832	7.2297	6.2205	44.972	23
24	18.788	.05323	136.831	.00731	.13731	7.2829	6.3431	46.196	24
25	21.231	.04710	155.620	.00643	.13643	7.3300	6.4566	47.326	25

COMPOUND INTEREST FACTOR TABLES

INTEREST RATE 14.000 PERCENT

N	F/P	P/F	F/A	A/F	A/P	P/A	A/G	P/G	N
1	1.140	.87719	1.000	1.00000	1.14000	.8772	.0000	.000	1
2	1.300	.76947	2.140	.46729	.60729	1.6467	.4673	.769	2
3	1.482	.67497	3.440	.29073	.43073	2.3216	.9129	2.119	3
4	1.689	.59208	4.921	.20320	.34320	2.9137	1.3370	3.896	4
5	1.925	.51937	6.610	.15128	.29128	3.4331	1.7399	5.973	5
6	2.195	.45559	8.536	.11716	.25716	3.8887	2.1218	8.251	6
7	2.502	.39964	10.730	.09319	.23319	4.2883	2.4832	10.649	7
8	2.853	.35056	13.233	.07557	.21557	4.6389	2.8246	13.103	8
9	3.252	.30751	16.085	.06217	.20217	4.9464	3.1463	15.563	9
10	3.707	.26974	19.337	.05171	.19171	5.2161	3.4490	17.991	10
11	4.226	.23662	23.045	.04339	.18339	5.4527	3.7333	20.357	11
12	4.818	.20756	27.271	.03667	.17667	5.6603	3.9998	22.640	12
13	5.492	.18207	32.089	.03116	.17116	5.8424	4.2491	24.825	13
14	6.261	.15971	37.581	.02661	.16661	6.0021	4.4819	26.901	14
15	7.138	.14010	43.842	.02281	.16281	6.1422	4.6990	28.862	15
16	8.137	.12289	50.980	.01962	.15962	6.2651	4.9011	30.706	16
17	9.276	.10780	59.118	.01692	.15692	6.3729	5.0888	32.430	17
18	10.575	.09456	68.394	.01462	.15462	6.4674	5.2630	34.038	18
19	12.056	.08295	78.969	.01266	.15266	6.5504	5.4243	35.531	19
20	13.743	.07276	91.025	.01099	.15099	6.6231	5.5734	36.914	20
21	15.668	.06383	104.768	.00954	.14954	6.6870	5.7111	38.190	21
22	17.861	.05599	120.436	.00830	.14830	6.7429	5.8381	39.366	22
23	20.362	.04911	138.297	.00723	.14723	6.7921	5.9549	40.446	23
24	23.212	.04308	158.659	.00630	.14630	6.8351	6.0624	41.437	24
25	26.462	.03779	181.871	.00550	.14550	6.8729	6.1610	42.344	25

COMPOUND INTEREST FACTOR TABLES

INTEREST RATE 15.000 PERCENT

N	F/P	P/F	F/A	A/F	A/P	P/A	A/G	P/G	N
1	1.150	.86957	1.000	1.00000	1.15000	.8696	.0000	.000	1
2	1.322	.75614	2.150	.46512	.61512	1.6257	.4651	.756	2
3	1.521	.65752	3.472	.28798	.43798	2.2832	.9071	2.071	3
4	1.749	.57175	4.993	.20027	.35027	2.8550	1.3263	3.786	4
5	2.011	.49718	6.742	.14832	.29832	3.3522	1.7228	5.775	5
6	2.313	.43233	8.754	.11424	.26424	3.7845	2.0972	7.937	6
7	2.660	.37594	11.067	.09036	.24036	4.1604	2.4498	10.192	7
8	3.059	.32690	13.727	.07285	.22285	4.4873	2.7813	12.481	8
9	3.518	.28426	16.786	.05957	.20957	4.7716	3.0922	14.755	9
10	4.046	.24718	20.304	.04925	.19925	5.0188	3.3832	16.979	10
11	4.652	.21494	24.349	.04107	.19107	5.2337	3.6549	19.129	11
12	5.350	.18691	29.002	.03448	.18448	5.4206	3.9082	21.185	12
13	6.153	.16253	34.352	.02911	.17911	5.5831	4.1438	23.135	13
14	7.076	.14133	40.505	.02469	.17469	5.7245	4.3624	24.972	14
15	8.137	.12289	47.580	.02102	.17102	5.8474	4.5650	26.693	15
16	9.358	.10686	55.717	.01795	.16795	5.9542	4.7522	28.296	16
17	10.761	.09293	65.075	.01537	.16537	6.0472	4.9251	29.783	17
18	12.375	.08081	75.836	.01319	.16319	6.1280	5.0843	31.156	18
19	14.232	.07027	88.212	.01134	.16134	6.1982	5.2307	32.421	19
20	16.367	.06110	102.444	.00976	.15976	6.2593	5.3651	33.582	20
21	18.822	.05313	118.810	.00842	.15842	6.3125	5.4883	34.645	21
22	21.645	.04620	137.632	.00727	.15727	6.3587	5.6010	35.615	22
23	24.891	.04017	159.276	.00628	.15628	6.3988	5.7040	36.499	23
24	28.625	.03493	184.168	.00543	.15543	6.4338	5.7979	37.302	24
25	32.919	.03038	212.793	.00470	.15470	6.4641	5.8834	38.031	25

COMPOUND INTEREST FACTOR TABLES

NUMBER OF PERIODS 5.

i	F/P	P/F	F/A	A/F	A/P	P/A	A/G	P/G	i
.010	1.051	.95147	5.101	.19604	.20604	4.8534	1.9800	9.610	.010
.020	1.104	.90573	5.204	.19216	.21216	4.7135	1.9604	9.240	.020
.030	1.159	.86261	5.309	.18835	.21835	4.5797	1.9409	8.889	.030
.040	1.217	.82193	5.416	.18463	.22463	4.4518	1.9216	8.555	.040
.050	1.276	.78353	5.526	.18097	.23097	4.3295	1.9025	8.237	.050
.060	1.338	.74726	5.637	.17740	.23740	4.2124	1.8836	7.935	.060
.070	1.403	.71299	5.751	.17389	.24389	4.1002	1.8649	7.647	.070
.080	1.469	.68058	5.867	.17046	.25046	3.9927	1.8465	7.372	.080
.090	1.539	.64993	5.985	.16709	.25709	3.8897	1.8282	7.111	.090
.100	1.611	.62092	6.105	.16380	.26380	3.7908	1.8101	6.862	.100
.110	1.685	.59345	6.228	.16057	.27057	3.6959	1.7923	6.624	.110
.120	1.762	.56743	6.353	.15741	.27741	3.6048	1.7746	6.397	.120
.130	1.842	.54276	6.480	.15431	.28431	3.5172	1.7571	6.180	.130
.140	1.925	.51937	6.610	.15128	.29128	3.4331	1.7399	5.973	.140
.150	2.011	.49718	6.742	.14832	.29832	3.3522	1.7228	5.775	.150

COMPOUND INTEREST FACTOR TABLES

NUMBER OF PERIODS 10.

i	F/P	P/F	F/A	A/F	A/P	P/A	A/G	P/G	i
.010	1.105	.90529	10.462	.09558	.10558	9.4713	4.4179	41.844	.010
.020	1.219	.82035	10.950	.09133	.11133	8.9826	4.3367	38.955	.020
.030	1.344	.74409	11.464	.08723	.11723	8.5302	4.2565	36.309	.030
.040	1.480	.67556	12.006	.08329	.12329	8.1109	4.1773	33.881	.040
.050	1.629	.61391	12.578	.07950	.12950	7.7217	4.0991	31.652	.050
.060	1.791	.55839	13.181	.07587	.13587	7.3601	4.0220	29.602	.060
.070	1.967	.50835	13.816	.07238	.14238	7.0236	3.9461	27.716	.070
.080	2.159	.46319	14.487	.06903	.14903	6.7101	3.8713	25.977	.080
.090	2.367	.42241	15.193	.06582	.15582	6.4177	3.7978	24.373	.090
.100	2.594	.38554	15.937	.06275	.16275	6.1446	3.7255	22.891	.100
.110	2.839	.35218	16.722	.05980	.16980	5.8892	3.6544	21.522	.110
.120	3.106	.32197	17.549	.05698	.17698	5.6502	3.5847	20.254	.120
.130	3.395	.29459	18.420	.05429	.18429	5.4262	3.5162	19.080	.130
.140	3.707	.26974	19.337	.05171	.19171	5.2161	3.4490	17.991	.140
.150	4.046	.24718	20.304	.04925	.19925	5.0188	3.3832	16.979	.150

COMPOUND INTEREST FACTOR TABLES

NUMBER OF PERIODS 15.

i	F/P	P/F	F/A	A/F	A/P	P/A	A/G	P/G	i
.010	1.161	.86135	16.097	.06212	.07212	13.8650	6.8143	94.481	.010
.020	1.346	.74301	17.293	.05783	.07783	12.8493	6.6309	85.202	.020
.030	1.558	.64186	18.599	.05377	.08377	11.9379	6.4500	77.000	.030
.040	1.801	.55526	20.024	.04994	.08994	11.1184	6.2721	69.736	.040
.050	2.079	.48102	21.579	.04634	.09634	10.3797	6.0973	63.288	.050
.060	2.397	.41727	23.276	.04296	.10296	9.7122	5.9260	57.555	.060
.070	2.759	.36245	25.129	.03979	.10979	9.1079	5.7583	52.446	.070
.080	3.172	.31524	27.152	.03683	.11683	8.5595	5.5945	47.886	.080
.090	3.642	.27454	29.361	.03406	.12406	8.0607	5.4346	43.807	.090
.100	4.177	.23939	31.772	.03147	.13147	7.6061	5.2789	40.152	.100
.110	4.785	.20900	34.405	.02907	.13907	7.1909	5.1275	36.871	.110
.120	5.474	.18270	37.280	.02682	.14682	6.8109	4.9803	33.920	.120
.130	6.254	.15989	40.417	.02474	.15474	6.4624	4.8375	31.262	.130
.140	7.138	.14010	43.842	.02281	.16281	6.1422	4.6990	28.862	.140
.150	8.137	.12289	47.580	.02102	.17102	5.8474	4.5650	26.693	.150

COMPOUND INTEREST FACTOR TABLES

NUMBER OF PERIODS 20.

i	F/P	P/F	F/A	A/F	A/P	P/A	A/G	P/G	i
.010	1.220	.81954	22.019	.04542	.05542	18.0456	9.1694	165.466	.010
.020	1.486	.67297	24.297	.04116	.06116	16.3514	8.8433	144.600	.020
.030	1.806	.55368	26.870	.03722	.06722	14.8775	8.5229	126.799	.030
.040	2.191	.45639	29.778	.03358	.07358	13.5903	8.2091	111.565	.040
.050	2.653	.37689	33.066	.03024	.08024	12.4622	7.9030	98.488	.050
.060	3.207	.31180	36.786	.02718	.08718	11.4699	7.6051	87.230	.060
.070	3.870	.25842	40.995	.02439	.09439	10.5940	7.3163	77.509	.070
.080	4.661	.21455	45.762	.02185	.10185	9.8181	7.0369	69.090	.080
.090	5.604	.17843	51.160	.01955	.10955	9.1285	6.7675	61.777	.090
.100	6.727	.14864	57.275	.01746	.11746	8.5136	6.5081	55.407	.100
.110	8.062	.12403	64.203	.01558	.12558	7.9633	6.2590	49.842	.110
.120	9.646	.10367	72.052	.01388	.13388	7.4694	6.0202	44.968	.120
.130	11.523	.08678	80.947	.01235	.14235	7.0248	5.7917	40.685	.130
.140	13.743	.07276	91.025	.01099	.15099	6.6231	5.5734	36.914	.140
.150	16.367	.06110	102.444	.00976	.15976	6.2593	5.3651	33.582	.150

COMPOUND INTEREST FACTOR TABLES

NUMBER OF PERIODS 25.

i	F/P	P/F	F/A	A/F	A/P	P/A	A/G	P/G	i
.010	1.282	.77977	28.243	.03541	.04541	22.0232	11.4831	252.894	.010
.020	1.641	.60953	32.030	.03122	.05122	19.5235	10.9744	214.259	.020
.030	2.094	.47761	36.459	.02743	.05743	17.4131	10.4768	182.434	.030
.040	2.666	.37512	41.646	.02401	.06401	15.6221	9.9925	156.104	.040
.050	3.386	.29530	47.727	.02095	.07095	14.0939	9.5238	134.228	.050
.060	4.292	.23300	54.865	.01823	.07823	12.7834	9.0722	115.973	.060
.070	5.427	.18425	63.249	.01581	.08581	11.6536	8.6391	100.676	.070
.080	6.848	.14602	73.106	.01368	.09368	10.6748	8.2254	87.804	.080
.090	8.623	.11597	84.701	.01181	.10181	9.8226	7.8316	76.926	.090
.100	10.835	.09230	98.347	.01017	.11017	9.0770	7.4580	67.696	.100
.110	13.585	.07361	114.413	.00874	.11874	8.4217	7.1045	59.832	.110
.120	17.000	.05882	133.334	.00750	.12750	7.8431	6.7708	53.105	.120
.130	21.231	.04710	155.620	.00643	.13643	7.3300	6.4566	47.326	.130
.140	26.462	.03779	181.871	.00550	.14550	6.8729	6.1610	42.344	.140
.150	32.919	.03038	212.793	.00470	.15470	6.4641	5.8834	38.031	.150

GEOMETRIC SERIES COMPOUND INTEREST TABLE
INTEREST RATE 4.00 PERCENT

GEOMETRIC GROWTH RATE 1.00 TO 6.00 PERCENT

n	(P/G,.010,i,n)	(P/G,.027,i,n)	(P/G,.043,i,n)	(P/G,.060,i,n)	n
1	.96154	.96154	.96154	.96154	1
2	1.89534	1.91075	1.92616	1.94157	2
3	2.80221	2.84779	2.89387	2.94044	3
4	3.68291	3.77282	3.86468	3.95853	4
5	4.53821	4.68599	4.83861	4.99619	5
6	5.36884	5.58745	5.81566	6.05381	6
7	6.17551	6.47735	6.79584	7.13177	7
8	6.95891	7.35585	7.77916	8.23046	8
9	7.71971	8.22308	8.76563	9.35028	9
10	8.45856	9.07920	9.75526	10.49163	10
11	9.17610	9.92434	10.74806	11.65493	11
12	9.87295	10.75864	11.74405	12.84060	12
13	10.54969	11.58225	12.74323	14.04907	13
14	11.20691	12.39529	13.74562	15.28078	14
15	11.84517	13.19792	14.75121	16.53618	15
16	12.46502	13.99025	15.76003	17.81573	16
17	13.06699	14.77243	16.77208	19.11987	17
18	13.65160	15.54458	17.78737	20.44910	18
19	14.21934	16.30683	18.80592	21.80389	19
20	14.77071	17.05930	19.82774	23.18474	20
21	15.30617	17.80213	20.85283	24.59214	21
22	15.82618	18.53544	21.88120	26.02660	22
23	16.33120	19.25934	22.91287	27.48865	23
24	16.82164	19.97397	23.94785	28.97882	24
25	17.29794	20.67943	24.98614	30.49764	25

Appendix C

Labor and Material Fractional Cost Analyses and Typical Commercial Shipbuilding Cost Account Structure

LABOR AND MATERIAL FRACTIONAL COST ANALYSIS

Labor and Material by Cost Accounts for Double-Hull Tanker, 825′ × 136′ × 71′, 119,054 dwt, 30,000 shp and for RO/RO Cargo Vessel, 643′ × 92′ × 60.135′, 14,090 dwt, 30,000 shp

Acct No.	*Account Description*	*RO/RO Material, Dollars, Fraction*	*RO/RO Labor-Hours, Fraction*	*Tanker Material, Dollars, Fraction*	*Tanker Labor-Hours, Fraction*
714	Total hull steel	0.405672	0.443480	0.242449	0.441251
713	Stem, stern frame, rudder	0.007065	0.008638	0.012266	0.007909
721	Outfitting structures	0.000758	0.000950	0.001265	0.003130
722	Hatches and side ports	0.001863	0.000351	0.008371	0.001381
723	Door-actuating equipment	0.000000	0.000000	0.000216	0.000816
724	Structural doors	0.001506	0.001239	0.001296	0.001703
725	Fixed lights and windows	0.000472	0.000169	0.000591	0.000143
726	Metal skylights	0.000031	0.000030	0.000000	0.000000
734	Foundation for auxiliary machine equipment	0.000364	0.003966	0.000445	0.001683
735	Stabilization system	0.000000	0.000000	0.000000	0.000000
744	Structural masts and booms	0.002684	0.002874	0.000649	0.003960
745	Blocks and rigging	0.000524	0.000108	0.001911	0.000358
747	Cargo cranes	0.000000	0.000000	0.000000	0.000000
748	Elevators, hoists, conveyors	0.002162	0.000101	0.012439	0.000873
750	Joiner work	0.000009	0.000873	0.000054	0.002229
751	Refrigerated compartments	0.000171	0.000049	0.000000	0.000000
752	Accommodation ladder	0.000366	0.000290	0.000665	0.000229
755	Metal sheathing	0.000225	0.000847	0.001240	0.004077
756	Light metal bulkheads	0.000137	0.000508	0.000103	0.000745
758	Deck machinery	0.015798	0.000317	0.017695	0.000249

Acct No.	*Account Description*	*RO/RO Material, Dollars, Fraction*	*RO/RO Labor-Hours, Fraction*	*Tanker Material, Dollars, Fraction*	*Tanker Labor-Hours, Fraction*
760	Carpenter work	0.000000	0.000000	0.000000	0.000000
765	Painting	0.016444	0.009416	0.027615	0.021686
770	Structural insulation	0.001565	0.000000	0.007702	0.000000
780	Scupper and drains	0.000086	0.000225	0.000058	0.000315
783	Mooring and towing gear	0.002676	0.001868	0.004041	0.004350
784	Lines, hawsers	0.000690	0.000112	0.000719	0.000091
785	Anchors, chains	0.007500	0.000986	0.006051	0.001284
788	Swimming pool, gymnasium, etc.	0.000000	0.000000	0.000000	0.000000
789	Rat proofing	0.000009	0.000049	0.000010	0.000217
790	Lifeboats, rafts	0.002863	0.000853	0.004720	0.000731
791	Hotel and galley equipment	0.000988	0.000477	0.005553	0.001051
792	Handrails and stanchions	0.000382	0.001162	0.000437	0.001805
793	Metal ladders	0.001296	0.004365	0.000319	0.001384
794	Miscellaneous hull fittings	0.001683	0.002617	0.003266	0.004829
795	Storeroom and fittings	0.000057	0.000331	0.000148	0.000526
796	Rigid awnings	0.000068	0.000076	0.000055	0.000163
797	Deck covers, permanent ballast	0.001706	0.000022	0.002574	0.000935
798	Canvas outfitting	0.000020	0.000000	0.000016	0.000000
799	Owner's items	0.000000	0.000002	0.000012	0.000003
	Total outfitting	0.072169	0.043871	0.122498	0.068855
331	Cargo securing gear	0.000000	0.000000	0.022714	0.006353
811	Main condenser	0.010260	0.000145	0.015207	0.000666
817	Main engine throttle	0.000046	0.000247	0.000076	0.000363
840	Propeller	0.009939	0.000761	0.008914	0.001044
841	Bow thruster	0.000000	0.000000	0.011989	0.001202
842	Shafts, bearings	0.006454	0.002953	0.017554	0.009973
844	Floors, ladders, grating	0.001954	0.004905	0.001504	0.004828
850	Mechanical insulation	0.006712	0.000000	0.006998	0.000004
853	Dredging pump	0.000000	0.000000	0.000000	0.000000
855	Refrigeration plant	0.001632	0.000204	0.001483	0.000148
856	Miscellaneous fittings and tanks	0.004272	0.006059	0.006219	0.009475
858	Steering gear, anchor windlass	0.005556	0.002408	0.015684	0.003860
867	Auxiliary boilers	0.000000	0.000000	0.000000	0.000000
868	Fuel calification system	0.000000	0.000000	0.000000	0.000000
869	Fuel oil system	0.003510	0.005254	0.010625	0.016232
877	Cargo hold ventilation	0.000005	0.000093	0.005769	0.017878
878	Miscellaneous ventilation	0.000427	0.001215	0.000000	0.000001
879	Accomodation vent and air-conditioning	0.002993	0.004378	0.004664	0.007433
880	Smokestacks and uptakes	0.002293	0.002611	0.004062	0.007059
882	Forced draft system	0.002938	0.001335	0.004800	0.003213

Acct No.	Account Description	*RO/RO Material, Dollars, Fraction*	*RO/RO Labor-Hours, Fraction*	*Tanker Material, Dollars, Fraction*	*Tanker Labor-Hours, Fraction*
890	Main engine	0.067488	0.003156	0.113911	0.005364
893	Auxiliary machinery	0.005200	0.000154	0.008339	0.000307
894	Main boiler	0.033395	0.008866	0.056051	0.016769
900	Steam and exhaust piping in machinery	0.021708	0.014738	0.018460	0.015570
901	Cooling system	0.015364	0.009427	0.018853	0.011987
902	Compressed-air system	0.001121	0.000976	0.001538	0.001725
903	Auxiliary engine fuel oil system	0.000113	0.000464	0.000096	0.000401
904	Lube oil system	0.003285	0.003142	0.005202	0.004225
905	Auxiliary engine circulating water system	0.000000	0.000000	0.000000	0.000000
906	Feedwater and conditioning system	0.010970	0.005647	0.016530	0.008452
907	Steam drains in machinery space	0.002858	0.003710	0.002565	0.004401
908	Steam and exhaust system, outside machinery	0.000064	0.000331	0.000005	0.000005
914	Heating system	0.000174	0.000640	0.001267	0.001620
915	Ventilation system, engine spaces	0.002050	0.005528	0.002787	0.007613
916	Vents and sounding	0.000741	0.001290	0.000789	0.006072
917	Weather deck drains	0.000092	0.000319	0.000217	0.001199
918	Bilge and ballast system	0.007497	0.009333	0.008575	0.009292
919	Plumbing system	0.005985	0.005339	0.008087	0.009421
920	Carbon dioxide system	0.000271	0.000216	0.007336	0.009374
921	Salt water, steam, foam fire system	0.010506	0.005891	0.009762	0.009402
922	Cargo tank inerting system	0.021010	0.007093	0.000000	0.000000
923	Liquid cargo system	0.073377	0.031148	0.008754	0.004090
924	Engine room watertight doors	0.000000	0.000000	0.000000	0.000000
926	Machinery space skylight operating gear	0.000000	0.000000	0.000000	0.000000
927	Whistle pulls	0.000019	0.000045	0.000041	0.000051
929	Heavy lift system	0.000000	0.000000	0.000000	0.000000
	Total mechanical	0.342279	0.150022	0.404712	0.210720
930	Propulsion plant automatic control system	0.006695	0.000819	0.012345	0.001665
937	Electrical modifications	0.000000	0.000000	0.000000	0.000000
938	Electric generation	0.018700	0.000676	0.039023	0.001259
939	Wiring and cableways	0.012799	0.012360	0.022266	0.026361
940	Electric lighting	0.002086	0.002550	0.007223	0.008303

Acct No.	*Account Description*	*RO/RO Material, Dollars, Fraction*	*RO/RO Labor-Hours, Fraction*	*Tanker Material, Dollars, Fraction*	*Tanker Labor-Hours, Fraction*
941	Miscellaneous power distribution	0.001246	0.002577	0.005530	0.007144
942	Interior communication	0.002632	0.001503	0.009736	0.001730
943	Exterior communication	0.005496	0.000359	0.008991	0.000748
944	Electric motors	0.000000	0.000000	0.000000	0.000000
945	Motor controls	0.003424	0.000001	0.007361	0.000008
946	Cathodic protection	0.000029	0.000000	0.001117	0.000368
	Total electrical	0.053107	0.020845	0.113591	0.047586
950	Setting up and launch	0.005033	0.010825	0.002470	0.013683
952	Staging and blocking	0.018738	0.062806	0.013861	0.052330
957	Crane service	0.000458	0.029271	0.000000	0.028261
958	Cleaning	0.001295	0.026894	0.000832	0.019645
959	Temporary air, light, water	0.006396	0.012946	0.002827	0.011457
960	Miscellaneous construction	0.020376	0.030069	0.021294	0.027172
962	Testing, trials, delivery	0.006816	0.010378	0.009143	0.011943
963	Towing, port charges	0.000000	0.000000	0.000000	0.000000
964	Repairs after delivery	0.000000	0.000000	0.000000	0.000000
965	Drydocking	0.002291	0.000763	0.001914	0.000757
956	Special services	0.000000	0.015553	0.000037	0.015542
440	Defective material	0.002936	0.000545	0.002130	0.000523
441	Accident to product	0.000000	0.000000	0.000000	0.000000
442	Lost material	0.000000	0.000000	0.000000	0.000000
	Total service and quality cost	0.064339	0.200050	0.054507	0.181313
700	Naval architecture	0.016051	0.008994	0.002241	0.001970
701	Mold loft	0.012491	0.024847	0.000072	0.006588
702	Hull structural	0.003356	0.018551	0.000000	0.007783
703	Hull outfitting	0.000000	0.012930	0.000223	0.006199
731	Pattern making	0.000000	0.000000	0.000052	0.000030
800	Mechanical design	0.002145	0.041599	0.000935	0.010958
804	Machinery model	0.000000	0.000262	0.000000	0.000000
935	Electrical design	0.000615	0.009650	0.000311	0.004223
953	Jigs, tools, fixtures	0.000641	0.002998	0.000190	0.000947
975	Miscellaneous engineering	0.001243	0.010306	0.000069	0.004484
	Total engineering	0.036542	0.130139	0.004093	0.043182
200	Hull extra purchased parts	0.000915	0.001874	0.000521	0.000740
201	Spare parts	0.002776	0.001874	0.000000	0.000000
207	Owner outfit	0.000000	0.007844	0.000000	0.000000
	Total	1.000000	1.000000	1.000000	1.000000

TYPICAL COMMERCIAL SHIPYARD COST ACCOUNT STRUCTURE

700 *Naval architecture:* Includes naval architecture work, comprising design calculations and arrangement drawings together with associated material and services. Includes such technical areas as hydrostatic and hydrodynamic considerations, inclining experiments and stability data, speed and power calculations, model tests, and propeller design.

701 *Mold loft work:* Includes template work for structural steel, comprising body plans, layout templates, burning machine templates, bending templates, shop lofting data sheets and replacement, and repairs or alteration to templates.

702 *Hull structural design:* Includes hull structural engineering work, comprising design calculations, arrangement and detail drawings, material specifications, and associated services. Includes technical areas such as structural analysis, bending moment calculations, and compensation for hull penetrations.

703 *Hull outfitting design:* Includes hull outfitting engineering work, comprising arrangement and detail drawings, material specifications, allowance lists, and associated services.

713 *Stem, stern frame, rudder, propeller bossings:* Includes hull structural castings and forgings, or special weldments in lieu thereof, stem, paravane skeg, stern frame, stern post, stern skeg, structural stern tube, propeller shaft bossings and struts, spectacle frames, rudder, rudder stock, rudder trunk, special fairing devices in way of stem, stern and bow thruster, and antigalvanic anodes and passive devices for exterior shell. Also includes material and labor on all castings, forgings, and weldments up to the point of erection and attachment to the structural hull. Includes cutout for stern tube in the aft peak tank bulkhead.

714 *Structural steel:* Includes all material and labor for structural steel or nonferrous material used as plates, shapes, bars, and pipe used in such parts of the vessel as shell plating, framing system, center vertical keels, flat keels, bilge keels, docking keels, fenders, longitudinal hull girders, stringers, breast hooks, inner bottom plating and framing, deck plating and beams, deck platforms, deck flats, deck ramps, transverse and longitudinal main bulkheads, miscellaneous bulkheads, engine casing, boiler casing, pump room houses, trunks, deck girders, deck pillers, deck stanchions, continuous hatch coamings, shaft tunnels, engine and boiler foundations, auxiliary machin-

ery foundations, deckhouses, deck enclosures, and independent or built-in independent tanks. Includes also the testing and inspection of materials and workmanship and the stress relieving of all hull structural work. Also includes auxiliary and accessory materials such as welding rod, rivets, service bolts, washers, dogs, clips, strongbacks, and pads.

715 *National defense features:* Includes material and installation for gun foundations, emplacements, armor, armament, ammunition magazines, spar decks, platforms, and blackout screens and devices. Also includes emergency escapes, flexible ladders, special life-saving equipment, equipment for fueling/stores transfer at sea, protection of degaussing cables, and all other national defense feature items.

721 *Outfitting structures:* Includes fore and aft walkways, foundations and platforms for winches, capstans, and miscellaneous hull fittings, hinged or portable ramps, fixed or portable bulwarks, breakwaters, wind deflectors and dodgers, oil hose bolsters and gallows frames, stowages, companions and booby hatches, gratings in way of opening, structural protection of piping and cable, curbing and guards, grain divisional bulkheads and fittings, metal sparring, special fittings structures, portable decks, and container guides.

722 *Hatches and side ports:* Includes main cargo hatches, fuel bunker hatches, ballast tank and cofferdam hatches, miscellaneous hatches (stores, cordage, grain trimming), steel or nonferrous metal, structural fabrication or cast, raised coaming or flush, tight or nontight, portable, end or side rolling, sliding, hinged, pivoted, pontoon, bolted plate, sectional metal construction. Also includes operating and handling gear, but excludes power operations (for which see account 723), metal strongbeams for wood covers, tarpaulin cleats, batten bars, wedges, locks for wood covers, ullage covers, and screens, tank gauges, side ports, cargo, stores, embarkation, fueling, hinged or sliding, bow or stern doors, bulwark doors hinged or sliding.

723 *Hatch, port, and door-actuating equipment:* Includes hydraulic components, pumps, accumulators, controls and piping; electric motors, gears, controls, and other power applications.

724 *Structural doors, manholes, and scuttles:* Includes all structural doors, steel or nonferrous metal, fabricated or cast, watertight, gastight, nontight, fire screen, hinged, sliding, bolted plate, overhead roller type. Also includes operating gear for same, manholes, hinged or bolted plate, tight or nontight, special; scuttles, raised or flush, tight or nontight, special; freeing ports, bar or plate type.

725 *Air ports, fixed lights, windows:* Includes air ports, hinged, pivoted, patent fixed lights, deck lights; windows, hinged, sliding, casement, fixed, fabricated or cast frames; dead covers, guards, insect screens,

wind scoops, hangers, and hold-back devices; clear vision devices, manual or power operated.

726 *Metal skylights:* Includes metal skylights and coamings, metal fidley covers and coamings, insect screens, and operating gear.

734 *Foundation for auxiliary machinery and equipment:* Includes fabrication and installation of foundations that are sourced on drawings bearing the prefix 734.

735 *Stabilization systems:* Includes solid or liquid weights transfer systems, gyroscopic systems, actuated planes or fins, and gear and foundations.

744 *Structural masts, king posts, and booms:* Includes masts, topmasts, radar masts, king posts, derrick posts, towers of all kinds, gin poles, cross trusses, outriggers, struts, platforms, boom tables and foundations, posts, boom stowates, booms (tubular or lattice), yards, signal yards, equalizing booms, gaffs, flagstaffs, TV masts, antennae poles, cages, fittings, and wire. Also includes mast trucks, navigating light brackets, ladders, grabs, vang and topping winch foundations, all rigging fittings, pads, cleats, deadeyes, belaying pins, and stenciling or identification markings.

745 *Blocks and rigging*: Includes standing rigging, wire rope (manila or other), running rigging, wire rope, manila or other; slings, bridles, pendants, windsail spans, metal blocks, wooden blocks, cargo hooks, swivels, overhauling weights, turnbuckles, shackles, thimbles, and clips.

747 *Cargo cranes:* Includes all cargo cranes; controls, electric, hydraulic, electrohydraulic, electro-mechanical; revolving gantry, bridge, davit type, boom type, bull wheel; track, foundations, securing gear, boom stowages, and installation of cranes.

748 *Elevators, hoists, dumbwaiters, and conveyors:* Includes elevators; passenger, cargo, stores, car type, platform type, monorail type, continuous belt type; hoisting machinery, controls, fittings, trunks and foundations; hoists, ammunition, operating gear, trunks, dumbwaiters, manual or power operated, trunks, foundation controls and fittings. Also includes conveyors, stair type, bucket, roller, belt, chain, manual or power operated, operating machinery, trunks, foundations, installation of elevators, hoists, dumbwaiters, and conveyors.

750 *Joiner work:* Includes contract work performed by joiner subcontractor, joiner bulkheads, sheathing, panel work, ceiling, marinite, T & G wood, or other insulation grounds, furring, mouldings, joiner base, trim, cornices, pilasters, casings, joiner doors and trim, hardware, storm rails, shop and shipwork complete, installation only of joiner

windows. Also includes interior decorative arrangements, joiner ladders and stairways, balustrades, joiner rails, wind dodgers, name boards; furniture, fixed and movable, beds, berths, metal lockers, deck lockers, license and notice frames, racks and shelves, wood skylights, wood gratings, joiner decks, margins, carlins, canvas coverings and fastenings, shop and shipwork.

751 *Refrigerated compartments:* Includes refrigerated cargo spaces, cooling rooms, ship's walk-in stores refrigerators, insulation, sheathing and furring on boundary bulkheads and decks, insulated and divisional bulkheads, platforms, and decks, all internal rails, shelves, bins, fittings, refrigerator doors, battens, and gratings. Also includes metal linings, metal floor pans, composition walking surfaces (commercial refrigerators to be entered under account 791).

752 *Accommodation ladders and gangways:* Includes accommodation ladders, with platforms, fenders, bridles, lifting tackle, davits, fittings, storage racks, winches, gangways and stowage, and testing.

755 *Metal sheathing:* Includes material and installation for metal sheathing and furring, over insulation as well as where serving as interior finish other than as protective covering for insulation, but not as a nonwatertight divisional bulkhead.

756 *Light metal bulkheads:* Includes light metal bulkheads serving as nontight divisions in shower and toilet rooms, expanded metal or wire screen bulkheads, and corrugated metal bulkheads.

758 *Winches and capstans:* Includes winches and controlling appliances with spares, capstan towing machines or other warping machines, spares for same, and controlling appliances, including installation.

760 *Carpenter work and cargo battens:* Includes wood hatch covers, ceiling on tank tops, cargo battens, sparring, bossing of piping, cables, fenders fastened to hull, wood stanchions, gratings, miscellaneous spar decks, flats and foundations, cordage racks, all wood installed by carpenters; bosun's chairs, painters, and other staging. Also includes caulked wood decks, margins, and foundations, milling, laying, caulking, planing, and scraping; caulking materials and fastenings.

765 *Painting and surface preparation:* Includes labor and material for surface preparation and painting of the interior and exterior hulls, structural work, accommodation spaces, necessary joiner work, and machinery, including such processes as wire brushing, sand blasting, sand brushing, mineral shot blasting, pickling, various liquid cleaning methods, hand painting, spray painting, prime coating, finish coating, varnishing, and oiling. Also includes special interior decorative effects, such as textile finishes, murals, textile panels, and art effects generally. Also includes special protective finishes, such as

proprietary plastics, vinyls, epoxies, chemical flotation coatings, neoprene coatings, antifouling and antigalvanic coatings. Also includes the costs for operating the shot blasting and priming plant for steel plates and shapes. Includes special services connected with the work, such as rental of special equipment, special cleaning to remove residuals, special staging or scaffolding, special ventilation during operation of processes, and special testing and inspection procedures.

770 *Structural insulation:* Includes material and installation for heat, fire, and acoustical insulation, insulation materials on bulkheads and overhead areas covered by sheet metal, as well as deck insulation covered by floor covering.

777 *Cargo hold ventilation:* Includes natural cargo hold ventilation, supply and exhaust, mechanical cargo hold ventilation, supply and exhaust, dehumidification systems, fans, motors, controls, ducts, trunks, hangers, louvers, dampers, insulation, dehumidification machinery and controls, cowls, mushroom heads, goosenecks, vent head operating gear, covers, foundations, vibration dampers, materials, fabrication, and installation. Does not include machinery space ventilation.

778 *Miscellaneous ventilation:* Includes natural or mechanical ventilation, supply, and exhaust of all hull spaces not included in accounts 777 and 779, storerooms, work shops, steering gear room, deck machinery room, and lockers; fans, motors, controls, ducts, trunks, filters, coils, hangars, and fittings.

779 *Accommodation ventilation and air-conditioning:* Includes natural ventilation of accommodations, supply and exhaust, mechanical ventilation of accommodations, supply and exhaust, air-conditioning systems, fans, motors, controls, ducts, trunks hangers, louvers, dampers, registers, diffusers, grilles, filters, heating and cooling coils, thermostatic controls, insulation, air-conditioning machinery and controls, cowls, mushroom heads, goosenecks, vent head operating gear, covers, foundations, vibration dampers, materials, fabrication and installation. Does not include machinery space ventilation.

780 *Scuppers and drains:* Includes material and installation for deck scuppers in ship's refrigeration and refrigerated cargo spaces and for window drains. Also includes drains for special conditions, such as winch control stands and coamings for flush deck hatch covers on interior nonweather decks. Does not include drain piping for deck scuppers.

783 *Mooring and towing gear fittings:* Includes cast, forged, or weld fabricated fittings, bitts, chocks, cleats, pads, mooring pipes, fairleaders, sheaves, rollers, towing hook, towing knees, doubler plates, headers, shell mouldings, and foundations.

784 *Lines, hawsers, warps, and reels:* Includes stream line, tow line, mooring lines, spring lines, fire lines, hawsers and warps, reels for wire rope, foundations, and bracing.

785 *Anchors, chains, handling equipment, and stowage:* Includes anchors, bower, spare bower, stream, kedge, mushroom, anchor cable, cast, forged, patent; spare shackles and links, wire rope ground tackle, hawsepipe, cast or welded, hawse pipe bolsters at shell or deck, buckler plates, anchor recesses, chain pipe, chain pipe covers, and chain locker bellmouths. Also includes anchor cranes and davits, catheads, tripping devices, shell or deck, anchor lashing, chain locker fittings, gratings, concrete, anchorage buoy systems, anchorage spuds, frames and lifting gear, special anchorage systems, doublers, headers, and foundations.

788 *Swimming pools, gymnasium and recreational gear:* Includes built-in swimming pools, except steel work, tiling, beach, gratings, scum gutters, scuppers, ladders, rails, lifelines, diving boards, beach furniture, lockers, portable swimming pools, metal or other framework, pool envelope, ladders, securing gear, fittings, and stowage. Filling and draining, lighting systems not included in this account. Includes gymnasium equipment, recreational equipment, play equipment; dog kennels, pens, and fencing.

789 *Rat proofing:* Includes material and labor required to comply with public health rules and regulations.

790 *Boats, rafts, davits, and boat winches:* Includes lifeboats, oar-propelled, hand-propelled, motor-propelled; work boats, dinghies, prams, powerboats, launches, auxiliary sailing boats, whaleboats, other pulling boats; rafts, balsa, catamaran, inflatable, buoyant apparatus; davits, rotary, quadrant, boom, gravity, pivoting; boat stowage, chocks, cradles, platforms, skid beams, boat gripes, frapping lines, blocks, falls, and stays. Also includes boat winches, hand and power operated, controls; boat covers, canvas, wood or metal; access ladders, Jacob's ladders, foundations, doublers, headers, installation, and testing.

791 *Hotel and galley outfit:* Includes all outfits in galleys, bakeries, pantries, sculleries, butcher shops, vegetable rooms, steward's storerooms, ranges, ovens, broilers, griddles, fry kettles, incinerators, vegetable steamers, stock kettles, coffee and water urns, canopies over, galley smoke pipe, coat box, commercial type refrigerators, mixers, peelers, egg boilers, slicers, food choppers, toasters, proofers, dough troughs, steam tables, bainmaries, ice cream cabinets, dishwashers, garbage disposal units, garbage chutes, silver burnishers, and sharpeners. Also includes creels, baskets, trays, scales, ice makers, ice crushers, meat blocks, cutting boards, dressers, dish warmers,

dish racks, cabinets, lockers, tables, bins, shelves, pan racks, platforms, pots, pans, cooking and baking utensils, cutlery, flat tableware, hollow ware, chinaware, crockery, glassware, linens, bedding, mattresses, upholstery, curtains, drapes, blinds, roller shades, fittings, carpets, rugs, runners, mats, safes, and miscellaneous articles of furniture, equipment, and fixtures not included in joiner work. Also includes printing equipment, presses, duplicating machines, paper cutters, moving picture projection outfits, screens, darkroom equipment, medical and surgical outfits, hospital and dispensary equipment and furniture, operating tables, sterilizers, laundry equipment, washers, extractors, dryers, drying racks, ironers, irons, ironing boards, laundry trays, and tables. Also includes office equipment and furniture, typewriters, computers, business machines, cash registers, barber and beauty shops, equipment and furniture, chairs, mirror cases, lunch counters, bars, soda fountains, stools, vending machines, foundations, curbs, gratings, and installation of equipment.

792 *Handrails, rails, and awning stanchions:* Includes handrails and stanchions at deck boundaries, safety rails and stanchions in way of deck openings, walkways, lifeboats, deck machinery, rails, of pipe solid rods, chain, rope, wood cap rail, stanchions fixed or portable, braces, storm or grab rails, of pipe, wood, safety chains, rope, awning stanchions, ridge poles, spreaders, jack rods, and ridge ropes for nonrigid awnings. Also includes jack rods as required for lacing wind dodgers and weather cloths, portable gate sections in rails and bulwarks (excluding rails and stanchions in machinery spaces).

793 *Metal ladders:* Includes inclined and vertical metal ladders, exterior and interior ladders in way of accommodations, ladders to dry cargo holds, ore holds, reefer holds, cargo tanks, fuel bunkers, ballast tanks, peak tanks, cofferdams, void spaces, and double bottoms; portable ladders for access (other than step or extension ladders). Also includes grab rods or rungs as required, ladders, platforms, safety treads, and rails. Does not include ladders in machinery spaces.

794 *Miscellaneous hull fittings and equipment:* Includes hawser and cordage, racks, platforms, cargo hauling and handling pads, lashing gear, painter's staples, grounding pads, oil hose davits, stores davits, Suez davits, trolley rails, stores-handling winches (hand or power), docking plugs, blow-out plugs, ash and store chutes, spare propeller stowage, oil drum stowage, oil hose stowage, miscellaneous stowages, wheelhouse visor, side light boxes, emergency stop, station, and light recesses. Also includes cathodic protection in tank spaces (interior), name and hailing port, name boards, draft marks, freeboard marks, label plates and numbers, certification plates, builder's nameplates, net tonnage and official number marks, tank marks, caution and warning signs and marks, plaques, notice boards, sounding boards, deck outfit, carpenter's outfit, shop equipment, collision mats, fend-

ers, boat hooks, nets, rat guards, flexible ladders, and cooperage. Also includes navigating outfit such as compasses, logs, leads, binoculars, thermometers, barometers, rules, dividers, hydrometers, chronometers, clocks, line gun, gauges, sounding machine, compass platform, leadsman's platform, sounding boom, log bracket, line gun, and drum stowage. Also includes flags, pennants, day shapes, signals, and flares, pyrotechnic locker, small arms and stowage, damage control gear and stowage, fire extinguishers, fire axes, fire buckets, breathing apparatus, flame safety lamps, protective appliances, sand boxes, life preservers, life buoys, special buoys, stowages, ship's bells, belfries, fog gong, foghorn, lubrication guns and devices, and miscellaneous tools and equipment.

795 *Storerooms and fittings:* Includes deck department stores, bosun's stores, carpenter shop, paint room, lamp room, battery room, blacksmith shop, and welding shop; shelving, racks, bins, benches, battens, clips, small tanks,and miscellaneous fittings.

796 *Rigid awnings:* Includes rigid awnings of metal, wood, plastic, or other, sheathed over permanent framework and supported by stanchions and braces.

797 *Deck covering, fills, and permanent ballast:* Includes composition deck coverings, underlayment, cove base; terrazzo deck covering, cove base; latex type deck covering, or underlayment, brick or quarry tile, nonslip tile, cove base; ceramic or vitreous tile, cove base; border tile, rubber, asphalt, cork, vinyl, or other tile, cove base; linoleum, sheet rubber, or other deck covering. Also includes portland cement and sand deck covering, underlayment, plastic coatings with nonskid aggregate, special deck coverings, safety treads, bonding clips or mesh, bituminous coatings on deck, underdeck coverings, cleaning decks to receive deck coverings, and installation of deck coverings. Also includes concrete or cement chocks; concrete or cement fill in tanks and void spaces, cork and petroleum base fill in void spaces, chemical foaming fill in void spaces; permanent ballast, all kinds, including installation. Does not include canvas decking and wood gratings.

798 *Canvas outfitting:* Includes awnings, dodgers, weather cloths, windsails, hatch tarpaulins, covers for deck machinery, outfit, and gear exposed to weather, canvas or other suitable materials, plain or coated, grommets, lashing stuff, and sail equipments.

799 *Storage and installation of owner items:* Includes storage and installation aboard ship of all owner-furnished items, purchase of budget items, storage and installation aboard ship, and special purchases.

800 *Mechanical design:* Includes mechanical marine engineering work, comprising design calculation, arrangement and detail drawings,

material specifications, and allowance lists, together with associated material services. Includes such technical areas as heat balance calculations, propeller shaft calculations, piping system design, ventilation and air-conditioning system design, refrigeration application, steering gear design, anchor windlass and winch design and application.

801 *Mechanical design for diesel engine:* Includes finished plans, arrangements, and details, calculations, ordering material booklets and catalogs.

802 *Patterns for diesel engine*

803 *Jigs and fixtures for diesel engine*

804 *Arrangement models:* Includes models of machinery arrangements and machinery space, deckhouses, steering gear room, and similar space.

807 *Diesel engine bed plates:* Includes caps, liners, and main bearing. Excludes foundation bolts.

808 *Diesel engine cylinders:* Includes assistant cylinders and covers, covers and false faces, stuffing boxes and packing, valve chests and liners.

809 *Diesel engine housing:* Includes distant pieces, guides, housing ties, studs, brackets, bearing caps, and covers.

810 *Diesel engine receiver pipes (cast or copper)*

811 *Main condensers and air removal equipment:* Main condensers, including patterns. Does not include supporting brackets, which are charged to 807 (when condenser is attached to or built into the main engine) or 714 (when the condenser is independent of the main engine).

812 *Column bridge for diesel engine*

813 *Diesel engine pistons, rods, and crossheads:* Includes connecting rods, chunk ring, packing, pin, rings, springs, tail rods, and pistons.

814 *Diesel engine crankshaft:* Includes spares, pins, webs, and coupling bolts.

815 *Exhaust belt and silencer for diesel engine*

816 *Diesel engine valves and valve gear:* Includes eccentrics and rods, gibs, ring and packing, slide and piston valve bodies, valve stems, crossheads, and guides.

817 *Main engine throttle valve and handling gear:* Includes cylinder relief valves, governor, and gear, indicator gear, main engine drains, mechanical telltale gears, receiver safety valve.

818 *Diesel engine scavenger pump complete:* Includes connecting rod, guides, crossheads, slipper, piston and rod, air inlet silencer, beams, rack shaft and bearing. Also includes air compressor.

819 *Diesel engine turning and reversing gear:* Includes brackets and bearings, connecting rods, reversing and turning engines, reverse and rack shafts, turning wheels and worms.

820 *Diesel engine and shafting oil and water service:* Includes lubricators, sight feed, oil boxes, cups and covers, piping and fitting to oil tank.

822 *Diesel engine attached pumps, main air pumps:* Includes all pumps attached to main engine, including beam shaft and beams, with pins, distance pieces, and beam shaft bearings, pump links, crosshead links, relief valves, air vessel and overflow pipe and all studs, bolts, nuts, and screws used for the above pattern and installed on ship.

823 *Diesel main engine lagging and pipe covering*

824 *Diesel engine assembling in shops:* Includes miscellaneous material incidental to erection, scaffolding, temporary foundations, and disassembling.

825 *Diesel engine internal movement:* Includes all labor and material for internal movement of the main engine.

826 *Diesel engine repairs and renewals after delivery*

840 *Propeller:* Propellers complete with caps. Pattern and machine work are included.

841 *Bow thruster and installation*

842 *Shafts and bearings:* Includes line shaft, stern tube shaft, propeller shaft, inboard and outboard coupling and propeller nut, feather key for propeller, composition liners for stern tube shaft, all bolts, nuts, keys, adjusting wedges, holding down bolts, line shaft bearings, stern tube bearings with lignum vitae, phenolic resin bearings, oil-lubricated bearing seals, lubricating oil systems for white metal bearings, zinc protecting rings, stuffing boxes, strut bearings with lignum vitae and protecting rings, fairwaters and fillings for same, bulkhead stuffing boxes, all studs, bolts, nuts, keys, and white metal. Includes thrust shaft, thrust bearing, and sole plate when not part of the main engine. Includes boring of the stern frame boss, shaft struts, and stern tube. Also includes installation of the above.

843 *Diesel engine bearings and shafts:* Includes thrust shaft, thrust bearing complete, and sole plate for thrust bearing. Does not include installation.

844 *Floors, ladders, and gratings*: Includes floor plates, guard plates, and splash plates in the engine rooms, boiler rooms, and shaft alleys, including floor angles and supports, except bounding angles riveted to ship's structure. Includes ladders, gratings, grating supports, hand- and grab rails in engine rooms, boiler rooms, shaft alleys, and engine and boiler hatches, except gratings over hatch openings.

845 *Ladders, gratings, and casings for diesel engine:* Includes all engine casings, lights for engine casings, ladders and gratings, steps, grab rails, and floor plates which are attached to the engine.

850 *Mechanical insulation:* Includes all thermal insulation and lagging on pipes and equipment, including tanks and ventilation ducts throughout the ship.

853 *Dredging machinery:* Includes material and installation for dredging pump, suction and discharge piping, cutterhead and ladder, and suction heads, together with hoisting and handling winches and windlasses.

855 *Refrigeration plant:* Includes compressors, ice-making tanks, brine tanks, condensers, special valves, refrigeration coils and headers, and all other associated materials.

856 *Miscellaneous fittings and tanks:* Includes main feed tanks complete with gauge glasses and fittings, oil waste, tallow and miscellaneous tanks with drip pans and fittings. Includes all pressure and vacuum gauges, recording gauges, draft gauges, revolution counters, torsion meters, thermometers, pyrometers, gauge testers, gauge board piping and fittings, and associated installation. Includes all machine tools, with belting, shafting, and driving motors, carpenter's and machinist's tools, engine room trolleys, turbine and machinery-lifting gear, safety valve–easing gear, boiler stop valve operating gear, chronometer valve–operating gear, special wrenches, boiler-firing tools, eye bolts, assorted bolts, plates, and bars. Includes all installation labor.

857 *Miscellaneous fittings for diesel engine:* Includes tools, gauges, instruments, wrenches, and tools, wear-down gauges, engine indicators, gauges, counter-operating gear complete with shafting, brackets, gears, fuel valve testers, packing for engine valves, and engine name plates. Includes installation of the above.

858 *Steering gear and anchor windlasses:* Includes steering engine, steering engine motor, steering stand and wheels, telemotors, and auxil-

iary steering engine with spares. Also includes transmission and steering gear fittings such as control shafting, bevel gears and brackets, transmission ropes and drums, tiller ropes and sleeves, telemotor typing and fittings, rudder brake, rudder carrier, and stuffing box, quadrant crosshead for yoke, and spare tiller. Includes installation and the finish machining and fitting of the rudder stock and associated fittings. Includes material and installation for the anchor windlass and chain stopper.

867 *Auxiliary boilers:* Includes material and installation for auxiliary boilers, both oil-fired or gas waste heat types.

868 *Fuel clarification system*

869 *Fuel oil system:* Includes piping, pumps, strainers, manifolds, valves, and fittings connecting to main engine or to boilers.

870 *Fuel oil system for main engine:* Includes a high-pressure fuel oil system for a main diesel engine, complete with pipe and fittings, and installation.

877 *Cargo hold ventilation:* Includes natural cargo hold ventilation, supply and exhaust, mechanical cargo ventilation, supply and exhaust, dehumidification systems, fans, motors, controls, ducts, trunks, hangers, louvers, dampers, insulation, dehumidification machinery and controls, cowls, mushroom heads, goosenecks, vent head–operating gear, covers, foundations, and vibrations dampers. Includes installation.

878 *Miscellaneous ventilation:* Includes natural and mechanical ventilation, supply, and exhaust of all hull spaces not included in accounts 877 and 879, storerooms, work shops, steering gear room, deck machinery room, lockers; fans, motors, controls, ducts, trunks, filters, coils, hangars, and fittings.

879 *Accommodation ventilation and air-conditioning:* Includes natural ventilation of accommodations, supply and exhaust; mechanical ventilation of accommodations, supply and exhaust; air-conditioning systems, fans, motors, controls, ducts, trunks, hangers, louvers, dampers, registers, diffusers, and grilles. Also includes filters, heating and cooling coils, thermostatic controls, insulation, air-conditioning machinery and controls, cowls, vibration dampers, goosenecks, vent head–operating gear, covers, foundations, materials; fabrication and installation.

880 *Smokestacks and uptakes:* Includes all stacks and uptakes from main and auxiliary boilers, ladders and insignias on stacks, exhaust ducts and piping from main and auxiliary engines, expansion joints, piping fittings, sound attenuators, spark arrestors, guards, and rails.

882 *Forced draft system:* Includes air heaters, blowers, door, and uptakes up to and including top tube sheet of heater boxes; ducts and trunks, engines and motors, furnace fronts and shutters.

890 *Main propulsion engine:* Includes material and installation on ship for steam turbines, gas turbines, diesel engines, and electric propulsion generators, motors, and control equipments.

893 *Auxiliary machinery:* Includes material and installation of evaporators, distillers, auxiliary condensers, grease extractors, and other auxiliary machinery not otherwise designated.

894 *Main boilers*: Includes material, erection, and installation for main boilers, baffles, casings, downtakes, drums, manhole covers, thimbles, and tubes.

900 *Steam and exhaust systems in machinery spaces:* Includes pipe, valves, fittings, and hangers for all steam systems, safety valve escapes, and boiler blowdown systems.

901 *Cooling systems:* Includes material and installation for pumps, strainers, pipe, valves, fittings, and hangers for condensers, coolers, and seawater service systems.

902 *Compressed air system:* Includes main and auxiliary air storage tanks complete with all valves and air compressors; all piping and fittings between air compressors and air starting valves on main engine.

903 *Auxiliary engine fuel oil system:* Includes main and auxiliary diesel's fuel systems up to main engine manifold, daily service fuel oil tanks with heating coils, auxiliary fuel oil pump, all pipe, valves, and fittings between fuel oil transfer pump and suction manifold on main engine fuel pump and independent high pressure fuel oil pump. Does not include the transfer pump.

904 *Lubricating oil system:* Includes pumps, coolers, storage tanks with heating coils, and fittings for both cylinder and crankcase lubricating oil, and all necessary valves, pipe, and fittings connecting flange at engine bedplate on oil sumps to inlet manifold on main engine, and independent oil system to auxiliary engines.

905 *Auxiliary engine circulating water system:* Includes fresh-water system and salt-water system, exclusively for internal combustion engines for jacket- and piston-cooling services, salt-water circulating pumps, sump tanks, reserve supply tanks, heat exchangers, fresh-water circulating pumps, valves, strainers, pipe and pipe fittings connecting to engine manifold, or connection at main engine.

906 *Feedwater and condensate system:* Includes material and installation for pumps, strainers, feedwater heaters, pipe, valves, fittings, and hangers.

907 *Steam drain system in machinery spaces:* Includes all pipe, fittings, valves, traps, hangers, and strainers.

908 *Steam and exhaust system outside machinery spaces:* Includes all pipe, valves, traps, hangers, and fittings.

914 *Heating system:* Includes building and installing heating system complete with piping from connection to the auxiliary steam line in the machinery spaces, valves to radiators in rooms, or to thermo heating boxes, including heating coils, radiators, valves, hangers, air cocks, and pipe fittings. Also includes pipe, valves, and fittings for steam to galley, pantry, and fresh-water tanks on deck.

915 *Ventilating system, engine and boiler room:* Labor and material for all ventilator cowls and operating gear to provide engine and boiler room natural ventilation.

916 *Vents and soundings:* Includes all material and installation for all vents and soundings in hull tanks, such as inner bottom, fore and aft peak tanks, and deep tanks, including pipe, valves, and supports. Does not include cargo tank vents and sounding tubes.

917 *Weather deck drains*: Includes material and installation for all drains and scuppers on open decks, forecastle deck, poop deck, and superstructure decks. Also includes drain piping from deck scuppers in refrigeration spaces.

918 *Bilge and ballast systems:* Includes all material and installation for main and auxiliary drainage piping, sea chests, inner bottom and hold drainage pumps, strainers, pipe, valves, fittings, and hangers.

919 *Plumbing system:* Includes material and installation of plumbing and sanitary piping, scuttlebutts, toilet, galley and pantry drains, sanitary fixtures and accessories, pumps, strainers, pipe, valves, fittings, and hangers. Includes sewage system, holding tanks, and grey water and black water systems.

920 *CO_2 systems, fixed and portable:* Includes all piping, valves, fittings, racks, hose, and reels for fixed or portable systems.

921 *Salt-water fire mains, steam smothering and foam systems:* Includes material and installation for steam, water, and foam fire protection systems, pumps, strainers, pipe, valves, fittings, controls, and hangers. Also includes portable dry chemical extinguishers.

923 *Liquid cargo system:* Includes fill and discharge connections from account 921, all pumps, valves, controls, and fittings in pump room. Ladders, floor plates, gratings, and other pump room equipments, and cargo venting system.

924 *Engine room watertight doors:* Includes all material and installation.

926 *Machinery spaces skylight-operating gear:* Includes all material and installation.

927 *Mechanical telegraph, whistle pulls, and voice tubes*

930 *Propulsion plant automatic control system:* Includes material and labor for centralized control systems for throttle control, data acquisitions, data readout, data logging, acuators, alarms, and integral communications equipment. Also includes boiler control systems with combustion control, feedwater control, flame failure, and flame management systems.

935 *Electrical design:* Includes electrical engineering work comprising detail calculations, arrangement and detail drawings, material specifications, and allowance lists, together with associated material and services. Includes load analysis, fault current calculations, voltage drop calculations, lighting calculations, and communication systems design, planning, and application.

938 *Electric generation:* Includes material, and installation, mounting, and connection of ship's service power-generating equipments and all associated control and regulating equipments, from the generator to the first main distribution bus terminals. Does not include cable or wireways.

939 *Wiring and wireways:* Includes material, installation, mounting, and running cable, of all cable, cableways, hangers, and clips. Also includes wiring between system components, but does not include interconnecting, or local wiring of equipment components.

940 *Electric lighting:* Includes material, installation, mounting, and connecting of all electric lighting equipment, such as fixtures, searchlights, navigating lights, lighting panelboards, and lighting transformers. Does not include any cable.

941 *Miscellaneous power distribution:* Includes material, installation, mounting, and connecting, of power distribution panelboards, power transformers, power conversion apparatus not integral with other systems, and other miscellaneous power distribution equipments.

942 *Interior communications:* Includes material, installation, mounting, and connecting of electric communication and information systems used internally on the ship, such as engine order telegraph, shaft speed, rudder angle indicator, general alarm, telephones, intercom,

public address, annunciators, alarm panels, fire alarms, time systems, gryocompass, and steering control. Does not include cable.

943 *Exterior communications:* Includes material, installation, mounting, and connecting of electric communications and informations systems used externally to the ship, such as radio, radiotelephone, radiotelegraph, radar, loran, standard broadcast reception, depth sounding, sonar, radio direction finder, facsimile, and satellite communications. Also includes interconnecting wiring between equipment components. Does not include foundations and supports.

944 *Motors:* Includes material and installation for motors which are purchased separately from driven machinery units. Also includes installation of all other motors purchased integral with the driven unit, such as motors for axial fans, hermetically sealed motors, and special frame motors for close-coupled pumps. Does not include cable.

945 *Motor controls and drive systems:* Includes material and installation for motor controls, motor control centers, and drive systems which are purchased separately from the driven machinery units. Also includes electric drive systems and special control equipment in which the drive motor and the controller are matched, such as cargo winch drives, constant tension mooring winch drives, and emergency lube oil pump drives. Does not include cable.

946 *Cathodic protection systems:* Includes material and installation for impressed current type cathodic protection systems used for protection of the exterior hull.

950 *Setting up and launching:* Includes ground and sliding ways, setting up of ways, poppets, shoring, packing, wedging, crushing ships, masks, checking arrangement, temporary brackets, pads, wire rope, and fittings. Also includes launching lubricants, other friction-reducing devices, dressing of ship, launching platform, services of tow boats, and powerboats to retrieve launching ways. All work in connection with the above until the vessel is secured alongside the fitting-out dock. Includes special launching arrangements.

952 *Staging and blocking:* Includes installation and removal of staging, blocking, shoring, wood stanchions, false work, fairing battens, and leveling and plumbing of structural members.

953 *Jigs, tools, and fixtures:* Includes equipment used in the construction of the ship which does not become part of nor contribute to the value of the physical ship, but which are necessary during construction, comprising such items as boring bars, boiler erection frames, steel fabrication jigs, protection covers for equipment, lifting and handling rigs, and similar noncapital items.

957 *Crane service:* Includes all crane service in fabricating shops, shipway, and wet basin.

958 *Cleaning and caretaking:* Includes cleaning, watching, and caretaking of ship, removal of scrap materials, excess materials, debris, removal of ice and snow, pumping of bilges, and moving and securing ship.

959 *Temporary lighting, air, and water service:* Includes all labor and material in connection with these services.

960 *Miscellaneous construction charges:* Includes general expenses not chargeable to a particular account defined in this structure. Includes hull definite expenses such as insurance, bond, classification fees, royalties, commissions, travel expenses, progress photography, launching party, etc.

962 *Testing, trials, and delivery:* Includes testing of machinery and equipment, static and dynamic tests of rigging and ship's cargo gear, shaft and torsionmeter calibration, dock trials, preliminary and official sea trials, and delivery. Includes all supplies such as fuel oil, water, and commissary. Preparation of all reports, test memoranda, and trial reports.

963 *Towing, waterway, and port charges:* Includes towing charges in connection with delivery, wages and subsistence of crew for towing period, waterway and canal fees incurred during delivery, port charges, wharfage, etc.

964 *Repairs after delivery*

965 *Drydocking:* Includes material and labor charges for dock and staging, fees for suppling temporary steam, electricity, water, compressed air, towing charges in connection with docking, handling lines, and securing vessel, and for necessary crew, guards, and fire watch.

969 *Removals and replacement*

975 *Miscellaneous engineering charges:* Includes preliminary design and estimating, budgeting, preparation and printing of specifications, engineering planning, methods, and schedules.

Index

ISBN 0-87033-458-1
55000
9 780870 334580